Lecture Notes in Computer Science 12593

More information about this subseries at http://www.springer.com/series/7410

Deukjo Hong (Ed.)

Information Security and Cryptology – ICISC 2020

23rd International Conference
Seoul, South Korea, December 2–4, 2020
Proceedings

 Springer

Editor
Deukjo Hong 🆔
Jeonbuk National University
Jeonju-si, Korea (Republic of)

ISSN 0302-9743   ISSN 1611-3349   (electronic)
Lecture Notes in Computer Science
ISBN 978-3-030-68889-9   ISBN 978-3-030-68890-5   (eBook)
https://doi.org/10.1007/978-3-030-68890-5

LNCS Sublibrary: SL4 – Security and Cryptology

This Springer imprint is published by the registered company Springer Nature Switzerland AG
The registered company address is: Gewerbestrasse 11, 6330 Cham, Switzerland

Preface

The 23rd International Conference on Information Security and Cryptology (ICISC 2020) was held in virtual format during December 2–4, 2020. This year's conference was hosted by the KIISC (Korea Institute of Information Security and Cryptology).

The aim of this conference was to provide an international forum for the latest results of research, development, and applications within the field of information security and cryptology. This year, we received 51 submissions and were able to accept 15 papers, resulting in 15 presentations at the conference. The challenging review and selection processes were successfully conducted by Program Committee (PC) members and additional reviewers via the EasyChair review system. For transparency, it is worth noting that each paper underwent a blind review by three PC members. Furthermore, to aid in resolving conflicts concerning the reviewers' recommendations, individual review reports were open to all PC members, and detailed interactive discussions on each paper took place. For the LNCS post-proceedings, the authors of selcted papers had a few weeks to prepare their final versions, based on the comments received from the reviewers.

The conference featured three invited talks: "Tweakable Block Cipher-Based Cryptography" by Thomas Peyrin, "Designing the NIST post-quantum public-key candidate Saber" by Sujoy Sinha Roy, and "Next Generation Cryptography Standards" by Lily Chen. We thank the invited speakers for their kind acceptances and excellent presentations, all authors who submitted their papers to ICISC 2020, all PC members, and the additional reviewers. It was a truly wonderful experience to work with such talented and hard-working researchers.

Finally, we also thank all attendees for their active participation and the Organizing Committee (OC) members who successfully managed this conference.

December 2020

Deukjo Hong

Organization

General Chairs

Souhwan Jung   Soongsil University, South Korea
Hyojin Choi   National Security Research Institute, South Korea

Organizing Committee Chairs

Seungjoo Kim   Korea University, South Korea
HeeSeok Kim   Korea University, South Korea

Program Committee Chair

Deukjo Hong   Jeonbuk National University, South Korea

Program Committee

Joonsang Baek   University of Wollongong, Australia
Lynn Batten   Deakin University, Australia
Jonathan Bootle   IBM Research Zürich, Switzerland
Zhenfu Cao   East China Normal University, China
Donghoon Chang   IIIT-Delhi, India
Paolo D'Arco   University of Salerno, Italy
Keita Emura   NICT, Japan
Johann Groszschädl   University of Luxembourg, Luxembourg
Dong-Guk Han   Kookmin University, South Korea
Swee-Huay Heng   Multimedia University, Malaysia
David Jao   University of Waterloo, Canada
Seok Won Jung   Mokpo National University, South Korea
Dongchan Kim   Kookmin University, South Korea
Dongseong Kim   The University of Queensland, Australia
Howon Kim   Pusan National University, South Korea
Huy Kang Kim   Korea University, South Korea
Jihye Kim   Kookmin University, South Korea
Kee Sung Kim   Daegu Catholic University, South Korea
Minkyu Kim   National Security Research Institute, South Korea
Myungsun Kim   The University of Suwon, South Korea
Sungwook Kim   Seoul Women's University, South Korea
Young-Gab Kim   Sejong University, South Korea
Alptekin Küpçü   Koç University, Turkey
Jin Kwak   Ajou University, South Korea
Taekyoung Kwon   Yonsei University, South Korea

Changhoon Lee	SeoulTech, South Korea
Changmin Lee	École Normale Supérieure de Lyon, France
Hyung Tae Lee	Jeonbuk National University, South Korea
Jooyoung Lee	KAIST, South Korea
Kwangsu Lee	Sejong University, South Korea
Manhee Lee	Hannam University, South Korea
Mun-Kyu Lee	Inha University, South Korea
Iraklis Leontiadis	Inpher, Switzerland & USA
Jiqiang Lu	Beihang University, China
Sjouke Mauw	University of Luxembourg, Luxembourg
Atsuko Miyaji	JAIST, Japan
Nicky Mouha	National Institute of Standards and Technology, USA
Khoa Nguyen	Nanyang Technological University, Singapore
Katsuyuki Okeya	Hitachi High-Tech GLOBAL, Japan
Jong Hwan Park	Sangmyung University, South Korea
Ki-Woong Park	Sejong University, South Korea
Young-Ho Park	Sejong Cyber University, South Korea
Pedro Peris-Lopez	Universidad Carlos III de Madrid, Spain
Josef Pieprzyk	Queensland University of Technology, Australia
Dongyoung Roh	National Security Research Institute, South Korea
Bimal Roy	Indian Statistical Institute, India
Hwajeong Seo	Hansung University, South Korea
Jae Hong Seo	Hanyang University, South Korea
Seog Chung Seo	Kookmin University, South Korea
Taeshik Shon	Ajou University, South Korea
Daniel Slamanig	Austrian Institute of Technology, Austria
Hung-Min Sun	National Tsing Hua University, Taiwan
Shi-Feng Sun	Monash University, Australia
Jaechul Sung	University of Seoul, South Korea
Atsushi Takayasu	University of Tokyo, Japan
Qiang Tang	Ying Wu College of Computing, USA
Wenling Wu	ISCAS, China
Toshihiro Yamauchi	Okayama University, Japan
Okyeon Yi	Kookmin University, South Korea
Hyejeong Yoo	Sejong Cyber University, South Korea
Dae Hyun Yum	Myongji University, South Korea
Joobeom Yun	Sejong University, South Korea

Additional Reviewers

Behzad Abdolmaleki	Valerio Cini
Osman Biçer	Hanwen Feng
Jiageng Chen	Olga Gadyatskaya
Long Chen	Tomoaki Mimoto
Xihui Chen	Sebastian Ramacher
Seulki Choi	

Contents

Security Models

Security Definitions on Time-Lock Puzzles

Daiki Hiraga[1(✉)], Keisuke Hara[1,2], Masayuki Tezuka[1], Yusuke Yoshida[1], and Keisuke Tanaka[1]

[1] Tokyo Institute of Technology, Tokyo, Japan
hiraga.d.aa@m.titech.ac.jp
[2] National Institute of Advanced Industrial Science and Technology (AIST), Tokyo, Japan

Abstract. Time-lock puzzles allow one to encapsulate a message for a pre-determined amount of time. The message is required to be concealed from any algorithm running in parallel time less than the pre-determined amount of time. In the previous works, the security of time-lock puzzles was formalized in an indistinguishability manner. However, it is unclear whether it directly meets the security requirements of time-lock puzzles. In this work, we define semantic security for time-lock puzzles, which captures the security requirements of the time-lock puzzle more directly. We consider three computational restrictions of an adversary and see how the security relationship changes. At first, in the traditional setting, we observe that it is difficult to prove that the semantic security implies the indistinguishability, same as the opposite implication. Secondly, in a slightly relaxed setting, we show that it is possible to prove that the semantic security implies the indistinguishability. By contrast, we observe that it is difficult to prove the opposite implication. Thirdly, in the more relaxed setting, we show that it is possible to prove that semantic security is equivalent to the indistinguishability. This shows that an indistinguishability meets the security requirements of time-lock puzzles in a certain restriction.

1 Introduction

1.1 Background and Motivation

One of the important topics in cryptography is to design schemes that are secure against adversaries whose running time is bounded by some polynomial. In some real-world applications, it is more desirable to consider a more flexible quantification of an adversary's computational resources. In such cases, a fundamental primitive are *cryptographic puzzles* which can handle the flexible amount of computational time or space. As mentioned in [DN92, RSW96], these puzzles have a wide range of applications, such as combating junk mail and delayed digital cash payments.

As one of the novel cryptographic puzzles, Rivest, Shamir, and Wagner [RSW96] introduced the notion of *time-lock puzzles*, following May's work on

© Springer Nature Switzerland AG 2021
D. Hong (Ed.): ICISC 2020, LNCS 12593, pp. 3–15, 2021.
https://doi.org/10.1007/978-3-030-68890-5_1

timed-release cryptography [May93]. Informally, time-lock puzzles allow one to encapsulate messages for a precise amount of time. More specifically, a sender can generate a puzzle with a message m which is hidden until time t has elapsed. Regarding its efficiency requirement, the time needed to generate a puzzle must be much shorter than t. Moreover, as a security requirement, m should be hidden from adversaries running in time less than t. Here, we take into account parallel adversaries with many processors, or more widely, polynomial-sized circuits of depth less than t.

One of the obvious applications of time-lock puzzles is an e-voting protocol. An e-voting protocol consists of a voting phase and a counting phase. To explain the usefulness of time-lock puzzles, we consider the following naive e-voting protocol (without time-lock puzzles). In the voting phase, a voter decides his/her voting value and generates its commitment. Then, a voter posts this commitment to a public bulletin board which ensures that anyone cannot be deleted or changed. In the counting phase, voters are asked to open their commitments. By counting all open voting contents, everyone can know the voting result. At first glance, the above protocol seems to work well and make sense, but in fact, we have an important problem. The problem is that if some voters do not open their commitments in the counting phase, we cannot complete the protocol and obtain an exact result.

By utilizing time-lock puzzles in the above protocol, we can solve this problem. Instead of generating a commitment of a voting content in the voting phase, a voter generates a time-lock puzzle of his/her voting content with a precise period. Even if there are some voters who do not reveal their contents in the counting phase, we can solve their time-lock puzzles after the period and obtain their voting contents.

Currently, the security for time-lock puzzles is formalized in the indistinguishability setting by Bitansky, Goldwasser, Jain, Paneth, Vaikuntanathan, and Waters [BGJ+16]. Informally, this security captures that given a puzzle of one chosen by two messages at random, any circuit with a depth of t^ε or less cannot distinguish which one is encapsulated. However, this security is artificial and it is unclear whether it directly captures the security requirements of time-lock puzzles.

1.2 Our Contributions

We revisit the security definition of time-lock puzzles. In Sect. 3, we define new security for time-lock puzzles, semantic security, to see if indistinguishability is a practical security definition. In a nutshell, semantic security ensures that the information of a message that can be obtained by an adversary who knows some (partial) information (the "semantics") of the puzzle is only the information that can be obtained without knowing the puzzle. Intuitively, this security means that an adversary with a circuit with a depth less than a specified depth cannot extract any information about messages in a puzzle. Therefore, semantic security can directly capture the security requirement for time-lock puzzles. Note that the computational restriction of the adversary for indistinguishability and semantic security is $O(t^\varepsilon)$. This is because it is difficult to prove the security relationship in Sect. 4 if the computational restriction is t^ε.

In Sect. 4, we investigate the relationship between indistinguishability and semantic security for time-lock puzzles. In the existing security definition [BGJ+16], the adversary is modeled as a circuit family of depth t^ε or less ($0 < \varepsilon < 1$). It is difficult to show that indistinguishability and semantic security are equivalent. This is because, in a security proof, it is difficult to construct an adversary that breaks the security of one, while satisfying the calculation constraint, by using the adversary that breaks the security of the other. Therefore, we set the adversary's computational power in the indistinguishability and the semantic security to $O(t^\varepsilon)$ and prove the equivalence between indistinguishability and semantic security under this restriction. Since the adversary's computational power is changed to $O(t^\varepsilon)$, the security we improve is weaker than the security proposed by Bitansky et al. [BGJ+16], but the security proof is more strict.

In Sect. 5, we consider the adversary's computational power setting into three: $O(t^\varepsilon)$, $t^\varepsilon + O(1)$ or less, and t^ε or less, and consider how the security relationship changes. The security relationship is as follows.

1. $O(t^\varepsilon)$
   As above, we can prove that the semantic security implies the indistinguishability, same as the opposite implication.
2. less than $t^\varepsilon + O(1)$
   We can prove that the semantic security implies the indistinguishability. By contrast, we cannot prove the opposite implication.
3. less than t^ε
   This restriction is proposed by Bitansky et al. [BGJ+16]. We cannot prove that the semantic security implies the indistinguishability, same as the opposite implication.

1.3   Related Works

The concept of time-lock puzzles was proposed by Rivest et al. [RSW96]. They constructed a time-lock puzzle from the inherent sequentiality of repeated squaring in the RSA group. Bitansky et al. [BGJ+16] took a different approach and constructed (succinct/pre-processing/weak) time-lock puzzles, from non-parallelizable languages together with (succinct/reusable/non-succinct) randomized encodings [BGL+15], which can be instantiated from (indistinguishability obfuscation/sub-exponential LWE/one-way function). Liu, Jager, Kakvi, and Warinschi [LJKW18] constructed a time-lock puzzle using witness encryption [GGSW13] and blockchains. Also, Mahmoody, Moran, and Vadhan [MMV11] constructed a weak time-lock puzzle in the random oracle model.

Malavolta and Thyagarajan [MT19] constructed a fully homomorphic time-lock puzzle from indistinguishable obfuscators in addition to the inherent sequentiality of repeated squaring in RSA group. Brakerski, Döttlinng, Garg, and Malavolta [BDGM19] also constructed a fully homomorphic time-lock puzzle from LWE and the inherent sequentiality of repeated squaring in RSA group.

Ephraim, Freitag, Komargodski, and Pass [EFKP20] constructed a non-malleable time-lock puzzle in the random oracle model.

2 Preliminaries

2.1 Notations

Let λ be the security parameter and ε be a real number satisfying $0 < \varepsilon < 1$. PPT denotes the probabilistic polynomial time. $\mathsf{poly}(\lambda)$ denotes a polynomial in λ. For a circuit A, $\mathsf{dep}(\mathsf{A})$ denotes the depth of A, $y \leftarrow \mathsf{A}(x)$ denotes that A takes x as input and outputs y. For a finite set S, $s \leftarrow S$ denotes that s is sampled from S uniformly at random. For a distribution $\mathcal{M}$ over a space M, $m \leftarrow \mathcal{M}$ denotes that m is sampled from $\mathcal{M}$.

2.2 Time-Lock Puzzles

A formal definition of time-lock puzzles was introduced by Bitansky et al. [BGJ+16]. They proposed three definitions of time-lock puzzles depending on the efficiency requirement: weak, pre-processing, and succinct. Among them, we discusses succinct time-lock puzzles. In the following, unless otherwise noted, time-lock puzzles means succinct time-lock puzzles. In this section, we recall a definition of time-lock puzzles.

Definition 1 (Time-Lock Puzzles). *A time-lock puzzle scheme is a tuple of two PPT algorithms* (Gen, Solve) *defined as follows.*

$z \leftarrow \mathsf{Gen}(t, m)$: The puzzle generation algorithm takes a hardness-parameter t and a message $m \in M$, then outputs a puzzle z.
$m \leftarrow \mathsf{Solve}(z)$: The solve algorithm takes a puzzle z, then outputs a message m.

We require the following correctness and efficiency.

Correctness: For all $\lambda \in \mathbb{N}$, polynomials $t(\lambda)$, and $m \in M$, $m = \mathsf{Solve}(\mathsf{Gen}(t, m))$ holds.
Efficiency: The running time of Gen is bounded by $O(\log t)$ and the running time of Solve is $t \cdot \mathsf{poly}(\lambda)$.

Bitansky et al. [BGJ+16] define security for time-lock puzzles as follows.

Definition 2 (Security). *For a time-lock puzzle scheme* P = (Gen, Solve) *and an adversary* A = $(\mathsf{A}_1, \mathsf{A}_2)$, *we define the following experiment* $\mathsf{Exp}_{\mathsf{P},\mathsf{A}}^{\varepsilon\text{-sec-}b}(\lambda)$ *for* $b \in \{0, 1\}$.

- $\mathsf{Exp}_{\mathsf{P},\mathsf{A}}^{\varepsilon\text{-sec-}b}(\lambda)$:
    1. *The challenger executes* $z^* \leftarrow \mathsf{Gen}(t, m_b)$ *and sends* z^* *and* state *to* A_2.
    2. $\mathsf{A}_2(z^*, state)$ *outputs* $b' \in \{0, 1\}$.

A time-lock puzzle scheme P *is secure with gap* $\varepsilon < 1$ *if there exists a polynomial* $\tilde{t}(\lambda)$ *such that for all* $t \geq \tilde{t}$ *and every pair of messages* $m_0, m_1 \in M$, *it holds that*

$$\mathsf{Adv}_{P,A}^{\varepsilon\text{-sec}}(\lambda) = \left| \Pr[\mathsf{Exp}_{P,A}^{\varepsilon\text{-sec-}0}(\lambda) \to 1] - \Pr[\mathsf{Exp}_{P,A}^{\varepsilon\text{-sec-}1}(\lambda) \to 1] \right| < \mathsf{negl}(\lambda).$$

Here, the depth of A_1 *is* $\mathsf{poly}(\lambda)$ *and the depth of* A_2 *is* t^ε.

3 New Definition for Time-Lock Puzzles

In this section, we define new security for time-lock puzzles, the semantic security, to see if the indistinguishability is a practical security definition. Along with that, we redefine the indistinguishability for time-lock puzzles.

3.1 Indistinguishability for Time-Lock Puzzles

We redefine indistinguishability proposed by Bitansky et al. [BGJ+16] as the ε-indistinguishability (ε-IND). To do so, we change the security definition as follows.

The first is to relax an adversary's computational restriction to consider the relationship between indistinguishability and semantic security. Concretely, we restrict the depth of the adversary after receiving the puzzle to $O(t^\varepsilon)$. We discuss on this modification in Sect. 5. Note that the weaker the adversary's computational restriction, the more strict the security.

The second is that an adversary can decide the challenge messages (m_0, m_1) after it receives t.

Definition 3 (Indistinguishability (ε-IND)). *For a time-lock puzzle scheme* P = (Gen, Solve) *and an adversary* A = (A_1, A_2), *we define the following experiment* $\mathsf{Exp}_{P,A}^{\varepsilon\text{-IND-}b}(\lambda)$ *for* $b \in \{0, 1\}$.

- $\mathsf{Exp}_{P,A}^{\varepsilon\text{-IND-}b}(\lambda)$:
    1. $A_1(t)$ *outputs a pair of messages* (m_0, m_1) *and state information* state.
    2. *The challenger executes* $z^* \leftarrow \mathsf{Gen}(t, m_b)$ *and sends* z^* *and* state *to* A_2.
    3. $A_2(z^*, \mathsf{state})$ *outputs* $b' \in \{0, 1\}$.

A time-lock puzzle scheme P *satisfies the* ε*-indistinguishability (*ε*-IND) if there exists a polynomial* $\tilde{t}(\lambda)$ *such that for all* $t \geq \tilde{t}$ *and every polynomial-sized adversary* A = (A_1, A_2), *it holds that*

$$\mathsf{Adv}_{P,A}^{\varepsilon\text{-IND}}(\lambda) = \left| \Pr[\mathsf{Exp}_{P,A}^{\varepsilon\text{-IND-}0}(\lambda) \to 1] - \Pr[\mathsf{Exp}_{P,A}^{\varepsilon\text{-IND-}1}(\lambda) \to 1] \right| < \mathsf{negl}(\lambda).$$

Here, the depth of A_1 *is* $\mathsf{poly}(\lambda)$, *and the depth of* A_2 *is* $O(t^\varepsilon)$.

3.2   Semantic Security for Time-Lock Puzzles

We introduce semantic security for time-lock puzzles. This directly captures the intuition that no information about the message is leaked from the puzzle by the time. Informally, this security states that for enough large t, it is hard to distinguish the following two experiments by time $O(t^\varepsilon)$.

- Real: An adversary can get the puzzle generated from the message and tried to get information about the message in time $O(t^\varepsilon)$.
- Ideal: The simulator tried to get information about the message without the puzzle.

In conjunction with ε-IND, an adversary's computational restrictions are discussed in Sect. 5. Formally, the definition of semantic security for time-lock puzzles is given as follows.

Definition 4 (Semantic Security (ε-SS)). *For a time-lock puzzle* $\mathsf{P} = (\mathsf{Gen},$ $\mathsf{Solve})$, *an adversary* $\mathsf{A} = (\mathsf{A}_1, \mathsf{A}_2)$, *and a simulator* $\mathsf{Sim} = (\mathsf{Sim}_1, \mathsf{Sim}_2)$, *we define the following two experiments* $\mathsf{Exp}_{\mathsf{P},\mathsf{A}}^{\varepsilon\text{-SS-real}}(\lambda)$ *and* $\mathsf{Exp}_{\mathsf{P},\mathsf{Sim}}^{\varepsilon\text{-SS-ideal}}(\lambda)$.

- $\mathsf{Exp}_{\mathsf{P},\mathsf{A}}^{\varepsilon\text{-SS-real}}(\lambda)$:
  1. $\mathsf{A}_1(t)$ *outputs a distribution* $\mathcal{M}$ *over the message space* M *and state information* state.
  2. *The challenger samples* $m \leftarrow \mathcal{M}$, *computes* $z^* \leftarrow \mathsf{Gen}(t, m)$, *and sends* z^* *and* state *to* A_2.
  3. $\mathsf{A}_2(z^*, \mathsf{state})$ *outputs* out.
  4. *The experiment outputs* $(\mathcal{M}, m, \mathsf{out})$.

- $\mathsf{Exp}_{\mathsf{P},\mathsf{Sim}}^{\varepsilon\text{-SS-ideal}}(\lambda)$:
  1. $\mathsf{Sim}_1(t)$ *outputs a distribution* $\mathcal{M}$ *over the message space* M *and state information* state.
  2. *The challenger samples* $m \leftarrow \mathcal{M}$ *and sends* state *to* Sim_2.
  3. $\mathsf{Sim}_2(\mathsf{state})$ *outputs* out.
  4. *The experiment outputs* $(\mathcal{M}, m, \mathsf{out})$.

A time-lock puzzle scheme P *satisfies* ε-*semantic security (ε-SS) if there exists a polynomial* $\tilde{t}(\lambda)$, *for any* $t \geq \tilde{t}$, *for any adversary* $\mathsf{A} = (\mathsf{A}_1, \mathsf{A}_2)$, *exists a simulator* $\mathsf{Sim} = (\mathsf{Sim}_1, \mathsf{Sim}_2)$, *for any distinguisher* D,

$$\mathsf{Adv}_{\mathsf{P},\mathsf{A},\mathsf{Sim},\mathsf{D}}^{\varepsilon\text{-SS}}(\lambda) = \left| \Pr[\mathsf{D}(\mathsf{Exp}_{\mathsf{P},\mathsf{A}}^{\varepsilon\text{-SS-real}}(\lambda)) \to 1] - \Pr[\mathsf{D}(\mathsf{Exp}_{\mathsf{P},\mathsf{Sim}}^{\varepsilon\text{-SS-ideal}}(\lambda)) \to 1] \right|$$
$$< \mathsf{negl}(\lambda)$$

holds where $\mathsf{dep}(\mathsf{A}_1) = \mathsf{poly}(\lambda)$, $\mathsf{dep}(\mathsf{Sim}_1) = \mathsf{poly}(\lambda)$, $\mathsf{dep}(\mathsf{A}_2) = O(t^\varepsilon)$, $\mathsf{dep}(\mathsf{Sim}_2) = O(t^\varepsilon)$, *and* $\mathsf{dep}(\mathsf{D}) = O(t^\varepsilon)$.

4   The Equivalence of Indistinguishability and Semantic Security

In this section, we show that ε-IND is equivalent to ε-SS for time-lock puzzles. In this proof, the adversary's computational restriction in ε-IND and ε-SS is $O(t^\varepsilon)$. We stress that if the computational restrictions remain conventional, the proof will not work. Details on these changes are discussed in Sect. 5.

First, we show that ε-IND implies ε-SS.

Theorem 1. *If a time-lock puzzle* P *satisfies* ε-IND, *then* P *satisfies* ε-SS.

Proof. Let $A = (A_1, A_2)$ be an adversary against ε-SS of P where $\mathsf{dep}(A_1) = \mathsf{poly}(\lambda)$ and $\mathsf{dep}(A_2) = O(t^\varepsilon)$. We construct a simulator $\mathsf{Sim} = (\mathsf{Sim}_1, \mathsf{Sim}_2)$ as follows.

- $\mathsf{Sim}_1(t)$:
    1. Given a hardness-parameter t, execute $A_1(t)$ and receive a distribution $\mathcal{M}$ and information state.
    2. Sample a message m_1 according to the distribution $\mathcal{M}$.
    3. Execute $z^* \leftarrow \mathsf{Gen}(t, m_1)$ and send $(\mathcal{M}, \mathsf{state}')$ to the challenger where $\mathsf{state}' = (\mathcal{M}, m_1, z^*, \mathsf{state})$.

- $\mathsf{Sim}_2(\mathsf{state}')$:
    1. Execute $A_2(z^*, \mathsf{state})$ and receive out.
    2. Output $\mathsf{out}' = (\mathcal{M}, m_1, \mathsf{out})$.

Note that $\mathsf{dep}(\mathsf{Sim}_1) = \mathsf{poly}(\lambda)$ and $\mathsf{dep}(\mathsf{Sim}_2) = O(t^\varepsilon)$. For any adversary A and the simulator Sim described above, let D be an arbitrary distinguisher where $\mathsf{dep}(D) = O(t^\varepsilon)$. Since there is no difference between the information that Sim_1 obtained and the information that Sim_2 obtained, there is essentially no difference even if the calculation amount of Sim_2 is $O(t^\varepsilon)$.

Here, we introduce the following experiment sequence $\{\mathsf{Exp}_i\}_{i=0}^3$.

Exp_0: The original experiment $\mathsf{Exp}_{\mathsf{P},\mathsf{A}}^{\varepsilon\text{-SS-real}}(\lambda)$.

Exp_1: Same as Exp_0 except for the following:

In Step 3 of Exp_0, the challenger samples two messages m_0 and m_1 according to the distribution $\mathcal{M}$ over the message space, calculate a puzzle $z^* \leftarrow \mathsf{Gen}(t, m_0)$ and send z^* and state to A_2.

Exp_2: Same as Exp_1 except for the following:

In Step 3 of Exp_1, we change the message to be a puzzle from m_0 to m_1.

Exp_3: Same as Exp_2 except for the following:

We replace the adversary $A = (A_1, A_2)$ in Steps 2, 3, and 4 of Exp_2 with the simulator $\mathsf{Sim} := (\mathsf{Sim}_1, \mathsf{Sim}_2)$. This experiment is the same as $\mathsf{Exp}_{\mathsf{P},\mathsf{A}}^{\varepsilon\text{-SS-ideal}}(\lambda)$.

For each $i = 0, \ldots, 3$, let out'_i be the final output of the challenger in Exp_i and $\mathbf{T}_i$ be the event that the challenger outputs 1 in Exp_i.

To evaluate $\text{Adv}_{\mathsf{P},\mathsf{A}}^{\varepsilon\text{-IND}}(\lambda) = |\Pr[\mathbf{T}_0] - \Pr[\mathbf{T}_3]|$, we show the following lemmas.

Lemma 1. $|\Pr[\mathbf{T}_0] - \Pr[\mathbf{T}_1]| = 0$.

Proof. The difference between Exp_1 and Exp_0 is that, in Step 3, the number of messages sampled according to the distribution $\mathcal{M}$ and the puzzle output by the challenger. It does not change any information that the adversary can get about the message. Therefore, $|\Pr[\mathbf{T}_0] - \Pr[\mathbf{T}_1]| = 0$ holds.

$\square$ **(Lemma 1)**

Lemma 2. *There exists an adversary against ε-SS $\mathsf{B} = (\mathsf{B}_1, \mathsf{B}_2)$ where* $|\Pr[\mathbf{T}_1] - \Pr[\mathbf{T}_2]| = \text{Adv}_{\mathsf{P},\mathsf{B}}^{\varepsilon\text{-IND}}(\lambda)$, $\text{dep}(\mathsf{B}_1) = \text{poly}(\lambda)$, *and* $\text{dep}(\mathsf{B}_2) = O(t^\varepsilon)$.

Proof. We construct an adversary $\mathsf{B} = (\mathsf{B}_1, \mathsf{B}_2)$ against ε-IND for P as follows:

- $\mathsf{B}_1(t)$:
    1. Execute $\mathsf{A}_1(t)$ and receive $(\mathcal{M}, \text{state})$.
    2. Sample $m_0, m_1 \leftarrow \mathcal{M}$ and output $(m_0, m_1, \text{state}')$ where $\text{state}' = (m_0, \mathcal{M}, \text{state})$.

- $\mathsf{B}_2(z^*, \text{state}')$:
    1. Execute $\mathsf{A}_2(z^*, \text{state})$ and receive out.
    2. Execute $\mathsf{D}(\mathcal{M}, m_0, \text{out})$, receive b', and send b' to the challenger.

First, we evaluate the depth of B_1 and B_2. You can see that

$$\text{dep}(\mathsf{A}_1) = \text{poly}(\lambda) \tag{1}$$
$$\text{dep}(\mathsf{A}_2) = O(t^\varepsilon) \tag{2}$$
$$\text{dep}(\mathsf{D}) = O(t^\varepsilon) \tag{3}$$

holds. By (1),

$$\text{dep}(\mathsf{B}_1) = \text{poly}(\lambda)$$

holds. By (2) and (3),

$$\text{dep}(\mathsf{B}_2) = \text{dep}(\mathsf{A}_2) + \text{dep}(\mathsf{D}) = O(t^\varepsilon)$$

holds. If computational restriction of A_2 and B_2 is equal or less than t^ε, this construction of B does not satisfy the computational restriction. The case where the adversary's computational restriction is changed is explained in Sect. 5.

Next, we evaluate $|\Pr[\mathbf{T}_1] - \Pr[\mathbf{T}_2]|$. In the case of $\text{Exp}_{\mathsf{P},\mathsf{B}}^{\varepsilon\text{-IND-0}}(\lambda)$, A_2 have received a puzzle z^* of m_0. B completely simulates Exp_1 against A. Since the distribution of the input to D is the same as the distribution in Exp_1,

$\Pr[\mathbf{T}_1] = \Pr[\mathsf{Exp}_{\mathsf{P},\mathsf{B}}^{\varepsilon\text{-IND-0}}(\lambda) \to 1]$ holds. In the case of $\mathsf{Exp}_{\mathsf{P},\mathsf{B}}^{\varepsilon\text{-IND-1}}(\lambda)$, A_2 have received a puzzle z^* of m_1. B completely simulates Exp_2 against A. Since the distribution of the input to D is the same as the distribution in Exp_2, $\Pr[\mathbf{T}_2] = \Pr[\mathsf{Exp}_{\mathsf{P},\mathsf{B}}^{\varepsilon\text{-IND-1}}(\lambda) \to 1]$ holds. Therefore,

$$
\begin{aligned}
\left|\Pr[\mathbf{T}_1] - \Pr[\mathbf{T}_2]\right| &= \left|\Pr[\mathsf{Exp}_{\mathsf{P},\mathsf{B}}^{\varepsilon\text{-IND-0}}(\lambda) \to 1] - \Pr[\mathsf{Exp}_{\mathsf{P},\mathsf{B}}^{\varepsilon\text{-IND-1}}(\lambda) \to 1]\right| \\
&= \mathsf{Adv}_{\mathsf{P},\mathsf{B}}^{\varepsilon\text{-IND}}(\lambda)
\end{aligned}
$$

holds.   □ (**Lemma** 2)

Lemma 3. $\left|\Pr[\mathbf{T}_2] - \Pr[\mathbf{T}_3]\right| = 0$.

Proof. In the case of Exp_3, A_2 executed by Sim_2 has received a puzzle z^* of m_1. The output of Sim in Exp_3 has the same distribution as the output of A in Exp_2. Therefore, $\left|\Pr[\mathbf{T}_2] - \Pr[\mathbf{T}_3]\right| = 0$ holds.   □ (**Lemma** 3)

Now, we evaluate $\mathsf{Adv}_{\mathsf{P},\mathsf{A}}^{\varepsilon\text{-IND}}(\lambda) = \left|\Pr[\mathbf{T}_0] - \Pr[\mathbf{T}_3]\right|$. The advantage $\mathsf{Adv}_{\mathsf{P},\mathsf{A}}^{\varepsilon\text{-IND}}(\lambda)$ can be evaluated as follows.

$$
\mathsf{Adv}_{\mathsf{P},\mathsf{A}}^{\varepsilon\text{-SS}}(\lambda) = \left|\Pr[\mathbf{T}_0] - \Pr[\mathbf{T}_3]\right| \le \sum_{i=0}^{2}\left|\Pr[\mathbf{T}_i] - \Pr[\mathbf{T}_{i+1}]\right|
$$

By combining Lemma 1, 2, and 3,

$$
\mathsf{Adv}_{\mathsf{P},\mathsf{A},\mathsf{Sim},\mathsf{D}}^{\varepsilon\text{-SS}}(\lambda) \le \mathsf{Adv}_{\mathsf{P},\mathsf{B}}^{\varepsilon\text{-IND}}(\lambda)
$$

holds. Since, P is ε-IND secure, $\mathsf{Adv}_{\mathsf{P},\mathsf{B}}^{\varepsilon\text{-IND}}(\lambda) = \mathsf{negl}(\lambda)$ holds. Therefore, we can conclude Theorem 1.

□ (**Theorem** 1)

Next, we show that ε-SS implies ε-IND. The adversary's computational restrictions are the same as in Theorem 1.

Theorem 2. *If a time-lock puzzle* P *satisfies* ε-SS, *then* P *satisfies* ε-IND.

Proof. Take any ε-IND adversary $\mathsf{A} = (\mathsf{A}_1, \mathsf{A}_2)$, and construct an ε-SS adversary $\mathsf{B} := (\mathsf{B}_1, \mathsf{B}_2)$ and ε-SS distinguisher as follows.

– $\mathsf{B}_1(t)$:
  1. Execute $\mathsf{A}_1(t)$ and receive a pair of message (m_0, m_1) and information state.
  2. Set the distribution $\mathcal{M}$ over the message space as uniform distribution on $\{m_0, m_1\}$.
  3. Outputs $(\mathcal{M}, \mathsf{state}')$ where $\mathsf{state}' = (\mathcal{M}, m_0, m_1, \mathsf{state})$.

- $B_2(z^*, \text{state}')$:
  1. Execute $A_2(z^*, \text{state})$ and receive b'.
  2. Output out $= m_{b'}$.

- $D_i(\mathcal{M}, m, \text{out})$:
  1. If $\|\mathcal{M}\| \neq 2$, then output $1 - i$.
  2. If $\|\mathcal{M}\| = 2$ and out $= m$, output 0.
  3. Otherwise output 1.

In this construction, We can see that

$$\text{dep}(A_1) = \text{poly}(\lambda) \tag{4}$$

$$\text{dep}(A_2) = O(t^\varepsilon) \tag{5}$$

hold. By (4),

$$\text{dep}(B_1) = \text{poly}(\lambda)$$

holds. By (5),

$$B_2 = O(t^\varepsilon) \text{ and } D_i = O(t^\varepsilon)$$

holds. If the computational restrictions of A_2 and B_2 is equal or less than t^ε, this construction of B does not satisfy the computational restriction. It is because B_2 uses ε-IND adversary A_2 as a subroutine, and selects the message m_0, m_1 based on $b' \in \{0, 1\}$ received from A_2. The case where the adversary's computational restriction is changed is explained in Sect. 5.

Let $\|\mathcal{M}\|$ be the number of messages that can be taken when choosing according to the distribution $\mathcal{M}$. E_b the event that the challenger selects m_b when choosing according to the distribution $\mathcal{M}$.

First, we show that $\|\mathcal{M}\| \neq 2$ contradicts the security of ε-SS for the time-lock puzzle. Since puzzle P is ε-SS secure, there is a simulator Sim and the following holds for any $i = 0, 1$.

$$\left| \Pr[D_i(\text{Exp}_{P,B}^{\varepsilon\text{-SS-real}}(\lambda)) \rightarrow 1] - \Pr[D_i(\text{Exp}_{P,\text{Sim}}^{\varepsilon\text{-SS-ideal}}(\lambda)) \rightarrow 1] \right| < \text{negl}(\lambda) \tag{6}$$

If the distribution $\mathcal{M}$ output from Sim is $\|\mathcal{M}\| \neq 2$, the following holds.

$$\Pr[D_0(\text{Exp}_{P,\text{Sim}}^{\varepsilon\text{-SS-ideal}}(\lambda)) \rightarrow 1] = 1$$
$$\Pr[D_1(\text{Exp}_{P,\text{Sim}}^{\varepsilon\text{-SS-ideal}}(\lambda)) \rightarrow 1] = 0$$
$$\Pr[D_0(\text{Exp}_{P,B}^{\varepsilon\text{-SS-real}}(\lambda)) \rightarrow 1] = \Pr[D_1(\text{Exp}_{P,B}^{\varepsilon\text{-SS-real}}(\lambda)) \rightarrow 1] \tag{7}$$

From (6), the difference between the probability that the distinguisher of $\text{Exp}_{P,B}^{\varepsilon\text{-SS-real}}(\lambda)$ and $\text{Exp}_{P,\text{Sim}}^{\varepsilon\text{-SS-ideal}}(\lambda)$ outputs 1 is negligible. However, there does not exists $\Pr[D_0(\text{Exp}_{P,B}^{\varepsilon\text{-SS-real}}(\lambda)) \rightarrow 1]$ that simultaneously satisfies (6) and (7).

This contradicts the assumption that the time-lock puzzle P is ε-SS secure, and thus the output of Sim $\mathcal{M}$ must satisfy $\|\mathcal{M}\| = 2$. At this time, in $\mathsf{Exp}_{\mathsf{P,Sim}}^{\varepsilon\text{-SS-ideal}}(\lambda)$, the following holds because no information about the selection of m_0, m_1 is input in Sim.

$$\Pr[\mathsf{D}_i(\mathsf{Exp}_{\mathsf{P,Sim}}^{\varepsilon\text{-SS-ideal}}(\lambda)) \to 1] = \frac{1}{2} \tag{8}$$

For any $i \in \{0, 1\}$, we can see that the following equations hold.

$$\Pr[E_0] = \Pr[E_1] = \frac{1}{2}$$
$$\Pr[\mathsf{D}_i(\mathsf{Exp}_{\mathsf{P,B}}^{\varepsilon\text{-SS-real}}(\lambda)) \to 1 \mid E_0] = \Pr[\mathsf{Exp}_{\mathsf{P,A}}^{\varepsilon\text{-IND-0}}(\lambda) \to 0]$$
$$\Pr[\mathsf{D}_i(\mathsf{Exp}_{\mathsf{P,B}}^{\varepsilon\text{-SS-real}}(\lambda)) \to 1 \mid E_1] = \Pr[\mathsf{Exp}_{\mathsf{P,A}}^{\varepsilon\text{-IND-1}}(\lambda) \to 1]$$

Using these equations, we can evaluate the following. In the following proof, let p_b be the probability that 1 is output in $\mathsf{Exp}_{\mathsf{P,A}}^{\varepsilon\text{-IND-}b}(\lambda)$ for $b \in \{0, 1\}$ Next, we evaluate $\Pr[\mathsf{D}_i(\mathsf{Exp}_{\mathsf{P,B}}^{\varepsilon\text{-SS-real}}(\lambda)) \to 1]$ using p_0, p_1.

$$\Pr[\mathsf{D}_i(\mathsf{Exp}_{\mathsf{P,B}}^{\varepsilon\text{-SS-real}}(\lambda)) \to 1]$$
$$= \Pr[E_0] \cdot \Pr[\mathsf{D}_i(\mathsf{Exp}_{\mathsf{P,B}}^{\varepsilon\text{-SS-real}}(\lambda)) \to 1 \mid E_0]$$
$$\quad + \Pr[E_1] \cdot \Pr[\mathsf{D}_i(\mathsf{Exp}_{\mathsf{P,B}}^{\varepsilon\text{-SS-real}}(\lambda)) \to 1 \mid E_1]$$
$$= \frac{1}{2}(\Pr[\mathsf{Exp}_{\mathsf{P,A}}^{\varepsilon\text{-IND-0}}(\lambda) \to 0] + \Pr[\mathsf{Exp}_{\mathsf{P,A}}^{\varepsilon\text{-IND-1}}(\lambda) \to 1])$$
$$= \frac{1}{2}((1 - \Pr[\mathsf{Exp}_{\mathsf{P,A}}^{\varepsilon\text{-IND-0}}(\lambda) \to 1]) + \Pr[\mathsf{Exp}_{\mathsf{P,A}}^{\varepsilon\text{-IND-1}}(\lambda) \to 1])$$
$$= \frac{1}{2}((1 - p_0) + p_1)$$
$$= \frac{1}{2} + \frac{1}{2}(p_1 - p_0) \tag{9}$$

Therefore, from the equations (6), (8), (9), the following holds.

$$|p_1 - p_0| = 2\left|\frac{1}{2} + \frac{1}{2}(p_1 - p_0) - \frac{1}{2}\right|$$
$$= 2\left|\Pr[\mathsf{D}_i(\mathsf{Exp}_{\mathsf{P,B}}^{\varepsilon\text{-SS-real}}(\lambda)) \to 1] - \Pr[\mathsf{D}_i(\mathsf{Exp}_{\mathsf{P,Sim}}^{\varepsilon\text{-SS-ideal}}(\lambda)) \to 1]\right|$$
$$< \mathsf{negl}(\lambda)$$

Therefore, we can conclude Theorem 2. $\square$ **(Theorem 2)**

5   Reconsideration on Computational Power

In our security definition in Sect. 3, the depth of an adversary after receiving a puzzle and a distinguisher is limited to $O(t^\varepsilon)$. In Sect. 4, under this restriction, we show the equivalence of ε-IND and ε-SS. In this section, we consider the relation between ε-IND and ε-SS as the adversary's computational power changes. In the following, the discussion about the computational restriction of a distinguisher in the semantic security is omitted because it does not affect the implications.

1. $\mathsf{dep}(\mathsf{A}_2) = O(t^\varepsilon)$
   This is a restriction considered in Sect. 3 and 4. Under this restriction, ε-IND and ε-SS are equivalent, as shown in Sect. 4.

2. $\mathsf{dep}(\mathsf{A}_2) \leq t^\varepsilon + O(1)$
   This is a restriction that allows the depth of the adversary's circuit to be not only t^ε but also the calculation of a constant that is independent of the security parameter λ.
   In this setting, it cannot be shown that ε-IND implies ε-SS by the same proof method in Sect. 4. More precisely, in Lemma 2, the adversary against ε-IND B_2 after receiving the puzzle uses ε-SS adversary A_2 as a subroutine, and then uses the distinguisher D. The depth of the circuit B_2 is as follows.

   $$\mathsf{dep}(\mathsf{B}_2) = \mathsf{dep}(\mathsf{A}_2) + \mathsf{dep}(\mathsf{D}) \leq 2t^\varepsilon + O(1)$$

   This does not satisfy the computational power restriction.
   On the other hand, it can be shown that ε-SS implies ε-IND. In Theorem 2, the adversary against ε-SS B_2 uses the adversary against ε-IND A_2 as a subroutine, and selects the message m_0, m_1 based on $b' \in \{0,1\}$ received from A_2. Now, the depth of A_2 is $t^\varepsilon + O(1)$, and subsequent calculations can be performed with a constant depth. Therefore, B_2 satisfies the restriction of computational power and can be concluded.

3. $\mathsf{dep}(\mathsf{A}_2) \leq t^\varepsilon$
   This is a restriction that the depth of an adversary cannot exceed t^ε. Bitansky et al. [BGJ+16] use this restriction to define the indistinguishability of a time-lock puzzle.
   In this setting, it cannot be shown that ε-IND implies ε-SS by the same proof method in Sect. 4. This is because the depth of ε-IND adversary B_2 exceeds t^ε in Lemma 2. The opposite implication cannot be shown because the depth of the ε-SS adversary B_2 exceeds t^ε.

Previous definition proposed by Bitansky et al. [BGJ+16] use $\mathsf{dep}(\mathsf{A}_2) \leq t^\varepsilon$ as a restriction on an adversary's computational power. However, under this restriction, it is unclear the security relationship between the indistinguishability and the semantic security as described above. In this work, we considered changing the computational restrictions of the adversary, so we can see the relationship. Therefore, it can be said that the time-lock puzzle that satisfies ε-IND essentially meets security requirements of time-lock puzzles.

Acknowledgements. A part of this work was supported by NTT Secure Platform Laboratories, JST OPERA JPMJOP1612, JST CREST JPMJCR14D6, JSPS KAKENHI JP16H01705, JP17H01695, JP19J22363, JP20J14338.

References

[BDGM19] Brakerski, Z., Döttling, N., Garg, S., Malavolta, G.: Leveraging linear decryption: rate-1 fully-homomorphic encryption and time-lock puzzles. In: Hofheinz, D., Rosen, A. (eds.) TCC 2019. LNCS, vol. 11892, pp. 407–437. Springer, Cham (2019). https://doi.org/10.1007/978-3-030-36033-7_16

[BGJ+16] Bitansky, N., Goldwasser, S., Jain, A., Paneth, O., Vaikuntanathan, V., Waters, B.: Time-lock puzzles from randomized encodings. In: Proceedings of the 2016 ACM Conference on Innovations in Theoretical Computer Science, Cambridge, MA, USA, 14–16 January 2016, pp. 345–356 (2016)

[BGL+15] Bitansky, N., Garg, S., Lin, H., Pass, R., Telang, S.: Succinct randomized encodings and their applications. In: Proceedings of the Forty-Seventh Annual ACM on Symposium on Theory of Computing, STOC 2015, Portland, OR, USA, 14–17 June 2015, pp. 439–448 (2015)

[DN92] Dwork, C., Naor, M.: Pricing via processing or combatting junk mail. In: Brickell, E.F. (ed.) CRYPTO 1992. LNCS, vol. 740, pp. 139–147. Springer, Heidelberg (1993). https://doi.org/10.1007/3-540-48071-4_10

[EFKP20] Ephraim, N., Freitag, C., Komargodski, I., Pass, R.: Non-malleable time-lock puzzles and applications. IACR Cryptol. ePrint Arch. 2020:779 (2020)

[GGSW13] Garg, S., Gentry, C., Sahai, A., Waters, B.: Witness encryption and its applications. In: Symposium on Theory of Computing Conference, STOC 2013, Palo Alto, CA, USA, 1–4 June 2013, pp. 467–476 (2013)

[LJKW18] Liu, J., Jager, T., Kakvi, S.A., Warinschi, B.: How to build time-lock encryption. Des. Codes Cryptogr. **86**(11), 2549–2586 (2018). https://doi.org/10.1007/s10623-018-0461-x

[May93] May, T.C.: Timed-release crypto (1993)

[MMV11] Mahmoody, M., Moran, T., Vadhan, S.: Time-lock puzzles in the random oracle model. In: Rogaway, P. (ed.) CRYPTO 2011. LNCS, vol. 6841, pp. 39–50. Springer, Heidelberg (2011). https://doi.org/10.1007/978-3-642-22792-9_3

[MT19] Malavolta, G., Thyagarajan, S.A.K.: Homomorphic time-lock puzzles and applications. In: Boldyreva, A., Micciancio, D. (eds.) CRYPTO 2019. LNCS, vol. 11692, pp. 620–649. Springer, Cham (2019). https://doi.org/10.1007/978-3-030-26948-7_22

[RSW96] Rivest, R.L., Shamir, A., Wagner, D.A.: Time-lock puzzles and timed-release crypto. Technical report, Cambridge, MA, USA (1996)

Secret Sharing with Statistical Privacy and Computational Relaxed Non-malleability

Tasuku Narita[1(✉)], Fuyuki Kitagawa[2], Yusuke Yoshida[1], and Keisuke Tanaka[1]

[1] Tokyo Institute of Technology, Tokyo, Japan
`narita.t.ad@m.titech.ac.jp`
[2] NTT Secure Platform Laboratories, Tokyo, Japan

Abstract. Goyal and Kumar (STOC '18, CRYPTO '18) initiate the study of non-malleability for secret sharing and proposed the definition of information-theoretical non-malleability for secret sharing. Subsequently, Brian, Faonio, and Venturi (CRYPTO '19, TCC '19) proposed computational variants of non-malleability for secret sharing and showed that by focusing on computational non-malleability, it is possible to construct more efficient schemes compared to the existing ones. However, their schemes have a drawback that they do not satisfy statistical privacy.

In this paper, we propose a new definition of computational non-malleability for secret sharing in the public parameter model. Although our definition is relaxed compared to the one proposed by Brian et al., it captures a strong security notion called non-malleability against overlap-joint tampering. Then, we show how to transform any secret sharing scheme into the one satisfying our computational non-malleability with small efficiency overhead. This transformation has a nice property that it preserves the statistical privacy of the underlying secret sharing scheme. Thus, through our transformation, we can obtain efficient secret sharing schemes satisfying computational non-malleability and statistical privacy. We achieve this transformation using lossy encryption which satisfies IND-CCA security in the injective mode.

Keywords: Secret sharing · Non-malleability · Lossy encryption · Chosen ciphertext attack

1 Introduction

1.1 Background

Secret sharing was introduced by Shamir [Sha79] and Blakley [Bla79] as a tool that enables to securely store secret information. A secret sharing scheme divides a secret message into shares and distributes them to parties. Access structure of a secret sharing scheme describes access control to the shared message. If a set of parties is contained in the access structure, they can reconstruct the

D. Hong (Ed.): ICISC 2020, LNCS 12593, pp. 16–39, 2021.
https://doi.org/10.1007/978-3-030-68890-5_2

message from the distributed shares. On the other hand, if a set of parties is not contained in the access structure, they cannot learn any information on the message from their shares. Secret sharing is used as a fundamental building block for various cryptographic primitives such as multi-party computation [GMW87, CCD88, BGW88].

Classical secret sharing schemes are not tolerant against an adversary who tampers shares. To remedy this shortcoming, Goyal and Kumar [GK18a, GK18b] introduced the notion of non-malleable secret sharing (NMSS). Intuitively, their basic definition of non-malleability ensures that even if all shares are tampered once individually, the reconstructed message results in the original value or unrelated value. This tampering model is known as *individual tampering*. Also, they considered a more powerful tampering model called *joint tampering*. In the joint tampering, the adversary is allowed to partition the shares into any number of groups and tamper the shares in each group. Furthermore, they introduced the tampering model that allows overlap in the above partition. This model is called *overlap-joint tampering* and captures the most powerful adversarial tampering attacks.

Previous Works. Goyal and Kumar defined non-malleability for secret sharing for the first time. Goyal and Kumar focused on the complexity of the access structure and the partition of the adversary's input shares to evaluate the secret sharing scheme. We can say that their goal is to construct a scheme that recognizes the general access structure and has non-malleability against over-lap joint tampering.

Goyal and Kumar constructed a scheme that recognizes general access structure and has non-malleability against individual tampering. Also they constructed a scheme that recognizes an *n*-out-of-*n* access structure and has non-malleability against over-lap joint tampering. Since [GK18b], various studies of NMSS have been conducted. A line of works [BS19, SV19, ADN+19] considered multi-time tampering. In the multi-time tampering, the adversary tampers the shares several times. They considered individual tampering and general access structures. Kumar, Meka, and Sahai [KMS18] construct general access structure NMSS with leakage-resilience. In the leakage-resilient NMSS, the adversary can learn the part of information about all shares. The tampering model is individual tampering. Chattopadhyay and Li [CL18] construct threshold access structure non-malleable ramp secret-sharing schemes against joint tampering. Lin, Cheraghchi, Guruswami, Safavi-Naini, and Wang [LCG+19] studied affine tampering, which is a non-compartmentalized restriction on tampering. They proposed threshold NMSS against affine tampering.

The above non-malleable secret-sharing schemes satisfy the information-theoretical non-malleability. Recently, Faonio and Venturi [FV19] introduced computational non-malleable secret sharing. They introduced continuous NMSS and constructed it for threshold access structure. The continuous tampering is a variant of multi-time tampering. The tampering model is individual. Further, Brian, Faonio, and Venturi [BFV19] constructed a general access structure NMSS against individual tampering. Also, they constructed NMSS against selective k-partitioning joint tampering using a variant of non-interactive zero-knowledge proofs. Thus the latter scheme is constructed in the common reference strings model. In selective k-partitioning joint tampering, the size of the partitions must be smaller than k. Their NMSS has leakage-resilience and they considered continuous tampering. In addition, Brian, Faonio, Obremski, Simkin, and Venturi [BFO+20] proposes a scheme that satisfies the k-joint p-time non-malleability that recognizes the general access structure.

Brian et al. [BFO+20] proposed the compiler that obtains the scheme which recognizes the general access structure and has non-malleability against joint tampering. However, in their scheme, the size of the partition is limited to $k \in O(\sqrt{log\ n})$. We still do not have a secret sharing scheme for general access structure which is non-malleable against joint tampering without any limitation to the size of the partition.

Motivation. Goyal and Kumar [GK18a, GK18b] defined the non-malleability for secret sharing as an extension of non-malleable codes proposed by Dziembowski, Pietrzak and Wichs [DPW10]. NMSS studied in other prior works are also defined essentially as extensions of non-malleable codes. However, under those definitions, there is still no scheme that recognizes the general access structure and has non-malleability for over-lap joint tampering. Consider this situation, we give up to refer to the line of non-malleable codes once and explore the possibility of non-malleability definition by focusing on another non-malleability. Specifically, we focus on the definition of non-malleable commitment proposed by Crescenzo, Katz, Ostrovsky, and Smith [CKOS01], and we extend this definition to non-malleable secret sharing.

Focusing on a computational non-malleability opens up the possibility of constructing efficient schemes satisfying non-malleability against powerful tampering attacks. However, the existing computational non-malleable schemes have a drawback that they do not satisfy statistical privacy. We observe that in many cases it is desirable for secret sharing schemes to satisfy statistical privacy even if they satisfy only computational non-malleability. In some cases, it is realistic that an adversary may not able to spend enormous computational power on tampering shares that honest parties have rather than on extracting information from the shares the adversary has. Moreover, for some sort of information to be shared, we would wish that privacy holds for a much longer time period compared to non-malleability. Thus, it is an interesting question of whether we can construct efficient NMSS that is secure against powerful tampering attacks and satisfies statistical privacy.

1.2   Our Contribution

In this work, we introduce a new definition of computational non-malleability for secret sharing and propose a simple and efficient compiler that transforms any secret sharing scheme into the one satisfying our notion of computational non-malleability. Our compiler has an advantage that it preserves the statistical privacy of the underlying secret sharing scheme. Thus, through our transformation, we can obtain efficient secret sharing schemes satisfying computational non-malleability and statistical privacy.

Below, we first explain our new definition and then provide the details of our construction.

On Our Definition. We aimed at constructing a simple and efficient compiler that makes any secret sharing scheme non-malleable against overlap-joint tampering. For this goal, we employ non-malleable commitment, unlike existing non-malleable secret sharing schemes that essentially employ non-malleable codes.

To follow this approach, we reconsider the definition of NMSS and explore other possibilities. The existing definitions of non-malleable secret sharing have the same spirit as the definition of non-malleable codes proposed by Dziembowski et al. [DPW10]. In this work, we work with a somewhat relaxed definition that has the same spirit as the definition of the non-malleable commitment proposed by Crescenzo, Katz, Ostrovsky, and Smith [CKOS01].[1] The motivation of the work by Crescenzo et al. was to propose an efficient and simple non-malleable commitment scheme by working with a relaxed definition. This work shows a similar possibility in the context of non-malleable secret sharing.

Our definition adopts the simulation-based paradigm similar to the existing definitions. In our definition, a simulator, who is not given any information about the target shares, is allowed to succeed in tampering with significantly better probability than an adversary who receives shares. Recall that in many definitions of non-malleability, a simulator is required to succeed in tampering with essentially the same probability as the adversary.

As discussed in the work by Crescenzo et al. [CKOS01], even if we allow such relaxation for simulators, the definition still captures the intuition of non-malleability since it guarantees that the success probability of an adversary is not (significantly) greater than that of a simulator who is not given any information about the shares and does not have any chance to succeed in tampering.

At first glance, it seems unnatural that a simulator is more successful than an adversary because the adversary could use the simulator and be as successful as the simulator. However, the adversary has an option to make a difference. It can intentionally fail on tampering in a way that the simulator does not notice.

Due to this difference, our relaxed definition does not imply existing definitions. Intuitively, in our definition, a tampering attack is not considered to be successful if the result of the tampering is $\perp$ (it means reconstruction failed) though such an attack can be considered to be successful in the previous definitions. We note that whenever the reconstructed message is $\perp$, parties can notice that a tampering attack was occurred. Thus, by treating those cases appropriately as exceptions, we can exclude tampering attacks that result in $\perp$ from our focus.

---

[1] For the definition of non-malleable commitment proposed by Crescenzo et al., see the Sect. 3.

By working with this definition, we show that we can construct a relatively efficient and simple compiler that makes any secret sharing scheme non-malleable.

In Sect. 6, we show the gap between our definition and the conventional definition.

On Our Construction. Our construction strategy is simple. In the construction, essentially, we use a non-malleable commitment scheme to protect the secret message from tampering and use the underlying secret sharing scheme to share the decommitment. Thus, each of the resulting shares consists of the commitment of the secret message and a share of the decommitment. When reconstructing the message, we first gather the shares of the underlying secret sharing scheme to reconstruct the decommitment, then verify the consistency of all of the commitments attached to the shares.

However, it turns out that the above approach seems not to work if we use ordinary non-malleable commitment schemes such as those proposed by Crescenzo et al. [CKOS01]. The technical difficulty comes from our goal that our compiler preserves statistical privacy of secret sharing schemes. Specifically, in the above design strategy, a simulator somehow obtains a tampered message from outputs of an adversary which includes only an unauthorized set of shares. The only way to make this possible is to somehow extract the message from the commitment. However, without requiring any additional property to the commitment scheme, this is impossible because the commitment scheme needs to satisfy statistically-hiding property in order to make the resulting secret sharing scheme statistically private.

IND-CCA Secure Lossy Encryption. We solved this problem by introducing a primitive that we call IND-CCA secure lossy encryption (CCA-LE) and use it as an alternative to the non-malleable commitment. CCA-LE is a public key encryption where we can generate a public key in two modes: the injective mode and the lossy mode. In the lossy mode, CCA-LE guarantees that a ciphertext statistically loses the information of the encrypted message. Thus, CCA-LE in the lossy mode can play the role of the statistical-hiding commitment scheme in the security proof of the statistical privacy of the resulting secret sharing scheme. On the other hand, in the injective mode, CCA-LE satisfies the IND-CCA security that is equivalent to non-malleability.[2] Thus, CCA-LE in the injective mode can play the role of the non-malleable commitment that has an additional extractability since we can extract an encrypted message from a ciphertext in the injective mode. We construct such IND-CCA secure lossy encryption from the combination of IND-CCA secure public-key encryption and lossy encryption. Each of them can be instantiated from many standard assumptions such as the DDH assumption.

This approach preserves the statistical privacy of the underlying secret sharing schemes. This is in contrast to the compiler proposed by Brian et al. [BFV19]

---

[2] In the definition of ordinary lossy encryption [BHY09], it is required to satisfy only IND-CPA security.

with which the resulting scheme satisfies the computational non-malleability against joint tampering, but satisfies only computational privacy even if the underlying secret sharing scheme is statistically private.

On the Use of the Public Parameter Model. One caveat of this work is that we work in the public parameter model similar to the work by Brian et al. [BFV19]. This is because, in order to use the security notion of CCA-LE, we have to keep the public key of it intact in the security experiments. Note that Crescenzo et al. [CKOS01] showed any perfectly correct IND-CCA secure PKE scheme can be used as a non-malleable commitment scheme that is computationally private in the public parameter model. We extend this idea.

2 Preliminaries

In this section, we define some notations and cryptographic primitives.

2.1 Notations

In this paper, $x \xleftarrow{r} X$ denotes selecting an element from a finite set X uniformly at random, and $y \leftarrow \mathsf{A}(x; r)$ denotes that the algorithm A is performed on the input x and the randomness r and outputs y. If there is no need to specify the randomness used by A, just write $y \leftarrow \mathsf{A}(x)$. For strings x and y, $x\|y$ denotes the concatenation of x and y. λ denotes a security parameter. A function $f(\lambda)$ is negligible function if $f(\lambda)$ tends to 0 faster than $\frac{1}{\lambda^c}$ for every constant $c > 0$. We write $f(\lambda) = \mathsf{negl}(\lambda)$ to denote $f(\lambda)$ being negligible function. For an integer ℓ, $[\ell]$ denotes the set of integers $\{1, \ldots, \ell\}$. For a set A, $|A|$ denotes the number of elements of A. The statistical distance between two random variables X and Y over a finite common domain D is defined by $\Delta(X, Y) = \frac{1}{2} \sum_{z \in \mathsf{D}} |\Pr[X = z] - \Pr[Y = z]|$. We say that two families $X = (X_\lambda)_{\lambda \in \mathbb{N}}$ and $Y = (Y_\lambda)_{\lambda \in \mathbb{N}}$ of random variables are statistically close or statistically indistinguishable, denoted by $X \approx_s Y$, if $\Delta(X_\lambda, Y_\lambda)$ is negligible in λ.

2.2 Public Key Encryption

Here, we define public key encryption (PKE).

Definition 1 (Public key encryption). *A PKE scheme Π is a three tuple* (KG, Enc, Dec) *of PPT algorithms. Below, let $\mathcal{M}$ be the message space of Π.*

- *The key generation algorithm* KG, *given a security parameter 1^λ, outputs a public key pk and a secret key sk.*
- *The encryption algorithm* Enc, *given a public key pk and message $m \in \mathcal{M}$, outputs a ciphertext ct.*
- *The decryption algorithm* Dec, *given a secret key sk and ciphertext ct, outputs a message $\widehat{m} \in \{\bot\} \cup \mathcal{M}$.*

Correctness: We require $\mathsf{Dec}(sk, \mathsf{Enc}(pk, m)) = m$ *for every* $m \in \mathcal{M}$ *and* $(pk, sk) \leftarrow \mathsf{KG}(1^\lambda)$.

We introduce indistinguishability against chosen ciphertext attacks (IND-CCA security) for PKE.

Definition 2 (IND-CCA security). *Let* Π *be a PKE scheme. We define the IND-CCA game between a challenger and an adversary* $\mathcal{A}$ *as follows. We let* $\mathcal{M}$ *be the message space of* Π.

1. *First, the challenger chooses a challenge bit* $b \xleftarrow{r} \{0, 1\}$. *Next, the challenger generates a key pair* $(pk, sk) \leftarrow \mathsf{KG}(1^\lambda)$ *and sends* pk *to* $\mathcal{A}$.
2. $\mathcal{A}$ *sends* ct *to the challenger. The challenger returns* $m \leftarrow \mathsf{Dec}(sk, ct)$ *to* $\mathcal{A}$. $\mathcal{A}$ *can make this query repeatedly polynomially many times.*
3. $\mathcal{A}$ *sends* $(m_0, m_1) \in \mathcal{M}^2$ *to the challenger. We require that* $|m_0| = |m_1|$. *The challenger computes* $ct^* \leftarrow \mathsf{Enc}(pk, m_b)$ *and sends* ct^* *to* $\mathcal{A}$.
4. $\mathcal{A}$ *sends* ct *to the challenger, where* $ct \neq ct^*$. *The challenger returns* $m \leftarrow \mathsf{Dec}(sk, ct)$ *to* $\mathcal{A}$. $\mathcal{A}$ *can make this query repeatedly polynomially many times.*
5. $\mathcal{A}$ *outputs* $b' \in \{0, 1\}$.

In this game, we define the advantage of the adversary $\mathcal{A}$ *as*

$$\mathsf{Adv}_{\Pi, \mathcal{A}}^{indcca}(\lambda) = \left| \Pr[b' = b] - \frac{1}{2} \right|.$$

We say that Π *is* IND-CCA *secure if for any PPT adversary* $\mathcal{A}$, *we have* $\mathsf{Adv}_{\Pi, \mathcal{A}}^{indcca}(\lambda) = \mathsf{negl}(\lambda)$.

2.3   Lossy Encryption

Lossy encryption [BHY09] is a variant of PKE with two modes. The first one is called the injective mode. In this mode, lossy encryption works in the same way as ordinary PKE. The other one is called the lossy mode. In this mode, a ciphertext output by the encryption algorithm has no information about the underlying message. We can switch these two modes by switching key generation algorithms. For this, a lossy encryption scheme has a lossy key generation algorithm in addition to the ordinary algorithms consisting of a PKE scheme. The formal definition is as follows.

Definition 3 (Lossy encryption). *A lossy encryption scheme* LPKE *is a four tuple* (Gen, LGen, LEnc, LDec) *of PPT algorithms. Below, let* $\mathcal{M}$ *be the message space of* LPKE.

- *The key generation algorithm for injective mode* Gen, *given a security parameter* 1^λ, *outputs a public key* pk_{inj} *and a secret key* sk_{inj}. *We call* pk_{inj} *injective key.*

- *The key generation algorithm for lossy mode* LGen, *given a security parameter* 1^λ, *outputs a public key* pk_{los}. *We call* pk_{los} *lossy key. Note that* LGen *does not output a secret key.*
- *The encryption algorithm* LEnc, *given a public key* (pk_{inj} *or* pk_{los}) *and message* $m \in \mathcal{M}$, *outputs a ciphertext* ct.
- *The decryption algorithm* LDec, *given a secret key* sk_{inj} *and ciphertext* ct, *outputs a message* $\widehat{m} \in \{\bot\} \cup \mathcal{M}$.

We require LPKE *satisfies the following properties:*

1. *Correctness under injective keys: We require* $\mathsf{LDec}(sk_{inj}, \mathsf{LEnc}(pk_{inj}, m)) = m$ *for every* $m \in \mathcal{M}$ *and* $(pk_{inj}, sk_{inj}) \leftarrow \mathsf{Gen}(1^\lambda)$.
2. *Key indistinguishability: For any PPT distinguisher* $\mathcal{A}$ *it holds that the advantage*

$$\mathsf{Adv}^{ind\text{-}keys}_{\mathsf{LPKE},\mathcal{A}}(\lambda) := \left| \begin{array}{l} \Pr[1 \leftarrow \mathcal{A}(pk_{inj}) \mid (pk_{inj}, sk_{inj}) \leftarrow \mathsf{Gen}(1^\lambda)] \\ - \Pr[1 \leftarrow \mathcal{A}(pk_{los}) \mid pk_{los} \leftarrow \mathsf{LGen}(1^\lambda)] \end{array} \right|$$

is negligible in λ.
3. *Lossiness under lossy keys: For any* $pk_{los} \leftarrow \mathsf{LGen}(1^\lambda)$ *and pair of message* $(m_0, m_1) \in \mathcal{M}^2$, *it holds that*

$$\mathsf{LEnc}(pk_{los}, m_0) \approx_s \mathsf{LEnc}(pk_{los}, m_1).$$

2.4   Secret Sharing

Before defining a secret sharing scheme, we define the notion of access structures. An access structure is a family of parties that can reconstruct a message.

Definition 4 (Access structure). *Let* $\{1, \ldots, n\}$ *be a set of parties. A collection* $\mathsf{AS} \subseteq 2^{\{1,\ldots,n\}}$ *is monotone if* $B \in \mathsf{AS}$ *and* $B \subseteq C$ *imply that* $C \in \mathsf{AS}$. *An access structure over* $\{1, \ldots, n\}$ *is a monotone collection* $\mathsf{AS} \subseteq 2^{\{1,\ldots,n\}}$.

Next, we define the secret sharing scheme in the public parameter model.

Definition 5 (Secret sharing scheme). *Let* $\{1, \ldots, n\}$ *be a set of parties and* AS *be an access structure on* $\{1, \ldots, n\}$. *A secret sharing scheme* Σ *in the public parameter model realizing* AS *is three tuple* (Setup, Share, Rec) *of PPT algorithms. Below, let* $\mathcal{M}$ *be the message space of* Σ.

- *The setup algorithm* Setup, *given a security parameter* 1^λ, *outputs a public parameter* pp.
- *The share algorithm* Share, *given a public parameter* pp *and a message* $m \in \mathcal{M}$, *outputs a set of shares* $\{(i, s_i)\}_{i \in [n]}$.
- *The reconstruction algorithm* Rec, *given a public parameter* pp *and a set of shares* $\{(i, s_i)\}_{i \in B}$, *outputs a message* $\widehat{m} \in \{\bot\} \cup \mathcal{M}$, *where* $B \in 2^{\{1,\ldots,n\}}$ *is a set of parties.*

Correctness: We require $\mathsf{Rec}(pp, \{(i, s_i)\}_{i \in T}) = m$ *for every* $n \in \mathbb{N}, m \in \mathcal{M}$, *and* $T \in \mathsf{AS}$, *where* $pp \leftarrow \mathsf{Setup}(1^\lambda)$ *and* $\{(i, s_i)\}_{i \in [n]} \leftarrow \mathsf{Share}(pp, m)$.

Definition 6 (Statistical privacy). *Let* AS *be an access structure on* $\{1, \ldots, n\}$, *and* $\Sigma = (\mathsf{Setup}, \mathsf{Share}, \mathsf{Rec})$ *be a secret sharing scheme for* AS. *We say that* Σ *satisfies the statistical privacy if for every* $n \in \mathbb{N}$, *any* $B \notin \mathsf{AS}$, *and any* $(m_0, m_1) \in \mathcal{M}^2$, *we have* $\{s_i\}_{i \in B} \approx_s \{s_i'\}_{i \in B}$, *where* $pp \leftarrow \mathsf{Setup}(1^\lambda), \{(i, s_i)\}_{i \in [n]} \leftarrow \mathsf{Share}(pp, m_0)$, *and* $\{(i, s_i')\}_{i \in [n]} \leftarrow \mathsf{Share}(pp, m_1)$.

3   IND-CCA Secure Lossy Encryption in the Injective Mode

In this section, we construct a lossy encryption scheme that satisfies IND-CCA security in the injective mode. We call such a lossy encryption CCA-LE. Our construction is a fairly simple combination of a lossy encryption scheme and an IND-CCA secure PKE scheme. It encrypts a message by using a lossy encryption scheme firstly and then encrypt the ciphertext with an IND-CCA secure scheme.

3.1   Construction

Let $\Pi = (\mathsf{KG}, \mathsf{Enc}, \mathsf{Dec})$ be an IND-CCA secure PKE scheme and $\Lambda = (\mathsf{G}, \mathsf{LG}, \mathsf{LE}, \mathsf{LD})$ be a lossy encryption scheme. A CCA-LE scheme $\mathsf{LPKE} = (\mathsf{Gen}, \mathsf{LGen}, \mathsf{LEnc}, \mathsf{LDec})$ is constructed as follows.

$\mathsf{Gen}(1^\lambda)$: Compute $(pk_{inj}, sk_{inj}) \leftarrow \mathsf{G}(1^\lambda)$ and $(pk, sk) \leftarrow \mathsf{KG}(1^\lambda)$, then output (pk^*, sk^*), where $pk^* := (pk_{inj}, pk)$ and $sk^* := (sk_{inj}, sk)$.

$\mathsf{LGen}(1^\lambda)$: Compute $pk_{los} \leftarrow \mathsf{LG}(1^\lambda)$ and $(pk, sk) \leftarrow \mathsf{KG}(1^\lambda)$, then output $pk^* := (pk_{los}, pk)$.

$\mathsf{LEnc}(pk^*, m)$: Let $pk^* = (pk_\Lambda, pk)$. Compute $c \leftarrow \mathsf{LE}(pk_\Lambda, m)$ and $ct \leftarrow \mathsf{Enc}(pk, c)$, then output ct.

$\mathsf{LDec}(sk^*, ct)$: Let $sk^* = (sk_{inj}, sk)$. Compute $\widehat{c} \leftarrow \mathsf{Dec}(sk, ct)$ and $\widehat{m} \leftarrow \mathsf{LD}(sk_{inj}, \widehat{c})$, then output $\widehat{m}$.

The correctness in the injective mode of LPKE follows from the correctness of Π and the correctness in the injective mode of Λ. In the following subsections, we show that LPKE satisfies key indistinguishability and lossiness under lossy keys. Moreover, we show that LPKE satisfies IND-CCA security in the injective mode.

3.2   Key Indistinguishability

Here, we prove that LPKE satisfies the key indistinguishability. Specifically, we have the following theorem.

Theorem 1. *If* Λ *is a lossy encryption scheme, then* LPKE *satisfies the key indistinguishability.*

Proof of Theorem 1. A public key of LPKE consists of two components, that are public keys of Λ and Π. We see that the second component is always an honestly generated public key of Π thus distributes identically regardless of the mode in which the key of Λ is generated. Then, the key indistinguishability of LPKE follows from that of Λ. $\square$ (**Theorem** 1)

3.3 Lossiness Under Lossy Key

Here, we prove that LPKE satisfies the lossiness under lossy keys. Specifically, we have the following theorem.

Theorem 2. *If Λ is a lossy encryption scheme, then* LPKE *satisfies* lossiness *under lossy keys.*

Proof of Theorem 2. For any $(m_0, m_1) \in \mathcal{M}^2$, let $pk_{los} \leftarrow \mathsf{LG}(1^\lambda)$, $(pk, sk) \leftarrow \mathsf{KG}(1^\lambda)$, $pk^* := (pk_{los}, pk)$, $c_0 := \mathsf{LE}(pk_{los}, m_0; r_\ell)$, and $c_1 := \mathsf{LE}(pk_{los}, m_1; r'_\ell)$, where r_ℓ and r'_ℓ are randomness of LE.

We have to show $\mathsf{LEnc}(pk^*, m_0) \approx_s \mathsf{LEnc}(pk^*, m_1)$. That is, we have to show $\mathsf{Enc}(pk, c_0; r) \approx_s \mathsf{Enc}(pk, c_1; r')$, where r and r' are randomness of Enc. Since pk, r and r' are independent of c_0 and c_1, and $c_0 \approx_s c_1$ follows from the lossiness of Λ, we can say that $\mathsf{Enc}(pk, c_0; r) \approx_s \mathsf{Enc}(pk, c_1; r')$.

This means that LPKE satisfies the lossiness property. $\square$ (**Theorem** 2)

3.4 IND-CCA Security

Here, we prove that LPKE satisfies IND-CCA security. Specifically, we prove the following theorem.

Theorem 3. *If Π satisfies IND-CCA security, then* LPKE *satisfies IND-CCA security.*

Proof of Theorem 3. Let $\mathcal{A}$ be a PPT adversary that attacks the IND-CCA security of LPKE. Using $\mathcal{A}$, we construct a PPT adversary $\mathcal{B}$ that attacks the IND-CCA security of Π.

1. $\mathcal{B}$ receives pk as an input from the challenger, computes $(pk_{inj}, sk_{inj}) \leftarrow \mathsf{G}(1^\lambda)$ and sets $pk^* := (pk_{inj}, pk)$. $\mathcal{B}$ sends pk^* to $\mathcal{A}$.
2. $\mathcal{B}$ receives a pair of message (m_0, m_1) from $\mathcal{A}$. $\mathcal{B}$ sets $M_0 := \mathsf{LE}(pk_{inj}, m_0), M_1 := \mathsf{LE}(pk_{inj}, m_1)$ and sends (M_0, M_1) to the challenger. $\mathcal{B}$ receives ct^* from the challenger, and sends ct^* to $\mathcal{A}$.
3. When $\mathcal{A}$ queries decryption of a ciphertext ct, $\mathcal{B}$ sends it to the challenger. $\mathcal{B}$ receives c from the challenger, then computes $m \leftarrow \mathsf{LD}(sk_{inj}, c)$ and sends m to $\mathcal{A}$.
4. $\mathcal{B}$ receives bit b' from $\mathcal{A}$, and sends b' to the challenger.

Since $\mathcal{B}$ perfectly simulates the IND-CCA game of LPKE for $\mathcal{A}$, we can estimate advantage of $\mathcal{B}$ as

$$\mathsf{Adv}_{\Pi,\mathcal{B}}^{indcca}(\lambda) = \left| \Pr[b' = b] - \frac{1}{2} \right| = \mathsf{Adv}_{\mathsf{LPKE},\mathcal{A}}^{indcca}(\lambda).$$

Given Π satisfies the IND-CCA security, $\mathsf{Adv}_{\mathsf{LPKE},\mathcal{A}}^{indcca}(\lambda) = \mathsf{negl}(\lambda)$ holds. Thus LPKE satisfies the IND-CCA security.   □ (**Theorem** 3)

In this way, IND-CCA secure lossy encryption in the injective mode can be constructed. In this paper, we use the constructed lossy encryption as a non-malleable commitment. In the next subsection, we show that the above CCA-LE is a non-malleable commitment.

3.5 IND-CCA Secure Lossy Encryption is a Non-malleable Commitment

Commitment Schemes. A commitment scheme is a three tuple $Com = (G, Com, Decom)$ of PPT algorithm. $G(1^\lambda)$ outputs a public parameter pp. The commit algorithm $Com(pp, m)$ computes commitment to m com and its decommitment dec. The decommit algorithm $Decom(pp, com, dec)$ outputs the committed message m. Com satisfies statistically-hidng if commitments of m_0 and m_1 are statistically indistinguishable. Computational binding property demands that no PPT algorithm can find dec' such that $Decom(pp, com, dec') = m' \neq m$.

Next, we review the non-malleability for commitment. The following definition subtly differs from the one in [CKOS01], still they are essentially the same.

Definition 7 (Non-malleable commitment). *Let Com be a statistical-hiding commitment scheme. Let D be a distribution on message space $\mathcal{M}$ and R be a relation that can be computed in polynomial time, where for any message $m \in \mathcal{M}$, R satisfies $R(m, \perp) = 0$. The following two experiments are defined for a PPT adversary $\mathcal{A}$ and a PPT simulator Sim.*

$\mathrm{Exp}_{Com,\mathcal{A}}^{real}(\lambda):$	$\mathrm{Exp}_{Com,\mathsf{Sim}}^{sim}(\lambda):$
$pp \leftarrow G(1^\lambda)$	
$m \leftarrow D$	$m \leftarrow D$
$(com, dec) \leftarrow Com(pp, m)$	
$com' \leftarrow \mathcal{A}_1(pp, com)$	
$dec' \leftarrow \mathcal{A}_2(pp, com, dec)$	
$\widetilde{m} \leftarrow Decom(pp, com', dec')$	$\widetilde{m} \leftarrow \mathsf{Sim}(1^\lambda)$
Outputs: $com \neq com' \wedge R(m, \widetilde{m})$	Outputs: $R(m, \widetilde{m})$

We say Com satisfies the non-malleability with respect to opening, if for any $D, \mathcal{A}$ there exists a simulator Sim such that for any relation R

$$\Pr[\mathrm{Exp}_{Com,\mathcal{A}}^{real}(\lambda) = 1] - \Pr[\mathrm{Exp}_{Com,\mathsf{Sim}}^{sim}(\lambda) = 1] \leq \mathsf{negl}(\lambda)$$

holds.

Next, we show IND-CCA secure loosy encryption CCA-LE is a non-malleable commitment scheme.

We use the lossy key pk_{los} as a public parameter pp. A commitment to m, com is a ciphertext $ct \leftarrow \mathsf{LEnc}(pk_{los}, m; r)$. The decommitment dec consists of $m\|r$. The Decommit algorithm $Decom$ on input $(pk_{los}, ct, m\|r)$ outputs m if $ct = \mathsf{LEnc}(pk_{los}, m; r)$ holds.

Statistical-hiding follows from the lossiness of CCA-LE. Computational binding follows from the key indistinguishability of CCA-LE because if an adversary success in breaking the binding property, the adversary could sure that the key is pk_{los}, not pk_{inj}.

Theorem 4. *IND-CCA secure lossy encryption satisfies the non-malleability with respect to opening as a commitment scheme.*

Proof of Theorem 4. We define the following experiments.

Exp 0: This experiment is identical to $\mathrm{Exp}_{Com,\mathcal{A}}^{real}(\lambda)$.

1. Compute $pp := pk_{los} \leftarrow \mathsf{LGen}(1^\lambda)$.
2. Sample $m \leftarrow D$.
3. Compute $com := ct \leftarrow \mathsf{LEnc}(pk_{los}, m; r)$.
4. Run $\widetilde{ct} \leftarrow \mathcal{A}_1(pk_{los}, ct)$.
5. Run $\widetilde{dec} := \widetilde{m}\|\widetilde{r} \leftarrow \mathcal{A}_2(pk_{los}, ct, m\|r)$.
6. Set $\widetilde{m} := \bot$ if $\widetilde{ct} \neq \mathsf{LEnc}(pk_{los}, \widetilde{m}; \widetilde{r})$.
7. Output $ct \neq \widetilde{ct} \wedge R(m, \widetilde{m})$.

Exp 1: This experiment is identical to Exp 0 except that we generate $(pk_{inj}, sk_{inj}) \leftarrow \mathsf{Gen}(1^\lambda)$ in step 1 and use pk_{inj} instead of pk_{los} through the experiment.

Exp 2: In this experiment, we set $\widetilde{m} := \mathsf{LDec}(sk, \widetilde{ct})$ instead of the decommitment output by $\mathcal{A}_2$.

Exp 3: In this experiment, the commitment is computed as $ct \leftarrow \mathsf{LEnc}(pk_{inj}, 0; r)$.

Now we have the simulator Sim as follows.

1. Compute $(pk_{inj}, sk_{inj}) \leftarrow \mathsf{Gen}(1^\lambda)$.
2. Compute $ct \leftarrow \mathsf{LEnc}(pk_{inj}, 0; r)$.
3. Run $\widetilde{ct} \leftarrow \mathcal{A}_1(pk_{inj}, ct)$.
4. Output $\widetilde{m} \leftarrow \mathsf{LDec}(sk, \widetilde{ct})$.

Let R_t be the event in which Exp t outputs 1. To complete the proof, we show

$$\Pr[\mathrm{Exp}_{Com,\mathcal{A}}^{real}(\lambda) = 1] - \Pr[\mathrm{Exp}_{Com,\mathsf{Sim}}^{sim}(\lambda) = 1] = \Pr[\mathsf{R}_0] - \Pr[\mathsf{R}_3] \leq \mathsf{negl}(\lambda).$$

This is shown from the following lemmas.

Lemma 1. $|\Pr[R_0] - \Pr[R_1]| = \mathsf{negl}(\lambda)$ *holds by the key indistinguishability of* CCA-LE.

This hold because $|\Pr[R_0] - \Pr[R_1]|$ is the advantage of distinguishing the keys.

Lemma 2. $\Pr[R_1] \leq \Pr[R_2]$ *holds.*

This holds because if the output is 1 in Exp 1, then so is in Exp 2. Specifically, if the output of Exp 1 is 1, then $ct \neq \widetilde{ct}$, $R(m, \widetilde{m}) = 1$ and $\widetilde{ct} = \mathsf{LEnc}(pk_{inj}, \widetilde{m}; \widetilde{r})$ holds. By the correctness of CCA-LE, the same $\widetilde{m}$ is decrypted from $\widetilde{ct}$ in the Exp 2. Thus, the same condition $ct \neq \widetilde{ct} \wedge R(m, \widetilde{m})$ holds also in Exp 2. So the output of Exp 2 is 1.

Lemma 3. $|\Pr[R_2] - \Pr[R_3]| = \mathsf{negl}(\lambda)$ *holds by the* IND-CCA *security of* CCA-LE.

We show a reduction $\mathcal{B}$ which using $D, \mathcal{A}, R$, attacks the IND-CCA security of CCA-LE.

On receiving pk_{inj}, $\mathcal{B}$ samples $m \leftarrow D$ and sends $(m_0, m_1) = (m, 0)$ to the challenger. The challenger chooses a challenge bit $b \leftarrow \{0, 1\}$ and sends $ct := \mathsf{LEnc}(pk_{inj}, m_b)$ to $\mathcal{B}$. $\mathcal{B}$ runs $\widetilde{ct} \leftarrow \mathcal{A}(pk_{inj}, ct)$. If $ct = \widetilde{ct}$ then $\mathcal{B}$ outputs $b' := 0$. Otherwise $\mathcal{B}$ queries $\widetilde{ct}$ to the challenger and receives $\widetilde{m}$. $\mathcal{B}$ outputs $b' := R(m, \widetilde{m})$.

4   Definition of Computational Non-malleability

In this section, we introduce the definition of computational non-malleability for secret sharing that we work with. The non-malleability for secret sharing introduced by Goyal and Kumar [GK18a, GK18b] was defined in a similar way as the non-malleable codes [DPW10]. In this work, we define non-malleability in a similar way as the non-malleability for non-malleable commitment defined by Crescenzo et al. [CKOS01].

Definition 8 (Computational non-malleability).   *Let* AS *be an access structure on* $\{1, \ldots, n\}$ *and* $\Sigma = (\mathsf{Setup}, \mathsf{Share}, \mathsf{Rec})$ *be a secret sharing scheme for* AS. *Below, let* $\mathcal{M}$ *be a message space of* Σ.

We say Σ *satisfies the computational non-malleability, if for any PPT adversary* $\mathcal{F}$, *a distribution* D *on* $\mathcal{M}$, *there exists a simulator* Sim *such that for any relation* R,

$$\Pr[\mathsf{Exp}_{\Sigma, \mathcal{F}}^{real}(\lambda) = 1] - \Pr[\mathsf{Exp}_{\Sigma, \mathsf{Sim}}^{sim}(\lambda) = 1] \leq \mathsf{negl}(\lambda)$$

holds, where the two experiments are defined as follows.

$$\begin{array}{l|l}
\mathrm{Exp}^{real}_{\Sigma, \mathcal{F}}(\lambda): & \mathrm{Exp}^{sim}_{\Sigma, \mathsf{Sim}}(\lambda): \\
pp \leftarrow \mathsf{Setup}(1^\lambda) & pp \leftarrow \mathsf{Setup}(1^\lambda) \\
m \leftarrow D(pp) & m \leftarrow D(pp) \\
\{(i, s_i)\}_{i \in [n]} \leftarrow \mathsf{Share}(pp, m) & \\
\{(i, \tilde{s}_i)\}_{i \in T} \leftarrow \mathcal{F}(pp, \{(i, s_i)\}_{i \in [n]}) & \\
\tilde{m} \leftarrow \mathsf{Rec}(pp, \{(i, \tilde{s}_i)\}_{i \in T}) & \tilde{m} \leftarrow \mathsf{Sim}(1^\lambda) \\
\text{Outputs: } R(pp, m, \tilde{m}) & \text{Outputs: } R(pp, m, \tilde{m})
\end{array}$$

In the above experiment, we do not allow $\mathcal{F}$ to perform an obvious attack.[3] Concretely, $\mathcal{F}$ executes tampering in the following steps.

1. Depending on the public parameter pp, $\mathcal{F}$ selects a set $T \in \mathsf{AS}$ and functions $\{f_j\}_{j \in T}$.
2. Each function $f_j(pp, \{(i, s_i)\}_{i \in I_j \notin \mathsf{AS}})$ outputs $(j, \tilde{s}_j)$.
3. $\mathcal{F}$ outputs $\{(i, \tilde{s}_i)\}_{i \in T}$.

We also require the relation R to satisfy

$$R(pp, m, m) = R(pp, m, \bot) = 0$$

for any message $m \in \mathcal{M}$ and public parameter $pp \leftarrow \mathsf{Setup}(1^\lambda)$.

We give some notes on the above definition.

- Non-malleability for secret sharing cannot prevent the tampering performed by an adversary that once reconstructs the original message. This attack is called an *obvious attack* and not considered in the definition by Goyal and Kumar [GK18a, GK18b]. Similarly, we require that $\mathcal{F}$ does not perform the obvious attacks. More specifically, we pose the restriction that $\mathcal{F}$ consists of multiple sub-routine functions $\{f_j\}_{j \in T}$ each of whose inputs are not sufficient to reconstruct the original message. Since we do not require any restriction on $\mathcal{F}$ except prohibiting the obvious attacks, the above definition considers a powerful adversary who performs the overlap-joint tampering. Especially, we allow I_{j_1} and I_{j_2} ($j_1, j_2 \in T$) to include the same indicies. Furthermore, we allow T and $\{I_j\}_{j \in T}$ to be determined after seeing the public parameter.
- The adversary receives shares of a message m and aims to generate shares that are reconstructed to some message related to m. In the definition of non-malleability using a simulator, the adversary is simulated by a simulator that does not have any information on the message m. We consider the difference of tampering success probability between the adversary and the simulator. Many existing definitions of non-malleability including Goyal and Kumar [GK18a, GK18b] require that the difference in success probability between the adversary and the simulator be negligible.

---

[3] Intuitively, the obvious attack is the tampering performed by the adversary that once reconstructs the original message.

The definition of non-malleability for commitment defined by Crescenzo et al. [CKOS01] allows the simulator to satisfy the relation R with a probability significantly better than the adversary. We adopted such a definition because of the observation that even if the simulator without information about the message satisfies the relation R better than the adversary, the definition still captures the intuition of non-malleability. Similarly to the definition by Crescenzo et al. [CKOS01], our definition allows the simulator to satisfy the relation R better than the adversary.

– In most usage scenarios of secret sharing, it seems that we do not have to care about the cases in which the original message is reconstructed as expected even if the shares are tampered with. Thus, the relation R is required to satisfy $R(pp, m, m) = 0$ for any message m to exclude such (harmless) attacks in the above definition.

– Our definition is weaker than the conventional one. Our non-malleability ensures that an adversary can not tamper shares meaningfully. In other words, if the result of reconstruction is $\perp$, we do not regard the tampering is successful. By contrast, in the conventional definition, tampering which results in $\perp$ may be regarded as a success. This is because such tampering can help the adversary to leak partial information on m. The conventional non-malleability regards such attacks as a success.

Therefore, there is a gap between the two definitions. We give a detailed proof of this gap in Sect. 6.

5   Computationally Non-malleable Secret Sharing

In this section, we give a construction of a secret sharing scheme satisfying the above computational non-malleability. We use the CCA-LE as a building block.

5.1   Construction

Let LPKE = (Gen, LGen, LEnc, LDec) be a CCA-LE scheme, AS be an access structure on $\{1, \ldots, n\}$, and Σ = (Setup, Share, Rec) be a secret sharing scheme realizing AS. Let the message space of LPKE be $\{0, 1\}^k$, the randomness space of LEnc be $\{0, 1\}^\lambda$, and the message space of Σ be $\{0, 1\}^{k+\lambda}$. From these, the secret sharing scheme Σ_{NM} = (NMSetup, NMShare, NMRec) that realizes the access structure AS is constructed as follows. The message space of Σ_{NM} is $\{0, 1\}^k$.

NMSetup(1^λ): Compute $pk^* \leftarrow$ LGen(1^λ) and $pp \leftarrow$ Setup(1^λ), then output $pp_{\text{nm}} := (pk^*, pp)$.

NMShare(pp_{nm}, m): Compute $ct \leftarrow$ LEnc($pk^*, m; r$) and $\{(i, s_i)\}_{i \in [n]} \leftarrow$ Share($pp, m \| r$), then output $\{(i, share_i)\}_{i \in [n]}$, where $share_i := (ct, s_i)$ for all $i \in [n]$.

NMRec($pp_{\text{nm}}, \{(i, share_i)\}_{i \in B \subset [n]}$) :
1. For all $i \in B$, parse $share_i$ as (ct_i, s_i).
2. Let $B = \{i_1, \ldots, i_{|B|}\}$. If there exists $j \in B$ such that $ct_{i_1} \neq ct_j$, then output $\perp$ and terminate. Otherwise execute step 3, where i_1 is the smallest index in B.

3. Compute $\widehat{m}\|\widehat{r} \leftarrow \mathsf{Rec}(pp, \{(i, s_i)\}_{i \in B})$.
4. Compute $\widehat{ct} \leftarrow \mathsf{LEnc}(pk^*, \widehat{m}; \widehat{r})$. If $\widehat{ct} \neq ct_{i_1}$, then output $\perp$ and terminate. Otherwise output $\widehat{m}$ and terminate.

Correctness. The correctness of Σ_{NM} follows from that of LPKE and Σ.

In the following subsections, we show that Σ_{NM} satisfies the statistical privacy and the computational non-malleability.

5.2 Statistical Privacy

In this section, we prove that Σ_{NM} satisfies the statistical privacy. Specifically, we prove the following theorem.

Theorem 5. *If Σ satisfies the statistical privacy and LPKE is a CCA-LE, then Σ_{NM} satisfies the statistical privacy.*

Proof of Theorem 5. In the construction of Σ_{NM}, a key of CCA-LE is always generated in the lossy mode. In this case, we see that Σ_{NM} satisfies statistical privacy based on the statistical privacy of Σ and the lossieness under lossy keys of CCA-LE. $\qquad\square$ **(Theorem 5)**

5.3 Computational Non-malleability

In this section, we prove that Σ_{NM} satisfies the computational non-malleability. Specifically, we prove the following theorem.

Theorem 6. *If Σ satisfies the statistical privacy and LPKE is a CCA-LE, then Σ_{NM} satisfies the computational non-malleability.*

Proof of Theorem 6. Let $\mathcal{F}$ be an adversary for the computational non-malleability of Σ_{NM}, which determines subset $T = \{i_1, \ldots, i_\ell\} \in \mathsf{AS}$ and its sub-routines $(f_1, \ldots, f_\ell)$. I_j is the indices of the domain of each sub-routine f_j. Let D be any distribution on $\{0, 1\}^k$ and R be any relation that can be computed in polynomial time, where for any message $m \in \{0, 1\}^k$ and public parameter $pp \leftarrow \mathsf{Setup}(1^\lambda)$, R satisfies $R(pp, m, m) = R(pp, m, \perp) = 0$.

We define the following sequence of experiments.

Exp 0: This experiment is identical to $\mathrm{Exp}^{real}_{\Sigma_{\mathsf{NM}}, \mathcal{F}}(\lambda)$.

1. Compute $pk^* \leftarrow \mathsf{LGen}(1^\lambda)$ and $pp \leftarrow \mathsf{Setup}(1^\lambda)$, and set $pp_{\mathsf{nm}} := (pk^*, pp)$.
2. Sample the message $m \leftarrow D(pp)$, then compute $ct \leftarrow \mathsf{LEnc}(pk^*, m; r)$ and $\{(i, s_i)\}_{i \in [n]} \leftarrow \mathsf{Share}(pp, m\|r)$. Set $\{(i, share_i)\}_{i \in [n]}$, where for all $i \in [n]$, $share_i := (ct, s_i)$.
3. Run $\{(i, \widetilde{share_i})\}_{i \in T} \leftarrow \mathcal{F}(pp_{\mathsf{nm}}, \{(i, share_i)\}_{i \in [n]})$. In more detail, for $j \in [\ell]$, run $(i_j, \widetilde{share_{i_j}}) \leftarrow f_j(pp_{\mathsf{nm}}, \{(i, share_i)\}_{i \in I_j})$.

4. Reconstruct $\widetilde{m} \leftarrow \mathsf{NMRec}(pp_{\mathrm{nm}}, \{(i, \widetilde{share}_i)\}_{i \in T})$. In more detail, perform the following.
   (a) Let $\widetilde{share}_i = (\widetilde{ct}_i, \widetilde{s}_i)$ for all $i \in T$.
   (b) Let i_1 be the smallest index in T. If there exists $t \in T$ such that $\widetilde{ct}_t \neq \widetilde{ct}_{i_1}$, then let $\widetilde{m} := \bot$ and jump to step 5.
   (c) Compute $\widehat{m} \| \widehat{r} \leftarrow \mathsf{Rec}(pp, \{(i, \widetilde{s}_i)\}_{i \in T})$.
   (d) If $\mathsf{LEnc}(pk^*, \widehat{m}; \widehat{r}) \neq \widetilde{ct}_{i_1}$, then set $\widetilde{m} := \bot$. Otherwise set $\widetilde{m} := \widehat{m}$.
5. Output $R(pp_{\mathrm{nm}}, m, \widetilde{m})$.

Exp 1: In this experiment, we change step 1 from Exp 0 in the following way. We compute $(pk^*, sk^*) \leftarrow \mathsf{Gen}(1^\lambda)$ instead of $pk^* \leftarrow \mathsf{LGen}(1^\lambda)$. In other words, LPKE is switched from the lossy mode to the injective mode.

Exp 2: In this experiment, only step 4 is different from Exp 1. We set $\widetilde{m}$ as $\widetilde{m} := \mathsf{LDec}(sk^*, \widetilde{ct}_{i_1})$ at step 4.

Exp 3: In this experiment, only step 3 is different from Exp 2. We run only f_1 and not $f_2, \ldots, f_\ell$ in step 3.

Exp 4: In this experiment, we change step 2 from Exp 3. We change the input of the Share algorithm in step 2, from $m\|r$ to $0^{k+\lambda}$. That is, we compute $\{(i, s_i)\}_{i \in [n]} \leftarrow \mathsf{Share}(pp, 0^{k+\lambda})$ instead of $\{(i, s_i)\}_{i \in [n]} \leftarrow \mathsf{Share}(pp, m\|r)$.

Exp 5: In this experiment, we change step 4 from Exp 4. We set $\widetilde{m} := \bot$ if $ct = \widetilde{ct}_{i_1}$ holds. Otherwise, we set the value of $\widetilde{m}$ in the same way as Exp 4 (That is, we set $\widetilde{m}$ as the decryption result of $\widetilde{ct}_{i_1}$).

Exp 6: In this experiment, we change step 2 from Exp 5. We change the input of LEnc from m to 0^k. That is, we compute $ct \leftarrow \mathsf{LEnc}(pk^*, 0^k; r)$ instead of $ct \leftarrow \mathsf{LEnc}(pk^*, m; r)$.

Exp 6 is identical to the experiment $\mathsf{Exp}_{\Sigma_{\mathrm{NM}}, \mathsf{Sim}}^{sim}(\lambda)$ for Sim described below, except for a conceptual change.

$\mathsf{Sim}(1^\lambda)$:
1. Sim computes $(pk^*, sk^*) \leftarrow \mathsf{Gen}(1^\lambda)$ and $pp \leftarrow \mathsf{Setup}(1^\lambda)$, then sets $pp_{\mathrm{nm}} := (pk^*, pp)$.
2. Sim computes $ct \leftarrow \mathsf{LEnc}(pk^*, 0^k; r)$ and $\{(i, s_i)\}_{i \in [n]} \leftarrow \mathsf{Share}(pp, 0^{k+\lambda})$, then sets $\{(i, share_i)\}_{i \in [n]}$, where $share_i := (ct, s_i)$ for all $i \in [n]$.
3. Sim computes $(i_1, \widetilde{share}_{i_1}) \leftarrow f_1(pp_{\mathrm{nm}}, \{(i, share_i)\}_{i \in I_l})$. Below, we let $\widetilde{share}_{i_1} = (\widetilde{ct}_{i_1}, \widetilde{s}_{i_1})$.
4. If $ct = \widetilde{ct}_{i_1}$, then Sim sets $\widetilde{m} := \bot$. Otherwise, Sim sets $\widetilde{m} \leftarrow \mathsf{LDec}(sk^*, \widetilde{ct}_{i_1})$.
5. Sim outputs $\widetilde{m}$.

For every $t \in \{0, \ldots, 6\}$, we define R_t as the event that the output of Exp t is 1, that is, $R(pp_{\mathsf{nm}}, m, \tilde{m}) = 1$ holds.

Using the above events, we can estimate

$$\Pr[\mathrm{Exp}^{real}_{\Sigma_{\mathsf{NM}}, \mathcal{F}}(\lambda) = 1] - \Pr[\mathrm{Exp}^{sim}_{\Sigma_{\mathsf{NM}}, \mathsf{Sim}}(\lambda) = 1]$$

as

$$\Pr[\mathsf{R}_0] - \Pr[\mathsf{R}_6] = \sum_{t=0}^{5} \Pr[\mathsf{R}_t] - \Pr[\mathsf{R}_{t+1}].$$

In the following, we estimate $\Pr[\mathsf{R}_t] - \Pr[\mathsf{R}_{t+1}]$ for every $t \in \{0, \ldots, 5\}$.

Lemma 4. $\Pr[\mathsf{R}_0] - \Pr[\mathsf{R}_1] = \mathsf{negl}(\lambda)$ *holds by the key indistinguishability of* LPKE.

Proof of Lemma 4. Using the adversary $\mathcal{F}$, we construct an PPT adversary $\mathcal{B}$ that attacks the key indistinguishability of LPKE.

1. Given pk^* as an input, $\mathcal{B}$ computes $pp \leftarrow \mathsf{Setup}(1^\lambda)$. $\mathcal{B}$ then sets $pp_{\mathsf{nm}} := (pk^*, pp)$ and samples a message $m \leftarrow D(pp_{\mathsf{nm}})$. Then $\mathcal{B}$ computes $ct \leftarrow \mathsf{LEnc}(pk^*, m; r)$ and $\{(i, s_i)\}_{i \in [n]} \leftarrow \mathsf{Share}(pp, m \| r)$, and sets $share_i := (ct, s_i)$ for $i \in [n]$. $\mathcal{B}$ sends $\undertilde{pp_{\mathsf{nm}}}$ and $\{(i, share_i)\}_{i \in [n]}$ to $\mathcal{F}$.
2. $\mathcal{B}$ receives $\{(i, \widetilde{share_i})\}_{i \in T}$ from $\mathcal{F}$. $\mathcal{B}$ computes $\tilde{m} = \mathsf{NMRec}$ $(pp, \{\widetilde{share_i}\}_{i \in T})$. $\mathcal{B}$ sends $b' := 0$ to the challenger if $R(pp_{\mathsf{nm}}, m, \tilde{m}) = 0$ and $\mathcal{B}$ sets $b' := 1$ if $R(pp_{\mathsf{nm}}, m, \tilde{m}) = 1$.

We can estimate the advantage of $\mathcal{B}$ as $\mathsf{Adv}^{ind\text{-}keys}_{\mathsf{LPKE}, \mathcal{B}}(\lambda) = \frac{1}{2} |\Pr[\mathsf{R}_0] - \Pr[\mathsf{R}_1]|$. Since LPKE satisfies the key indistinguishability, we have $\Pr[\mathsf{R}_0] - \Pr[\mathsf{R}_1] = \mathsf{negl}(\lambda)$. $\qquad\square$ (**Lemma 4**)

Lemma 5. $\Pr[\mathsf{R}_1] \leq \Pr[\mathsf{R}_2]$ *holds.*

Proof of Lemma 5. We define the following events X_t for $t \in \{1, 2\}$.

X_t: In Exp t, $R(pp_{\mathsf{nm}}, m, \mathsf{LDec}(sk^*, \widetilde{ct}_{i_1})) = 1$ holds, where m is the message sampled in step 2, sk^* is the secret key corresponding to pk^*.

Consider the relation between events X_t and R_t. Since in Exp 1, event X_1 always occurs when event R_1 occurs, $\Pr[\mathsf{R}_1] \leq \Pr[\mathsf{X}_1]$ holds. Furthermore, the difference between Exp 1 and Exp 2 does not affect the probability that X_t occurs. Thus, we have $\Pr[\mathsf{X}_1] = \Pr[\mathsf{X}_2]$. Finally, in Exp 2, since we determine the value of $\tilde{m}$ as the message obtained by decrypting $\widetilde{ct}_{i_1}$, $\Pr[\mathsf{X}_2] = \Pr[\mathsf{R}_2]$ holds.

From the above, we have $\Pr[\mathsf{R}_1] \leq \Pr[\mathsf{X}_1] = \Pr[\mathsf{X}_2] = \Pr[\mathsf{R}_2]$. $\square$ (**Lemma 5**)

Lemma 6. $\Pr[\mathsf{R}_2] = \Pr[\mathsf{R}_3]$ *holds.*

Proof of Lemma 6. By the change from Exp 1 to Exp 2, the output distribution of Exp 2 is no longer affected by the outputs of $f_2, \ldots, f_\ell$. Thus, the change from Exp 2 to Exp 3 is only conceptual, and we have $\Pr[\mathsf{R}_2] = \Pr[\mathsf{R}_3]$. $\square$ (**Lemma 6**)

Lemma 7. $\Pr[\mathsf{R}_3] - \Pr[\mathsf{R}_4] = \mathsf{negl}(\lambda)$ *holds by the statistical privacy of* Σ.

Proof of Lemma 7. In Exp 3 and Exp 4, we run only f_1 and not $f_2, \ldots, f_\ell$. In other words, we can simulate these experiments with shares $\{(i, share_i)\}_{i \in I_1}$ for an unauthorized set I_1. Therefore, we can rely on the statistical privacy of Σ in this step, and we obtain $\Pr[\mathsf{R}_3] - \Pr[\mathsf{R}_4] = \mathsf{negl}(\lambda)$. $\qquad\square$ (**Lemma 7**)

Lemma 8. $\Pr[\mathsf{R}_4] = \Pr[\mathsf{R}_5]$ *holds.*

Proof of Lemma 8. In Exp 4 and Exp 5, if $ct = \widetilde{ct}_{i_1}$ holds, then $R(pp_{\mathsf{nm}}, m, \widetilde{m}) = 0$ holds. This is because Π has the (perfect) correctness and R has the condition that $R(pp_{\mathsf{nm}}, m', m') = 0$. Therefore, since we also have $R(pp_{\mathsf{nm}}, m', \perp) = 0$ for any m', the transition from Exp 4 to Exp 5 does not affect the output distribution of experiments. Thus, we have $\Pr[\mathsf{R}_4] = \Pr[\mathsf{R}_5]$. $\qquad\square$ (**Lemma 8**)

Lemma 9. $\Pr[\mathsf{R}_5] - \Pr[\mathsf{R}_6] = \mathsf{negl}(\lambda)$ *holds by the* IND-CCA *security of* Π.

Proof of Lemma 9. Using the adversary f_1, we construct an PPT adversary $\mathcal{B}$ that attacks the IND-CCA security of Π.

1. Given pk^* as an input, $\mathcal{B}$ computes $pp \leftarrow \mathsf{Setup}(1^\lambda)$ and sets $pp_{\mathsf{nm}} := (pk^*, pp)$.
2. $\mathcal{B}$ samples message $m \leftarrow D(pp_{\mathsf{nm}})$, sends $(M_0 := m, M_1 := 0^k)$ to the challenger, and receives ct^*. $\mathcal{B}$ also computes $\{(i, s_i)\}_{i \in [n]} \leftarrow \mathsf{Share}(pp, 0^{k+\lambda})$ and sets $share_i := (ct^*, s_i)$ for $i \in I_1$. Moreover, $\mathcal{B}$ sends pp_{nm} and $\{(i, share_i)\}_{i \in I_1}$ to f_1 and obtains $(i_1, \widetilde{share}_{i_1})$.
3. Letting $\widetilde{share}_{i_1} = (\widetilde{ct}_{i_1}, \widetilde{s}_{i_1})$, $\mathcal{B}$ sets $\widetilde{m}$ as follows. If $ct^* = \widetilde{ct}_{i_1}$, $\mathcal{B}$ sets $\widetilde{m} := \perp$. Otherwise, $\mathcal{B}$ sends $\widetilde{ct}_{i_1}$ as an decryption query and obtain the decryption result $\widetilde{m}$.
4. $\mathcal{B}$ sends $\beta' := R(pp_{\mathsf{nm}}, m, \widetilde{m})$ to the challenger.

Let β be the challenge bit of IND-CCA game for Π played by $\mathcal{B}$. Then, we have

$$\mathsf{Adv}_{\Pi, \mathcal{B}}^{indcca}(\lambda) = \frac{1}{2} | \Pr[\beta' = 1 | \beta = 0] - \Pr[\beta' = 1 | \beta = 1]|. \qquad (1)$$

In this game, $\mathcal{B}$ simulates Exp 5 for f_1 when $\beta = 0$, and B simulates Exp 6 for f_1 when $\beta = 1$. Moreover, $\beta' = 1$ occurs if and only if $R(pp_{\mathsf{nm}}, m, \widetilde{m}) = 1$ holds. Therefore, the Eq. 1 is described as

$$\mathsf{Adv}_{\Pi, \mathcal{B}}^{indcca}(\lambda) = \frac{1}{2} | \Pr[\mathsf{R}_5] - \Pr[\mathsf{R}_6]|.$$

Since Π satisfies the IND-CCA security, $\Pr[\mathsf{R}_5] - \Pr[\mathsf{R}_6] = \mathsf{negl}(\lambda)$ holds. $\qquad\square$ (**Lemma 9**)

From Lemma 4 to Lemma 9, we have

$$\Pr[\mathsf{Exp}_{\Sigma_{\mathsf{NM}}, \mathcal{F}}^{real}(\lambda) = 1] - \Pr[\mathsf{Exp}_{\Sigma_{\mathsf{NM}}, \mathsf{Sim}}^{sim}(\lambda) = 1] \leq \mathsf{negl}(\lambda).$$

From these arguments, we can conclude that Σ_{NM} satisfies the computational non-malleability. $\qquad\square$ (**Theorem 6**)

6   Gap with Conventional Definition

In this section, we describe the gap between our non-malleability and the conventional non-malleability proposed by Goyal and Kumar [GK18a, GK18b]. Let us consider a situation where the result of reconstruction is $\perp$ in the real experiment, but the simulator outputs $\widetilde{m}$. In this situation, success probability in the real experiment is smaller than the success probability of the simulator because the relation R satisfies $R(pp, m, \perp) = 0$ in our definition. On the other hand, in the conventional definition, this difference is distinguished. Thus, there is a gap between the two definitions.

First, we extend the conventional definition to the definition in the public parameter model. Next, we show that our non-malleability is weaker than the conventional non-malleability.

Definition 9. *Let* AS *be an access structure and* $\mathcal{F}$ *be a family of tampering functions. The following two experiments are defined for* $f \in \mathcal{F}$, *distribution* D, *distinguisher* Dis, *authorized set* $T \in$ AS *and simulator* Sim.

$$
\begin{array}{l|l}
\text{STamper}^{f,T,D}(\lambda): & \text{SSim}^{f,T,D}(\lambda): \\
pp \leftarrow \text{Setup}(1^\lambda) & pp \leftarrow \text{Setup}(1^\lambda) \\
m \leftarrow D(pp) & m \leftarrow D(pp) \\
\{s_i\}_{i\in[n]} \leftarrow \text{Share}(pp, m) & \\
\{\tilde{s}_i\}_{i\in[n]} \leftarrow f(pp, \{s_i\}_{i\in[n]}) & \\
\widetilde{m} \leftarrow \text{Rec}(pp, \{\tilde{s}_i\}_{i\in T}) & \widetilde{m} \leftarrow \text{Sim}(1^\lambda) \\
\text{Output: Dis}(pp, m, \widetilde{m}) & \text{Output: Dis}(pp, m, m) \; (if \; \widetilde{m} = same^*) \\
& \text{Dis}(pp, m, \widetilde{m}) \; (otherwise)
\end{array}
$$

A secret sharing scheme Σ *is non-malleable if for all* $f \in \mathcal{F}$, *and* $T \in$ AS, *there exists a simulator* Sim *such that for any distinguisher* Dis *and distribution* D,

$$
\left| \Pr[\text{STamper}^{f,T,D}(\lambda) = 1] - \Pr[\text{SSim}^{f,T,D}(\lambda) = 1] \right| = \text{negl}(\lambda)
$$

holds.

When simulating a tampering function f that makes the shares reconstructed to the original message, Sim outputs the special symbol $same^*$. We show that the conventional definition is truly stronger than our definition by showing the following two theorems. First, we show that if a scheme satisfies the conventional definition, then it also satisfies our definition.

Theorem 7. *If an NMSS scheme* $\Sigma = (\text{Setup}, \text{Share}, \text{Rec})$ *satisfies Definition 9, Σ satisfies Definition 8.*

Proof of Theorem 7. We prove that if a simulator exists which satisfies the Definition 9, then we can use it to construct a simulator that meets our definition.

Using the simulator Sim for f, T, we can construct the simulator $\mathcal{S}$ and using the relation R, we can construct a distinguisher Dis as follows.

$\mathcal{S}(1^\lambda):$	$\mathtt{Dis}(pp, m, \widetilde{m}):$
$\widetilde{m} \leftarrow \mathsf{Sim}(1^\lambda)$	Output: $R(pp, m, m)$ $(\widetilde{m} = same^*)$
Output: $\widetilde{m}$	$R(pp, m, \widetilde{m})$ (otherwise)

Since Sim exists for any f and T, and

$$\left| \Pr[\mathrm{STamper}^{f,T,D}(\lambda) = 1] - \Pr[\mathrm{SSim}^{f,T,D}(\lambda) = 1] \right| = \mathsf{negl}(\lambda)$$

holds for any Dis and D, thus the above equation holds even if Dis is constructed as above.

Also, at this time, the following holds.

$$\Pr[\mathrm{Exp}_{\Sigma,\mathcal{F}}^{real}(\lambda) = 1] = \Pr[\mathrm{STamper}^{f,T,D}(\lambda) = 1]$$
$$\Pr[\mathrm{Exp}_{\Sigma,\mathsf{Sim}}^{sim}(\lambda) = 1] = \Pr[\mathrm{SSim}^{f,T,D}(\lambda) = 1]$$

From the above and the construction of Dis,

$$\Pr[\mathrm{Exp}_{\Sigma,\mathcal{F}}^{real}(\lambda) = 1] - \Pr[\mathrm{Exp}_{\Sigma,\mathsf{Sim}}^{sim}(\lambda) = 1]$$
$$= \Pr[\mathrm{STamper}^{f,T,D}(\lambda) = 1] - \Pr[\mathrm{SSim}^{f,T,D}(\lambda) = 1]$$
$$= \mathsf{negl}(\lambda)$$

holds for arbitrary R and D. Therefore, we can say that the PPT algorithm $\mathcal{S}$ exists for any f' and T', and $\Pr[\mathrm{Exp}_{\Sigma,\mathcal{F}}^{real}(\lambda) = 1] - \Pr[\mathrm{Exp}_{\Sigma,\mathsf{Sim}}^{sim}(\lambda) = 1] \leq \mathsf{negl}(\lambda)$ holds for any R and D'.   □ (**Theorem 7**)

Next, we prove that even if a scheme satisfies our definition, it does not necessarily satisfy the conventional definition.

Theorem 8. *Even if an NMSS scheme* $\Sigma = (\mathsf{Setup}, \mathsf{Share}, \mathsf{Rec})$ *satisfies Definition 8, it does not necessarily satisfy Definition 9.*

Proof of Theorem 8. We construct $f, T, \mathtt{Dis}$ and D and show that

$$\left| \Pr[\mathrm{STamper}^{f,T,D}(\lambda) = 1] - \Pr[\mathrm{SSim}^{f,T,D}(\lambda) = 1] \right| \neq \mathsf{negl}(\lambda)$$

Let $\sigma = (\mathsf{setup}, \mathsf{share}, \mathsf{rec})$ be a secret sharing scheme that has homomorphism in the OR operation. We can compute the share of $m \vee m'$ using the share s of the message m and the share s' of m'. We express this calculation as $s \vee s'$. Let LPKE $= (\mathsf{Gen}, \mathsf{LGen}, \mathsf{LEnc}, \mathsf{LDec})$ be a CCA-LE scheme, $\Sigma = (\mathsf{Setup}, \mathsf{Share}, \mathsf{Rec})$ be the NMSS scheme obtained by applying our compiler to σ and LPKE. Let $\mathcal{M}$ be a message space of share and $\mathcal{R}$ a random space of LEnc.

We construct D, f, T and Dis as follows.

$$D(pp):\\ \text{Output: } m_0 = 0^{|\mathcal{M}|}$$

$$f(pp, \{s_i\}_{i\in[n]}):\\ M' = 10^{|\mathcal{M}|+|\mathcal{R}|-1}\\ \{s'_i\}_{i\in[n]} \leftarrow \mathsf{share}(pp, M')\\ \widetilde{s}_i := s_i \vee s'_i (i \in [n])\\ \text{Output: } \{\widetilde{s}_i\}_{i\in T}$$

$$T:\\ \textit{one of authorized sets}$$

$\mathtt{Dis}(pp, m, \widetilde{m}):$
if((the first bit of $\widetilde{m}$ = the first bit of m) $\wedge$ (the others of $\widetilde{m}$ = the others of m))
   Output: 0
else
   Output: 1

We define the event that the simulator outputs $same^*$ in SSim as Same. Thus, we can estimate about $\left|\Pr[\mathrm{STamper}^{f,T,D}(\lambda) = 1] - \Pr[\mathrm{SSim}^{f,T,D}(\lambda) = 1]\right|$ as follows.

$$\left|\Pr[\mathrm{STamper}^{f,T,D}(\lambda) = 1] - \Pr[\mathrm{SSim}^{f,T,D}(\lambda) = 1]\right|$$
$$= \left|0 - \left(\Pr[\mathrm{SSim}^{f,T,D}(\lambda) = 1 \wedge \mathsf{Same}] + \Pr[\mathrm{SSim}^{f,T,D}(\lambda) = 1 \wedge \neg\mathsf{Same}]\right)\right|$$
$$= \left|\Pr[\mathsf{Same}] \Pr[\mathrm{SSim}^{f,T,D}(\lambda) = 1|\mathsf{Same}] + \Pr[\neg\mathsf{Same}] \Pr[\mathrm{SSim}^{f,T,D}(\lambda) = 1|\neg\mathsf{Same}]\right|$$
$$= \left|\Pr[\mathsf{Same}] + \Pr[\neg\mathsf{Same}] \left(1 - \frac{1}{|\mathcal{M}|}\right)\right|$$
$$= \left|1 - \frac{1}{|\mathcal{M}|} \Pr[\neg\mathsf{Same}]\right|$$
$$\neq \mathsf{negl}(\lambda)$$

☐ (**Theorem** 8)

From Theorem 7 and Theorem 8, it is clear that our definition is truly weaker than the conventional definition.

7 Conclusion

In this work, we proposed a new definition of non-malleability for secret sharing in the public parameter model. Although our definition is relaxed compared to the one proposed by Brian et al. [BFV19], it captures the strong security notion called non-malleability against overlap-joint tampering.

Using a CCA secure lossy encryption scheme, we showed a simple and efficient transformation that makes any secret sharing scheme to the one satisfying non-malleability without sacrificing statistical privacy. One interesting open question is to propose a similar transformation in the plain model (i.e. model where there is no public parameter).

Acknowledgments. A part of this work was supported by NTT Secure Platform Laboratories, JST OPERA JPMJOP1612, JST CREST JPMJCR14D6, JSPS KAKENHI JP16H01705, JP17H01695, JP19J22363.

References

[ADN+19] Aggarwal, D., et al.: Stronger leakage-resilient and non-malleable secret sharing schemes for general access structures. In: Boldyreva, A., Micciancio, D. (eds.) CRYPTO 2019. LNCS, vol. 11693, pp. 510–539. Springer, Cham (2019). https://doi.org/10.1007/978-3-030-26951-7_18

[BFO+20] Brian, G., Faonio, A., Obremski, M., Simkin, M., Venturi, D.: Non-malleable secret sharing against bounded joint-tampering attacks in the plain model. In: Micciancio, D., Ristenpart, T. (eds.) CRYPTO 2020. LNCS, vol. 12172, pp. 127–155. Springer, Cham (2020). https://doi.org/10.1007/978-3-030-56877-1_5

[BFV19] Brian, G., Faonio, A., Venturi, D.: Continuously non-malleable secret sharing for general access structures. In: Hofheinz, D., Rosen, A. (eds.) TCC 2019. LNCS, vol. 11892, pp. 211–232. Springer, Cham (2019). https://doi.org/10.1007/978-3-030-36033-7_8

[BGW88] Ben-Or, M., Goldwasser, S., Wigderson, A.: Completeness theorems for non-cryptographic fault-tolerant distributed computation (extended abstract). In: 20th Annual ACM Symposium on Theory of Computing, Chicago, IL, USA, 2–4 May, pp. 1–10. ACM Press (1988)

[BHY09] Bellare, M., Hofheinz, D., Yilek, S.: Possibility and impossibility results for encryption and commitment secure under selective opening. In: Joux, A. (ed.) EUROCRYPT 2009. LNCS, vol. 5479, pp. 1–35. Springer, Heidelberg (2009). https://doi.org/10.1007/978-3-642-01001-9_1

[Bla79] Blakley, G.R.: Safeguarding cryptographic keys. In: Proceedings of AFIPS 1979 National Computer Conference, vol. 48, pp. 313–317 (1979)

[BS19] Badrinarayanan, S., Srinivasan, A.: Revisiting non-malleable secret sharing. In: Ishai, Y., Rijmen, V. (eds.) EUROCRYPT 2019. LNCS, vol. 11476, pp. 593–622. Springer, Cham (2019). https://doi.org/10.1007/978-3-030-17653-2_20

[CCD88] Chaum, D., Crépeau, C., Damgård, I.: Multiparty unconditionally secure protocols (extended abstract). In: 20th Annual ACM Symposium on Theory of Computing, Chicago, IL, USA, 2–4 May, pp. 11–19. ACM Press (1988)

[CKOS01] Di Crescenzo, G., Katz, J., Ostrovsky, R., Smith, A.: Efficient and non-interactive non-malleable commitment. In: Pfitzmann, B. (ed.) EUROCRYPT 2001. LNCS, vol. 2045, pp. 40–59. Springer, Heidelberg (2001). https://doi.org/10.1007/3-540-44987-6_4

[CL18] Chattopadhyay, E., Li, X.: Non-malleable codes, extractors and secret sharing for interleaved tampering and composition of tampering. Cryptology ePrint Archive, Report 2018/1069 (2018)

[DPW10] Dziembowski, S., Pietrzak, K., Wichs, D.: Non-malleable codes. In: Yao, A.C.-C. (ed.) ICS 2010: 1st Innovations in Computer Science, Tsinghua University, Beijing, China, 5–7 January, pp. 434–452. Tsinghua University Press (2010)

[FV19] Faonio, A., Venturi, D.: Non-malleable secret sharing in the computational setting: adaptive tampering, noisy-leakage resilience, and improved rate. In: Boldyreva, A., Micciancio, D. (eds.) CRYPTO 2019. LNCS, vol. 11693, pp. 448–479. Springer, Cham (2019). https://doi.org/10.1007/978-3-030-26951-7_16

[GK18a] Goyal, V., Kumar, A.: Non-malleable secret sharing. In: Diakonikolas, I., Kempe, D., Henzinger, M. (eds.) 50th Annual ACM Symposium on Theory of Computing, Los Angeles, CA, USA, 25–29 June, pp. 685–698. ACM Press (2018)

[GK18b] Goyal, V., Kumar, A.: Non-malleable secret sharing for general access structures. In: Shacham, H., Boldyreva, A. (eds.) CRYPTO 2018. LNCS, vol. 10991, pp. 501–530. Springer, Cham (2018). https://doi.org/10.1007/978-3-319-96884-1_17

[GMW87] Goldreich, O., Micali, S., Wigderson, A.: How to play any mental game or A completeness theorem for protocols with honest majority. In: Aho, A. (ed.) 19th Annual ACM Symposium on Theory of Computing, New York City, NY, USA, 25–27 May, pp. 218–229. ACM Press (1987)

[KMS18] Kumar, A., Meka, R., Sahai, A.: Leakage-resilient secret sharing. In: Electronic Colloquium on Computational Complexity (ECCC), vol. 25, p. 200 (2018)

[LCG+19] Lin, F., Cheraghchi, M., Guruswami, V., Safavi-Naini, R., Wang, H.: Non-malleable secret sharing against affine tampering. CoRR, abs/1902.06195 (2019)

[Sha79] Shamir, A.: How to share a secret. Commun. ACM **22**(11), 612–613 (1979)

[SV19] Srinivasan, A., Vasudevan, P.N.: Leakage resilient secret sharing and applications. In: Boldyreva, A., Micciancio, D. (eds.) CRYPTO 2019. LNCS, vol. 11693, pp. 480–509. Springer, Cham (2019). https://doi.org/10.1007/978-3-030-26951-7_17

Cryptography in Quantum Computer Age

(Quantum) Cryptanalysis of Misty Schemes

Aline Gouget[1], Jacques Patarin[2], and Ambre Toulemonde[1,2(✉)]

[1] Thales DIS, Meudon, France
{aline.gouget,ambre.toulemonde}@thalesgroup.com
[2] Université de Versailles Saint-Quentin-en-Yvelines, Versailles, France
jpatarin@club-internet.fr

Abstract. In this paper, we review the best known cryptanalysis results on the variants of Misty schemes and we provide new (quantum) cryptanalysis results. First, we describe a non-adaptive quantum chosen plaintext attack (QCPA) against 4-round Misty L and Misty LKF schemes, and a QCPA against 3-round Misty R and Misty RKF schemes. We extend the QCPA attack against 3-round Misty RKF schemes to recover the keys of d-round Misty RKF schemes with complexity $\tilde{\mathcal{O}}(2^{(d-3)n/2})$. We then provide a security proof for Misty R schemes with 3 rounds against chosen plaintext attacks using the *H coefficients technique*. This shows that the best known non-quantum attack against Misty R schemes with 3 rounds is optimal.

Keywords: Misty permutations · Pseudo-random permutation · Cryptanalysis · Quantum cryptanalysis · H coefficients

1 Introduction

The most studied way to build pseudo-random permutations from random function or random permutation is the d-round Feistel construction. However, there exist other well-known constructions such as the Misty constructions that we analyze in this paper. We study generic attacks on Misty schemes where we assume that the internal permutations $f_1, \ldots, f_d$ are randomly chosen. The Misty construction is important from a practical point of view since it has been used as a generic construction to design Kasumi [2] algorithm that has been adopted as the standard blockcipher in the third generation mobile systems.

The plaintext message of a Misty scheme is denoted by $[L, R]$ that stands for *Left* and *Right* and the ciphertext message, after applying d rounds, is denoted by $[S, T]$. Misty L and Misty R schemes are two different variants of Misty schemes. Indeed, the first round of a Misty L scheme takes as input $[L, R]$ and it outputs $[R, R \oplus f_1(L)]$ with f_1 a secret permutation from n bits to n bits whereas the first round of a Misty R scheme takes as input $[L, R]$ and it outputs $[R \oplus f_1(L), f_1(L)]$ with f_1 a secret permutation from n bits to n bits. We also consider in this paper a particular case of Misty L and Misty R constructions such that each round

© Springer Nature Switzerland AG 2021
D. Hong (Ed.): ICISC 2020, LNCS 12593, pp. 43–57, 2021.
https://doi.org/10.1007/978-3-030-68890-5_3

function f_i is defined by $f_i(x) = F_i(K_i \oplus x)$ with a public function F_i and a round secret key K_i. These constructions are named, respectively, *d-round Misty LKF scheme* and *d-round Misty RKF scheme*. To simplify the notation, the public functions F_i in each round are all denoted by F. These four variants of Misty schemes are studied in this paper.

Related Work. Cryptanalysis of Misty schemes have been studied by Nachef, Patarin and Treger in [9,10]. They described Known Plaintext Attack (KPA), Chosen Plaintext Attack (CPA) and Chosen Ciphertext Attack (CCA) against Misty L and Misty R schemes. In particular, they showed that there exists CPA and KPA attacks for $d = 5$ with complexity strictly less than 2^{2n}. They also studied some generic properties of Misty L and Misty R schemes such as the *inversion* property. They showed that the inverse of a Misty L function is a Misty R function, after composition by a permutation μ and μ^{-1} on the inputs and outputs, where μ is a permutation on $2n$ bits such that $\mu([L, R]) = [R, L \oplus R]$. They then showed that the security of Misty L and Misty R schemes are the same for all attacks where the inputs and outputs have the same possibilities which is the case for example in KPA attack and CCA attack. However, the security of Misty L and Misty R schemes may differ regarding CPA attacks as we will see in this paper for 3 rounds.

Quantum cryptanalysis has received much more attention in the last past years. It is known that Grover's algorithm [3] could speed up brute force search. Given a n-bit key, Grover's algorithm allows to recover the key using $\mathcal{O}(2^{n/2})$ quantum steps. It seems that doubling the key-length of one block cipher could achieve the same security against quantum attackers. However, Kuwakado and Morii [6] introduced a new family of quantum attacks using Simon's algorithm [12] which could find the period of a periodic function in polynomial time in a quantum computer. Indeed, they describe a quantum distinguishing CPA attack on the 3-round Feistel scheme. This work has been then extended by Ito *et al.* [5] to a quantum CCA distinguisher against the 4-round Feistel cipher.

Luo *et al.* [8] present quantum attacks on 3-round Misty L and Misty R schemes using Simon's algorithm. We describe a similar quantum attack on the 3-round Misty R structure. In this paper, we provide additional (quantum) cryptanalysis on variants of Misty L and Misty R schemes as explained in the "Our Contribution" paragraph.

Our Contribution. In this paper, we describe a non-adaptive quantum chosen plaintext attack (QCPA) against 4-round Misty L and Misty LKF schemes, and a non-adaptive quantum chosen plaintext attack (QCPA) against 3-round Misty R and Misty RKF schemes. These attacks enable to distinguish these Misty schemes from random permutations in polynomial time. We extend the quantum distinguishing attack against 3-round Misty RKF schemes to obtain a quantum key recovery attack against d-round Misty RKF schemes with complexity $\tilde{\mathcal{O}}(2^{(d-3)n/2})$. Then, we show that security of Misty L and Misty R schemes with 3 rounds differs regarding CPA attacks. The best known attack against Misty L schemes with 3 rounds has complexity 4 operations with 4 distinct messages. The best known attack against Misty R schemes has complexity $2^{n/2}$ operations with $2^{n/2}$ messages. In this paper, we provide a security proof with the same bound $2^{n/2}$ which shows that the best known cryptanalysis against Misty R schemes is optimal.

Organization. Section 2 describes the four variants of Misty schemes. Section 3 gives an overview of previous works and the new results provided in this paper. In Sect. 4, we present our QCPA against the four variants of Misty schemes and the quantum key recovery attack on Misty RKF schemes. Section 5 provides the security proof of Misty R schemes with 3 rounds against adaptive Chosen Plaintext attack (CPA-2). Finally, we conclude in Sect. 6.

2   Misty Constructions

In this section, we describe the four variants of Misty schemes. The set of all functions from $\{0,1\}^n$ to $\{0,1\}^n$ is denoted by F_n and the set of all permutations from $\{0,1\}^n$ to $\{0,1\}^n$ is denoted by B_n. We have $B_n \subset F_n$. We denote by M^d a Misty scheme of d rounds: $f = M^d(f_1, \ldots, f_d)$, where $f_1, \ldots, f_d$ are permutations from n bits to n bits, and f is a permutation from $2n$ bits to $2n$ bits.

2.1   Misty L Scheme

Let f_1 be a permutation of B_n. Let L, R, S and T be elements in $\{0,1\}^n$. Then by definition we have:

$$M_L(f_1)([L,R]) = [S,T] \Leftrightarrow S = R \text{ and } T = R \oplus f_1(L)$$

Let $f_1, \ldots, f_d$ be d bijections of B_n. Then by definition we have:

$$M_L^d(f_1, \ldots, f_d) = M_L(f_d) \circ \ldots M_L(f_2) \circ M_L(f_1)$$

The permutation $M_L^d(f_1, \ldots, f_d)$ is called a *Misty L scheme* with d rounds. We describe in detail the equations of Misty L for the first four rounds.

1 round: $\begin{cases} S = R \\ T = R \oplus f_1(L) = X^1 \end{cases}$ 2 rounds: $\begin{cases} S = X^1 \\ T = X^1 \oplus f_2(R) = X^2 \end{cases}$

3 rounds: $\begin{cases} S = X^2 \\ T = X^2 \oplus f_3(X^1) = X^3 \end{cases}$ 4 rounds: $\begin{cases} S = X^3 \\ T = X^3 \oplus f_4(X^2) = X^4 \end{cases}$

The figure of Misty L schemes for the first round is given in Fig. 1.

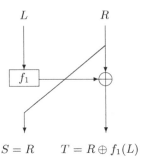

$$S = R \qquad T = R \oplus f_1(L)$$

Fig. 1. First round of Misty L

Misty LKF Scheme. Let F be a public function of F_n and K_1 be a key chosen in $\{0,1\}^n$. Let L, R, S and T be elements in $\{0,1\}^n$. Then, we define:

$$M_{LKF}(F, K_1)([L, R]) = [S, T] \Leftrightarrow S = R \text{ and } T = R \oplus F(K_1 \oplus L)$$

Let $K_1, \ldots, K_d$ be d keys chosen in $\{0,1\}^n$. Then we have:

$$M_{LKF}^d(F, K_1, \ldots, K_d) = M_{LKF}(F, K_d) \circ \ldots M_{LKF}(F, K_2) \circ M_{LKF}(F, K_1)$$

In this paper, we call $M_{LKF}^d(F, K_1, \ldots, K_d)$ a *Misty LKF scheme* with d rounds. The equations of the first four rounds of Misty LKF are as follows.

1 round: $\begin{cases} S = R \\ T = R \oplus F(K_1 \oplus L) = A^1 \end{cases}$ 2 rounds: $\begin{cases} S = A^1 \\ T = A^1 \oplus F(K_2 \oplus R) = A^2 \end{cases}$

3 rounds: $\begin{cases} S = A^2 \\ T = A^2 \oplus F(K_3 \oplus A^1) = A^3 \end{cases}$ 4 rounds: $\begin{cases} S = A^3 \\ T = A^3 \oplus F(K_4 \oplus A^2) = A^4 \end{cases}$

The figure of Misty LKF schemes for the first round is given in Fig. 2.

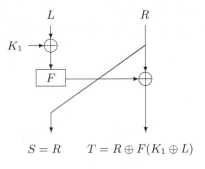

$$S = R \qquad T = R \oplus F(K_1 \oplus L)$$

Fig. 2. First round of Misty LKF

2.2 Misty R Scheme

Let f_1 be a permutation of B_n. Let L, R, S and T be elements in $\{0,1\}^n$. Then by definition we have:

$$M_R(f_1)([L, R]) = [S, T] \Leftrightarrow S = R \oplus f_1(L) \text{ and } T = f_1(L)$$

Let $f_1, \ldots, f_d$ be d bijections of B_n. Then by definition we have:

$$M_R^d(f_1, \ldots, f_d) = M_R(f_d) \circ \ldots M_R(f_2) \circ M_R(f_1)$$

The permutation $M_R^d(f_1, \ldots, f_d)$ is called a *Misty R scheme* with d rounds. We describe in detail the equations of Misty R for the first four rounds.

1 round: $\begin{cases} S = R \oplus f_1(L) = Y^1 \\ T = f_1(L) \end{cases}$ 2 rounds: $\begin{cases} S = f_1(L) \oplus f_2(Y^1) = Y^2 \\ T = f_2(Y^1) \end{cases}$

3 rounds: $\begin{cases} S = f_2(Y^1) \oplus f_3(Y^2) = Y^3 \\ T = f_3(Y^2) \end{cases}$ 4 rounds: $\begin{cases} S = f_3(Y^2) \oplus f_4(Y^3) = Y^4 \\ T = f_4(Y^3) \end{cases}$

The figure of Misty R schemes for the first round is given in Fig. 3.

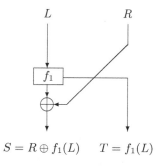

$$S = R \oplus f_1(L) \qquad T = f_1(L)$$

Fig. 3. First round of Misty R

Misty RKF Scheme. Let F be a public function of F_n and K_1 be a key chosen in $\{0,1\}^n$. Let L, R, S and T be elements in $\{0,1\}^n$. Then, we define:

$$M_{RKF}(F, K_1)([L, R]) = [S, T] \Leftrightarrow S = R \oplus F(K_1 \oplus L) \text{ and } T = F(K_1 \oplus L)$$

Let $K_1, \ldots, K_d$ be d keys chosen in $\{0,1\}^n$. Then we have:

$$M_{RKF}^d(F, K_1, \ldots, K_d) = M_{RKF}(F, K_d) \circ \ldots M_{RKF}(F, K_2) \circ M_{RKF}(F, K_1)$$

In this paper, we call $M_{RKF}^{d}(F, K_1, \ldots, K_d)$ a *Misty RKF scheme* with d rounds. The equations of Misty RKF for the first four rounds are as follows:

1 round:
$$\begin{cases} S = R \oplus F(K_1 \oplus L) = B^1 \\ T = F(K_1 \oplus L) \end{cases}$$

2 rounds:
$$\begin{cases} S = F(K_1 \oplus L) \oplus F(K_2 \oplus B^1) = B^2 \\ T = F(K_2 \oplus B^1) \end{cases}$$

3 rounds:
$$\begin{cases} S = F(K_2 \oplus B^1) \oplus F(K_3 \oplus B^2) = B^3 \\ T = F(K_3 \oplus B^2) \end{cases}$$

4 rounds:
$$\begin{cases} S = F(K_3 \oplus B^2) \oplus F(K_4 \oplus B^3) = B^4 \\ T = F(K_4 \oplus B^3) \end{cases}$$

The figure of Misty RKF schemes for the first round is given in Fig. 4.

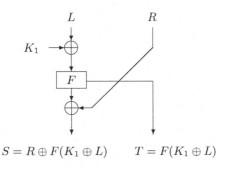

$$S = R \oplus F(K_1 \oplus L) \qquad T = F(K_1 \oplus L)$$

Fig. 4. First round of Misty RKF

3   Overview of (Quantum) Cryptanalysis on Misty Schemes

In this section, we review the cryptanalysis results of the state of the art on the Misty L and Misty R schemes and we point out the new results provided in this paper.

3.1   Misty L Schemes with Few Rounds

In Fig. 5, we summarize the cryptanalysis results on few rounds of Misty L schemes based on the state of the art distinguishing attacks presented in [9,10] together with our new contributions.

	KPA	CPA	CCA	QCPA	QCCA
M_L^1	1	1	1	1	1
M_L^2	$2^{n/2}$	2	2	2	2
M_L^3	2^n	4	3	4	3
M_L^4	2^n	$2^{n/2}$	4	**This paper :** n (distinguishing attack)	4

Fig. 5. Number of computations to distinguish Misty L schemes (with 1, 2, 3 and 4 rounds) from random permutations

On Misty L schemes with 1 round, we have $S = R$ which gives an attack with one message in all security models. We only have to check whether S is equal to R. For a Misty L scheme, this happens with probability 1 whereas for a random permutation it happens with probability $\frac{1}{2^n}$.

On Misty L schemes with 2 rounds, we have 2 cases depending on the security model. For CPA attack, we can choose 2 messages $[L_1, R_1]$ and $[L_2, R_2]$ such that $L_1 = L_2$. Then, we can check whether $S_1 \oplus S_2$ is equal to $R_1 \oplus R_2$. For a Misty L scheme, this happens with probability 1 whereas for a random permutation it happens with probability $\frac{1}{2^n}$. This cryptanalysis result is valid for other security models CCA, QCPA and QCCA. For KPA model, the CPA attack can be transformed into a KPA attack using $2^{n/2}$ messages and the birthday paradox bound to find a collision such that $L_i = L_j$.

On Misty L schemes with 3 rounds, there is a CPA attack with 4 messages [9] that can be transformed into a KPA attack with approximately 2^n messages and a CCA attack with 3 messages [10]. These two attacks also apply in the quantum model.

On Misty L schemes with 4 rounds, there is a CCA attack with 4 messages [10] that can be transformed into KPA attack or CPA attack. The same attacks in the quantum models hold. However, in this paper we describe a QCPA attack that enables to distinguish a Misty L permutation from a random permutation using only n computations instead of $2^{n/2}$ computations.

Misty LKF with Few Rounds. The KPA, CPA and CCA attacks against Misty L schemes of [9,10] can be applied on Misty LKF schemes. Therefore, we describe in Sect. 4 the QCPA attack that distinguishes a 4-round Misty LKF scheme from a random permutation using n computations.

3.2 Misty R Schemes with Few Rounds

On Misty R schemes, the results on 1 and 2 rounds are similar to the case of Misty L schemes. On Misty R schemes with 3 rounds and with 4 rounds, the results of the KPA, CCA and QCCA attacks are similar to those of Misty L schemes since a Misty R scheme is the inverse of a Misty L scheme [10].

On Misty R schemes with 3 rounds, the best known attack has a complexity in $2^{n/2}$ computations with $2^{n/2}$ messages [10]. In this paper, we provide the security proof of Misty R schemes with 3 rounds against CPA-2 with the same bound $2^{n/2}$. We describe also a QCPA attack that distinguishes a Misty R scheme from a random permutation by using n computations.

Figure 6 summarizes the cryptanalysis results that are distinguishing attacks on Misty R schemes based on [10] and our new contributions.

	KPA	CPA	CCA	QCPA	QCCA
M_R^1	1	1	1	1	1
M_R^2	$2^{n/2}$	2	2	2	2
M_R^3	2^n	**This paper:** $2^{n/2}$ (security proof)	3	**This paper:** n (distinguishing attack)	3
M_R^4	2^n	$2^{n/2}$	4	$2^{n/2}$	4

Fig. 6. Number of computations to distinguish Misty R schemes (with 1, 2, 3 and 4 rounds) from random permutations

Misty RKF Schemes. The state of the art distinguishing attacks on Misty R schemes are similar for Misty RKF schemes and are summarized in Fig. 7 together with our new contribution. In this paper, we provide first a QCPA attack that distinguishes a 3-round Misty RKF scheme from a random permutation by using n computations. Then, we describe a QCPA attack that uses this quantum distinguishing attack on 3-round Misty RKF schemes to recover the keys of d-round Misty RKF schemes, for $d > 3$, in time $2^{(d-3)n/2}$.

	KPA	CPA	CCA	QCPA	QCCA
M_{RKF}^3	2^n	$2^{n/2}$	3	**This paper:** n (distinguishing attack)	3
M_{RKF}^6	2^{2n}	2^{2n}	2^{2n}	**This paper:** $2^{3n/2}$ (key recovery)	2^{2n}
M_{RKF}^7	2^{4n}	2^{4n}	2^{4n}	**This paper:** 2^{2n} (key recovery)	2^{4n}
M_{RKF}^8	2^{4n}	2^{4n}	2^{4n}	**This paper:** $2^{5n/2}$ (key recovery)	2^{4n}
M_{RKF}^9	2^{6n}	2^{6n}	2^{6n}	**This paper:** 2^{3n} (key recovery)	2^{6n}
M_{RKF}^{10}	2^{6n}	2^{6n}	2^{6n}	**This paper:** $2^{7n/2}$ (key recovery)	2^{6n}
M_{RKF}^d, d odd $d \geq 9$	$2^{(d-3)n}$	$2^{(d-3)n}$	$2^{(d-3)n}$	**This paper:** $2^{(d-3)n/2}$ (key recovery)	$2^{(d-3)n}$
M_{RKF}^d, d even $d \geq 8$	$2^{(d-4)n}$	$2^{(d-4)n}$	$2^{(d-4)n}$	**This paper:** $2^{(d-3)n/2}$ (key recovery)	$2^{(d-4)n}$

Fig. 7. Number of computations to distinguish Misty RKF schemes from random permutations and number of computations to recover the keys when explicitly specified

4   Quantum Cryptanalysis on Misty

In this section, we recall the results of the two quantum algorithms that we use in our quantum cryptanalysis. The full details on how the algorithms work can be found in [3,12]. Then, we describe our QCPA attacks against the four variants of Misty schemes and the key recovery attack against Misty RKF schemes.

4.1   Simon's and Grover's Algorithms

Simon's Problem. Given a Boolean function, $f : \{0,1\}^n \mapsto \{0,1\}^n$, that is observed to be invariant under some n-bit XOR period a, find a.

Simon presents a quantum algorithm [12] that provides exponential speedup and requires only $\mathcal{O}(n)$ quantum queries to find a.

Grover's Problem. Given a Boolean function $f : \{0,1\}^n \to \{0,1\}$ and suppose that there exists a unique $x_0 \in \{0,1\}^n$ such that $f(x_0) = 1$. Given an oracle access to f, find x_0.

Grover presents a quantum algorithm [3] that requires $\mathcal{O}(2^{n/2})$ quantum queries to find x_0.

4.2   Quantum Distinguishing Attack on 4-Round Misty L Schemes

In this section, we describe a quantum chosen plaintext attack that distinguishes a 4-round Misty L scheme from a $2n$-bit random permutation in polynomial time. We also apply this attack on Misty LKF schemes to obtain a quantum distinguishing attack on 4-round Misty LKF schemes.

Let $[L_1, R_1], [L_2, R_2], [L_3, R_3], [L_4, R_4]$ be four messages such that $L_1 \neq L_2$, $R_1 \neq R_2$, $L_3 = L_1$, $R_3 = R_2$, $L_4 = L_2$ and $R_1 = R_4$. As it has been shown in [9], for such four messages, we have:

$$X_1^3 \oplus X_2^3 \oplus X_3^3 \oplus X_4^3 = f_3(X_1^1) \oplus f_3(X_2^1) \oplus f_3(X_3^1) \oplus f_3(X_4^1)$$

where X_i^3 is the left half of $M_L^4([L_i, R_i])$ as denoted in Sect. 2. Then, we have:

$$\begin{aligned} X_1^3 \oplus X_2^3 \oplus X_3^3 \oplus X_4^3 &= f_3(X_1^1) \oplus f_3(X_2^1) \oplus f_3(X_3^1) \oplus f_3(X_4^1) \\ &= f_3(R_1 \oplus f_1(L_1)) \oplus f_3(R_2 \oplus f_1(L_2)) \oplus f_3(R_2 \oplus f_1(L_1)) \\ &\quad \oplus f_3(R_1 \oplus f_1(L_2)) \end{aligned}$$

We set $R_1 = x$ and we define the function

$$g(x) = f_3(x \oplus f_1(L_1)) \oplus f_3(R_2 \oplus f_1(L_2)) \oplus f_3(R_2 \oplus f_1(L_1)) \oplus f_3(x \oplus f_1(L_2))$$

We observe that we have $g(x) = g(x \oplus f_1(L_1) \oplus f_1(L_2))$. Thus, the function g is periodic and the period is $f_1(L_1) \oplus f_1(L_2)$. Note that, this period works even if $x = R_2$. We can use the Simon's algorithm on g to get the period $s = f_1(L_1) \oplus f_1(L_2)$ in polynomial time.

In the case where g is constructed with a $2n$-bit random permutation instead of a 4-round Misty L scheme, g is not periodic with overwhelming probability. If we apply Simon's algorithm on g, the algorithm fails to find a period. Therefore, we can distinguish a 4-round Misty L scheme from a random permutation in polynomial time by using Simon's algorithm to check if g has a period.

Quantum Distinguishing Attack on 4-Round Misty LKF Schemes. In the same way as for 4-round Misty L schemes, we have a quantum distinguishing attack on 4-round Misty LKF schemes.

Let $[L_1, R_1], [L_2, R_2], [L_3, R_3], [L_4, R_4]$ four messages such that $L_1 \neq L_2$, $R_1 \neq R_2$, $L_3 = L_1$, $R_3 = R_2$, $L_4 = L_2$ and $R_1 = R_4$. We have also for Misty LKF:

$$
\begin{aligned}
A_1^3 \oplus A_2^3 \oplus A_3^3 \oplus A_4^3 &= F(K_3 \oplus A_1^1) \oplus F(K_3 \oplus A_2^1) \oplus F(K_3 \oplus A_3^1) \oplus F(K_3 \oplus A_4^1) \\
&= F(K_3 \oplus R_1 \oplus F(K_1 \oplus L_1)) \oplus F(K_3 \oplus R_2 \oplus F(K_1 \oplus L_2)) \\
&\quad \oplus F(K_3 \oplus R_2 \oplus F(K_1 \oplus L_1)) \oplus F(K_3 \oplus R_1 \oplus F(K_1 \oplus L_2))
\end{aligned}
$$

where A_i^3 is the left half of $M_{LKF}^4([L_i, R_i])$ as denoted in Sect. 2. We set $R_1 = x$ and we define the function g by

$$
\begin{aligned}
g(x) = {}&F(K_3 \oplus x \oplus F(K_1 \oplus L_1)) \oplus F(K_3 \oplus R_2 \oplus F(K_1 \oplus L_2)) \\
&\oplus F(K_3 \oplus R_2 \oplus F(K_1 \oplus L_1)) \oplus F(K_3 \oplus x \oplus F(K_1 \oplus L_2))
\end{aligned}
$$

We observe that $g(x) = g(x \oplus F(K_1 \oplus L_1) \oplus F(K_1 \oplus L_2))$. Thus, the function g is periodic and the period is $F(K_1 \oplus L_1) \oplus F(K_1 \oplus L_2)$. We can use the Simon's algorithm on g to get the period $s = F(K_1 \oplus L_1) \oplus F(K_1 \oplus L_2)$ in polynomial time. Thus, we obtain a quantum distinguishing attack on a 4-round Misty LKF scheme by checking with the Simon's algorithm if g has a period.

4.3   Quantum Distinguishing Attack on 3-Round Misty R Schemes

In this section, we describe a quantum chosen plaintext attack that distinguishes a 3-round Misty R scheme from a $2n$-bit random permutation in polynomial time that is already known [8]. We also apply this attack on Misty RKF schemes to obtain a quantum distinguishing attack on 3-round Misty RKF schemes.

We consider the value $S \oplus T = f_2(Y^1) = f_2(R \oplus f_1(L))$ where $[S, T] = M_R^3([L, R])$ as described in Sect. 2. Let $[L_1, R], [L_2, R]$ two messages such that $L_1 \neq L_2$. We set $R = x$ and we define the function

$$
\begin{aligned}
g(x) &= S_1 \oplus T_1 \oplus S_2 \oplus T_2 \\
&= f_2(x \oplus f_1(L_1)) \oplus f_2(x \oplus f_1(L_2))
\end{aligned}
$$

where $[S_i, T_i] = M_R^3([L_i, R])$. We observe that $g(x) = g(x \oplus f_1(L_1) \oplus f_1(L_2))$. Thus, g is a periodic function and the period is $f_1(L_1) \oplus f_1(L_2)$. We can use the Simon's algorithm on g to get the period $s = f_1(L_1) \oplus f_1(L_2)$ in polynomial time.

In the case where we apply Simon's algorithm on g that is constructed with a $2n$-bit random permutation, the algorithm fails to find a period with over-whelming probability. Thus, we can distinguish a 3-round Misty R scheme from a random permutation by checking with the Simon's algorithm if g has a period.

Quantum Distinguishing Attack on 3-Round Misty RKF Schemes. In the same way as for 3-round Misty R scheme, we have a quantum distinguishing attack on 3-round Misty RKF scheme. We can also consider the value $S \oplus T = F(K_2 \oplus B^1) = F(K_2 \oplus R \oplus F(K_1 \oplus L))$ where $[S, T] = M^3_{RKF}([L, R])$ as described in Sect. 2. Let $[L_1, R], [L_2, R]$ two messages such that $L_1 \neq L_2$. Thus, we set $R = x$ and we define the function g by

$$g(x) = S_1 \oplus T_1 \oplus S_2 \oplus T_2$$
$$= F(K_2 \oplus x \oplus F(K_1 \oplus L_1)) \oplus F(K_2 \oplus x \oplus F(K_1 \oplus L_2))$$

where $[S_i, T_i] = M^3_{RKF}([L_i, R])$. We observe that $g(x) = g(x \oplus F(K_1 \oplus L_1) \oplus F(K_1 \oplus L_2))$. The function g is periodic and the period of the function is $F(K_1 \oplus L_1) \oplus F(K_1 \oplus L_2)$. We can use the Simon's algorithm on g to get the period $s = F(K_1 \oplus L_1) \oplus F(K_1 \oplus L_2)$ in polynomial time.

Thus, we obtain a quantum distinguishing attack on 3-round Misty RKF scheme by using Simon's algorithm on g to check if g has a period.

4.4   Key Recovery Attack Against Misty RKF Schemes

Based on [1,4,7], we combine the quantum distinguishing attack on the 3-round Misty RKF scheme (Sect. 4.3) with the Grover search to obtain a key recovery attack against a d-round Misty RKF scheme. The attack recovers the keys of the d-round Misty RKF scheme $(K_1, \ldots, K_d)$. We apply the technique of [4] recalled in Proposition 1.

Proposition 1 (Proposition 3 in [4]). *Let $\Psi : F_m \times F_n \to F_n$ be a function such that $\Psi(k, \cdot) : F_n \to F_n$ is a random function for any fixed $k \in F_m$. Let $\Phi : F_m \times F_n \to F_n$ be a function such that $\Phi(k, \cdot) : F_n \to F_n$ is a random function for any fixed $k \in F_m \setminus \{k_0\}$ and $\Phi(k_0, x) = \Psi(k_0, x \oplus k_1)$. Then, given a quantum oracle access to $\Phi(\cdot, \cdot)$ and $\Psi(\cdot, \cdot)$, we can recover (k_0, k_1) with a constant probability and $\mathcal{O}((m + n^2)2^{m/2})$ queries, using $\mathcal{O}(m + n^2)$ qubits.*

For our attack, the key k_0 in Proposition 1 corresponds to the keys of the last $(d - 3)$-round of a d-round Misty RKF scheme $K_4, \ldots, K_d$ and k_1 corresponds to the period s recovered in the quantum distinguishing attack on the 3-round Misty RKF scheme described in Sect. 4.3. The idea is to search for the correct key $k_0 = (K_4, \ldots, K_d)$ with the Grover search and check if $\Phi(\cdot, \cdot) \oplus \Psi(\cdot, \cdot)$ is periodic or not for the candidate key $k = (K'_4, \ldots, K'_d)$ by running the Simon's algorithm in parallel.

The attack is the following. Assume that we have a quantum encryption oracle of a d-round Misty RKF scheme $\mathcal{O} : \{0, 1\}^{2n} \to \{0, 1\}^{2n}$. For $k = (K'_4, \ldots, K'_d) \in \{0, 1\}^{(d-3)n}$, let $D_k : \{0, 1\}^{2n} \to \{0, 1\}^{2n}$ denotes the partial decryption of the last $(d-3)$-round of Misty RKF with the key candidate k. Let $W : \{0, 1\}^{(d-3)n} \times \{0, 1\}^n \times \{0, 1\}^n \to \{0, 1\}^n$ be the function that is the sum of the right part and the left part obtained after the 3-round of the Misty RKF scheme. W is defined by

$$W(k, L, R) := \text{the sum of the left and right halves of } D_k \circ \mathcal{O}(L, R)$$

We implement a quantum circuit of W using the quantum encryption oracle $\mathcal{O}$. In the case where $k = k_0$, then $W(k_0, L, R) = F(K_2 \oplus R \oplus F(K_1 \oplus L))$.

Then, we choose two different n-bits string α, β and define $\Psi : \{0,1\}^{(d-3)n} \times \{0,1\}^n \rightarrow \{0,1\}^n$ and $\Phi : \{0,1\}^{(d-3)n} \times \{0,1\}^n \rightarrow \{0,1\}^n$ by $\Psi(k,x) := W(k,\alpha,x)$ and $\Phi(k,x) := W(k,\beta,x)$. The function $\Psi(k,\cdot)$ is an almost random function for each k and $\Phi(k,\cdot)$ is also an almost random function for each $k \neq k_0$. In the case where $k = k_0$, we have $\Phi(k_0, x) = \Psi(k_0, x \oplus k_1)$ where $k_1 = F(K_1 \oplus \alpha) \oplus F(K_1 \oplus \beta)$. Indeed, we have:

$$\Psi(k_0, x \oplus k_1) = W(k, \alpha, x \oplus k_1)$$
$$= F(K_2 \oplus x \oplus F(K_1 \oplus \alpha) \oplus F(K_1 \oplus \beta) \oplus F(K_1 \oplus \alpha))$$
$$= F(K_2 \oplus x \oplus F(K_1 \oplus \beta)) = W(k, \beta, x) = \Phi(k_0, x)$$

Thus, we can apply Proposition 1 and recover the keys $K_4, \ldots, K_d$. Then, we can recover K_1. To this end, we construct a quantum circuit that calculates the first 3 rounds of the Misty RKF scheme. Then, we compute the period $s = F(K_1 \oplus \alpha) \oplus F(K_1 \oplus \beta)$ with the quantum distinguishing attack on the 3-round Misty RKF scheme with two arbitrary messages $[\alpha, x], [\beta, x]$ such that $x, \alpha, \beta \in \{0,1\}^n$ and $\alpha \neq \beta$. Thus, we can recover K_1 by using the Grover search. Finally, we can easily recover K_2 and K_3 using the Grover search and the recovered key K_1.

Attack Complexity. By Proposition 1, we can recover $(K_4, \ldots, K_d)$ in time $\mathcal{O}(2^{(d-3)n/2})$[1]. Since the last keys K_1, K_2 and K_3 are recovered by using the Grover search in time $\mathcal{O}(2^{n/2})$, the complexity of the key recovery attack against a Misty RKF scheme is $\tilde{\mathcal{O}}(2^{(d-3)n/2})$.

5   Security Proof on Misty R Scheme with 3 Rounds

The best known CPA-1 attack against a Misty R scheme with 3 rounds is in $\mathcal{O}(2^{n/2})$ messages and computations [10]. In this section, we prove the security of the 3-round Misty R scheme against adaptive Chosen Plaintext CPA-2 attacks when the number of queries q is significantly smaller than $2^{n/2}$. Since this proof and the best known attack have the same bound $2^{n/2}$, the cryptanalysis of the 3-round Misty R scheme is optimal. For this proof, we use the result on H *coefficients technique* provided in [11].

5.1   H Coefficient Technique

Let N be a positive integer. Let I_N be the set $\{0,1\}^N$ and F_N be the set of all applications from I_N to I_N. Let B_N be the set of permutations from I_N to I_N. Let K denotes a set of k-uples of functions $(f_1, \ldots, f_k)$ of F_N. We define G as an application of $K \rightarrow F_N$.

---

[1] Taking into account the required numbers of qubits and operations, the complexity is in $\mathcal{O}(n^3 2^{(d-3)n/2})$ as explained in [4].

Definition 1 (H coefficient). *Let q be a positive integer. Let $(a_1, \ldots, a_q)$ with $a_i \in I_N$ for $i = 1, \ldots, q$ be a sequence of pairwise distinct elements of I_N . Let $(b_1, \ldots, b_q)$ with $b_i \in I_N$ for $i = 1, \ldots, q$. The H coefficient denoted by $H(a,b)$ or simply by H is the number of $(f_1, \ldots, f_k) \in K$ such that:*

$$\forall i, 1 \leq i \leq q, G(f_1, \ldots, f_k)(a_i) = b_i$$

5.2   Application to Misty R Scheme with 3 Rounds

Theorem 1 (Adaptive Chosen Plaintext attack with q queries) [11]. *Let ε and β be positive real numbers. Let E be a subset of I_N^q such that $|E| \geq (1 - \beta)2^{Nq}$. If for all $(a_1, \ldots, a_q)$ with $a_i \in I_N$ for $i = 1, \ldots, q$ such that $a_i \neq a_j$ when $i \neq j$ and for all $\beta \in E$ we have:*

$$H \geq \frac{|k|}{2^{Nq}}(1 - \varepsilon)$$

Then, the advantage Adv^{CPA-2} to distinguish $G(f_1, \ldots, f_k)$ with $(f_1, \ldots, f_k) \in_R K$ from a random function $f \in_R F_N$ fulfills:

$$Adv^{CPA-2} \leq \beta + \varepsilon.$$

Theorem 2 (CPA-2 security on 3 rounds Misty R). *The advantage of an attacker in an adaptive chosen plaintext attack against the construction Misty R with 3 rounds is upper bounded by:*

$$Adv^{CPA-2} \leq \frac{3}{2} \frac{q(q-1)}{2} \frac{1}{2^n}$$

Proof. On Misty R schemes with 3 rounds, the set of keys K is equal to B_N^3 with $N = 2n$.

The transformation M_R sends $[L_i, R_i]$ to $[U_i, T_i]$ such that:

$$\begin{cases} U_i = T_i \oplus S_i = f_2(R_i \oplus f_1(L_i)) \\ T_i = f_3(f_1(L_i) \oplus U_i) \end{cases}$$

We are looking to $H = \{(f_1, f_2, f_3) \in B_n^3$ such that $\forall i, 1 \leq i \leq q, M_R[L_i, R_i] = [U_i, T_i]\}$.

Let E be the set defined as follows: $E = \{[U_i, T_i], 1 \leq i \leq q, U_i \neq U_j$ when $i \neq j\}$. We have:

$$|E| \geq 2^{Nq}\left(1 - \frac{q(q-1)}{2 \cdot 2^n}\right)$$

and we deduce that we have $\beta = \frac{q(q-1)}{2 \cdot 2^n}$.

We select f_1 such that the values $R_i \oplus f_1(L_i)$ are pairwise distinct and the values $U_i \oplus f_1(L_i)$ are pairwise distinct with $[U_i, T_i] \in E$.

– $R_i \oplus f_1(L_i) = R_j \oplus f_1(L_j)$ implies that $L_i \neq L_j$ or $R_i \neq R_j$ since $i \neq j$. Then we have to remove at most $\frac{q(q-1)}{2 \cdot 2^n}|B_n|$ permutations f_1.

- $f_1(L_i) \oplus U_i = f_1(L_j) \oplus U_j$ implies $L_i \neq L_j$ since we have $U_i \neq U_j$. Then we have to remove at most $\frac{q(q-1)}{2 \cdot 2^n}|B_n|$ permutations f_1.

Now, the function f_1 is chosen and both f_2 and f_3 are fixed in q points pairwise distinct. Then we have:

$$H \geq \frac{|B_n|^3}{2^{2nq}}\left(1 - \frac{q(q-1)}{2^n}\right) = \frac{|K|}{2^{Nq}}\left(1 - \frac{q(q-1)}{2^n}\right)$$

Then, by applying Theorem 1, we have $\varepsilon = \frac{q(q-1)}{2^n}$, $\beta = \frac{q(q-1)}{2 \cdot 2^n}$ and

$$Adv^{\text{CPA}-2} \leq \left(\frac{3}{2}\right)\frac{q(q-1)}{2}\frac{1}{2^n}$$

This concludes the proof.

6   Conclusion

In this paper, we provide a quantum cryptanalysis of four variants of Misty schemes. Indeed, we describe QCPA attacks that enable to distinguish 4-round Misty L and Misty LKF schemes, and 3-round Misty R and Misty RKF schemes, from random permutations in complexity $\mathcal{O}(n)$ instead of $\mathcal{O}(2^{n/2})$. Note that the QCPA attack on 3-round Misty R schemes is already known in [8]. Moreover, we extend the quantum distinguishing attack on 3-round Misty RKF schemes to obtain a key recovery attack against Misty RKF schemes which recovers the keys of d-round Misty RKF schemes in time $\mathcal{O}(2^{(d-3)n/2})$. Then, we provide the security proof of 3-round Misty R schemes against CPA-2 attack with a complexity in $\mathcal{O}(2^{n/2})$. Since the best known attack against the 3-round Misty R schemes has the same bound, this shows that the state of the art attack is then optimal.

References

1. Dong, X., Wang, X.: Quantum key-recovery attack on Feistel structures. Sci. China Inf. Sci. **61**(10), 1–7 (2018). https://doi.org/10.1007/s11432-017-9468-y
2. ETSI: Specification of the 3GPP Confidentiality and Integrity Algorithm KASUMI. Document http://www.etsi.org/
3. Grover, L.K.: A fast quantum mechanical algorithm for database search. In: Proceedings of the Twenty-Eighth Annual ACM Symposium on Theory of Computing, STOC 1996, pp. 212–219 (1996)
4. Hosoyamada, A., Sasaki, Yu.: Quantum Demiric-Selçuk meet-in-the-middle attacks: applications to 6-round generic Feistel constructions. In: Catalano, D., De Prisco, R. (eds.) SCN 2018. LNCS, vol. 11035, pp. 386–403. Springer, Cham (2018). https://doi.org/10.1007/978-3-319-98113-0_21
5. Ito, G., Hosoyamada, A., Matsumoto, R., Sasaki, Yu., Iwata, T.: Quantum chosen-ciphertext attacks against Feistel ciphers. In: Matsui, M. (ed.) CT-RSA 2019. LNCS, vol. 11405, pp. 391–411. Springer, Cham (2019). https://doi.org/10.1007/978-3-030-12612-4_20

6. Kuwakado, H., Morii, M.: Quantum distinguisher between the 3-round Feistel cipher and the random permutation. In: Proceedings of the IEEE International Symposium on Information Theory, ISIT 2010, pp. 2682–2685. IEEE (2010)

7. Leander, G., May, A.: Grover meets Simon – quantumly attacking the FX-construction. In: Takagi, T., Peyrin, T. (eds.) ASIACRYPT 2017. LNCS, vol. 10625, pp. 161–178. Springer, Cham (2017). https://doi.org/10.1007/978-3-319-70697-9_6

8. Luo, Y.Y., Yan, H.L., Wang, L., Hu, H.G., Lai, X.J.: Study on block cipher structures against Simon's quantum algorithm. J. Cryptol. Res. **6**(5), 561 (2019)

9. Nachef, V., Patarin, J., Treger, J.: Generic attacks on Misty schemes -5 rounds is not enough. IACR Cryptology ePrint Archive **2009**, 405 (2009)

10. Nachef, V., Patarin, J., Treger, J.: Generic attacks on Misty schemes. In: Abdalla, M., Barreto, P.S.L.M. (eds.) LATINCRYPT 2010. LNCS, vol. 6212, pp. 222–240. Springer, Heidelberg (2010). https://doi.org/10.1007/978-3-642-14712-8_14

11. Patarin, J.: The "coefficients H" technique. In: Avanzi, R.M., Keliher, L., Sica, F. (eds.) SAC 2008. LNCS, vol. 5381, pp. 328–345. Springer, Heidelberg (2009). https://doi.org/10.1007/978-3-642-04159-4_21

12. Simon, D.R.: On the power of quantum computation. SIAM J. Comput. **26**(5), 1474–1483 (1997)

An Efficient Authenticated Key Exchange from Random Self-reducibility on CSIDH

Tomoki Kawashima[1(✉)], Katsuyuki Takashima[2], Yusuke Aikawa[2],
and Tsuyoshi Takagi[1]

[1] Department of Mathematical Informatics, The University of Tokyo, Tokyo, Japan
{tomoki_kawashima,takagi}@mist.i.u-tokyo.ac.jp
[2] Mitsubishi Electric Corporation, Kamakura, Kanagawa, Japan
Takashima.Katsuyuki@aj.MitsubishiElectric.co.jp,
Aikawa.Yusuke@bc.MitsubishiElectric.co.jp

Abstract. SIDH and CSIDH are key exchange protocols based on isogenies and conjectured to be quantum-resistant. Since the protocols are
similar to the classical Diffie–Hellman, they are vulnerable to the man-
in-the-middle attack. A key exchange which is resistant to such an attack
is called an authenticated key exchange (AKE), and many isogeny-based
AKEs have been proposed. However, the parameter sizes of the existing
schemes should be large since they all have relatively large security losses
in security proofs. This is partially because the random self-reducibility
of isogeny-based decisional problems has not been proved yet.

In this paper, we show that the computational problem and the gap
problem of CSIDH are random self-reducible. A gap problem is a computational problem given access to the corresponding decision oracle.
Moreover, we propose a CSIDH-based AKE with small security loss, following the construction of Cohn-Gordon et al. in CRYPTO 2019, as an
application of the random self-reducibility of the gap problem of CSIDH.
Our AKE is proved to be the fastest CSIDH-based AKE when we aim
at 110-bit security level.

Keywords: Post-quantum · Tight security · Authenticated key
exchange · Isogeny-based cryptography · CSIDH

1 Introduction

1.1 Backgrounds

Most of the public key cryptosystems currently used depend on the difficulty
of Discrete Logarithm Problem (DLP) or factorization, so will be broken by
Shor's algorithm [22] with quantum computers. To prepare for the appearance
of quantum computers, quantum-resistant cryptosystems are needed. Isogeny-
based cryptosystems, such as SIDH [16] and CSIDH [5], are expected to be
quantum-resistant. CSIDH can be considered to be an instantiation of Hard
Homogeneous Spaces (HHS) proposed in [8], and its protocol is similar to the

© Springer Nature Switzerland AG 2021
D. Hong (Ed.): ICISC 2020, LNCS 12593, pp. 58–84, 2021.
https://doi.org/10.1007/978-3-030-68890-5_4

classical Diffie–Hellman. SIDH also has a very similar protocol to the classical Diffie–Hellman, but we cannot regard it as an instantiation of HHS.

Since SIDH and CSIDH are subject to the man-in-the-middle attack, we cannot use them under insecure channels. Although the classical Diffie–Hellman is also vulnerable to the man-in-the-middle attack, there are many Diffie–Hellman-based key exchange protocols which are secure under insecure channels, e.g., [17,18]. Such key exchange protocols are called authenticated key exchanges (AKE).

As for isogeny-based cryptosystems, there are several SIDH-based AKEs [14, 19,23]. There also exist CSIDH-based AKEs [13], though [13] mainly focuses on a group AKE. However, these existing isogeny-based AKEs are not so efficient in that they all have large security losses. When we have a small security loss, we can use smaller parameters to achieve a specific security level, e.g., 110-bit security, which means the protocol becomes faster.

To achieve an AKE with a small security loss, "random self-reducibility" of the underlying problem is useful. Informally, we say that a problem is random self-reducible when we can produce multiple independent instances from a single instance such that we can restore the original answer from the answer of any one of the generated instances. In other words, a problem is random self-reducible if we can reduce the problem tightly to the corresponding multi-instance problem. It is well-known that the decisional Diffie–Hellman (DDH) problem and the computational Diffie–Hellman (CDH) problem are random self-reducible, see the left part of the Fig. 1.

Since the security of an AKE is often defined in the decisional game, it is plausible to use a decisional problem for the security proof. However, even though the (classical) DDH problem is random self-reducible, the decisional problem of SIDH and that of CSIDH (denoted as SI-DDH and CSI-DDH respectively) have not been proved so, see the middle and right part of the Fig. 1. Thus it seems difficult to construct an AKE with a small security loss from the decisional isogeny-based problems. Here, the next candidates are isogeny-based computational problems. However, the security proof of an AKE from a computational problem is often difficult, since we have to reduce the computational problem to the decisional game. Thus, other kinds of problems are needed to construct an AKE with a small security loss.

For that reason, we use gap problems introduced by Okamoto and Pointcheval [21]. A gap problem is a problem to solve a computational problem given access to the corresponding decisional oracle, and often used in the security proof of an AKE [13,18]. Though not quantum-resistant, the gap problem of the classical Diffie–Hellman (the GDH problem in short) is also random self-reducible, see the left part of the Fig. 1. As for SIDH, no gap problem which is suitable for cryptosystems has been found. Related discussions are in [10,15]. So we put our hope on the gap problem of CSIDH, CSI-GDH problem.

Following the discussion above, a natural question which arises is:

Can we construct a (almost) tightly secure quantum-resistant key exchange which is secure against man-in-the-middle attack from isogenies?

1.2   Contributions

Random Self-Reducibility of the Gap Problem of CSIDH. Our first contribution is that we prove the random self-reducibility of the CSI-GDH problem. Though the classical Diffie–Hellman, CSIDH, and SIDH have very similar structures, the situations about random self-reducibility are different. As for the classical Diffie–Hellman, the decisional problem and the gap problem are both random self-reducible (see the left part of the Fig. 1). On the contrary, neither the decisional problem and the gap problem of SIDH has been proved to be random self-reducible (see the right part of the Fig. 1). Here, our contribution is that we prove that the gap problem of CSIDH, CSI-GDH problem, is random self-reducible (see the middle part of the Fig. 1). So, among the classical Diffie–Hellman, SIDH, and CSIDH, CSIDH is the only quantum-resistant cryptosystem which has random self-reducibility.

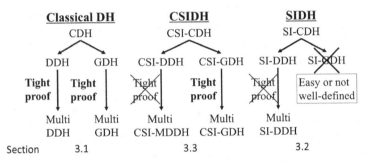

Fig. 1. The comparison of our result with existing result. We prove that the random self-reducibility of the CSI-GDH problem. As for the gap problems of SIDH, no suitable problems has been proposed, see the Sect. 3.2 for more details.

Efficient CSIDH-Based AKE. Our second contribution is that we construct a quantum-resistant efficient AKE based on CSIDH, following the construction of Cohn-Gordon et al. [7]. Our AKE has a security loss of $O(\mu)$, where μ is the number of users and Cohn-Gordon et al. showed that it is optimal among "simple" reductions.[1] So, even though the loss of $O(\mu)$ is not tight, i.e., the security loss is not constant, it is *optimally tight*, and our AKE is efficient in this sense. Moreover, since our AKE is based on CSIDH, it is quantum-resistant. As far as we know, our AKE has the smallest security loss among isogeny-based AKEs as shown in Table 1.

We also compare our AKE with other existing CSIDH-based AKEs. As far as we know, [13] is the only study that proposes other CSIDH-based AKEs. [13] proposes two CSIDH-based AKEs and we show that our AKE is faster than these two AKEs if we aim at 110-bit security.

---

[1] Informally, a reduction is simple if the reduction runs the adversary only once.

Table 1. AKEs proposed in the literature. "Assumption" shows which problem is assumed to be hard and "Model" represents which AKE security model was used. The number of hash queries and RevealSessKey are denoted as h and q. Also, l and n_s denote the maximum number of sessions per user and the number of sessions activated by the adversary.

Protocol	Assumption	Model	Loss
the classical Diffie–Hellman-based			
HMQV [17]	CDH	CK	$\mu^2 l^2$
NAXOS [18]	GDH	strong AKE	$\mu^2 l^2$
Protocol Π [7]	stDH	CCGJJ	μ
SIDH-based			
SIDH UM [12]	SI-DDH	CK	μl
SIDH-AKE [14]	SI-CDH	CK	$\mu^2 l^2$
SIGMA-SIDH [19]	SI-DDH	SK	μn_s
SIAKE$_2$ [23]	SI-DDH	CK$^+$	$\mu^2 l q$
SIAKE$_3$ [23]	1-OSIDH	CK$^+$	$\mu^2 l q$
CSIDH-based			
CSIDH UM [13]	2DDH	G-CK	$\mu n_s (h + q)^2$
CSIDH Biclique [13]	2GDH	G-CK$^+$	$\max(\mu^2, n_s^2)$
Proposed Protocol	CSI-stDH	CCGJJ	μ

1.3   Key Techniques

Relationship Between AKE and Random Self-reducibility. The security of an AKE is defined with the following game between a challenger $\mathcal{C}$ and an adversary $\mathcal{A}$. First, $\mathcal{C}$ gives public information of the AKE to $\mathcal{A}$, such as public keys of users and public parameters. Second, $\mathcal{A}$ carries out some attacks allowed in the security model. Finally, $\mathcal{A}$ chooses some sessions to get each person's shared keys or random keys. Such sessions are called test sessions. Here, a session is a single execution of the protocol. If $\mathcal{A}$ cannot decide whether the given keys are real-or-random with noticeable advantage, we say the AKE is secure. For further details, see Appendix A.

When we want to prove the security of an AKE from the difficulty of a problem, we construct a simulator of the game above. Given an instance of the problem, the simulator embeds the instance to the AKE. For example, the simulator sets a user's public key to the instance. The simulator embeds the problem deliberately so that if the adversary wins the game, then the simulator can answer the problem with high probability. In this case, the security loss is often equal to the reciprocal of the probability of the event that the adversary chooses one of the embedded sessions as a test session.

To lower the security loss, embedding the instance of the underlying problem to multiple sessions is helpful. This is because the probability that $\mathcal{A}$ "hits" the embedded sessions becomes higher and the security loss becomes smaller. In this case, the simulator has to embed instances so that the simulator can solve the underlying problem if the adversary hits the embedded sessions. However, while embedding the instance to AKE-settings, the simulator has to set the embedded keys independently. So, the simulator must generate multiple keys independently while the embedded keys are related to the original (single) instance.

Though it sounds difficult, we can do it easily if the underlying problem is random self-reducible. In the proof of the random self-reducibility, we make multiple and independent instances from one instance, which is analogous to the proof technique in AKE above.

Difference Between the Classical Diffie–Hellman and Isogeny-Based Cryptosystems. As for the classical Diffie–Hellman, the DDH problem and the CDH problem are both random self-reducible. The random self-reducibility of the GDH problem follows from that of the CDH problem.

Here, we describe the proof technique. Let $\mathbb{G}$ be a cyclic group of prime order p. Given $(X = g^x, Y = g^y, Z = g^z)$, the DDH problem is to decide whether $Z = g^{xy}$, where g is a generator of $\mathbb{G}$. In the proof of the random self-reducibility of the DDH problem, we take random elements $u_i, v_i, w_i \in \mathbb{Z}/p\mathbb{Z}$ and generate random instances $(X_i, Y_i, Z_i) = (X^{w_i} g^{u_i}, Y g^{v_i}, Z^{w_i} X^{v_i w_i} Y^{u_i} g^{u_i v_i})$ for $i = 1, 2, \cdots$. Note that in this rerandomization, we use the operations in $\mathbb{Z}/p\mathbb{Z}$ and $\mathbb{G}$. In other words, we use the operations in the set of secret keys and the set of public keys. On the contrary, in CDH case, we only use the operation in $\mathbb{Z}/p\mathbb{Z}$, the set of secret keys as discussed in Sect. 3.1.

As for SIDH, the key exchange is modeled as a random walk in isogeny graphs. Thus, we have poor algebraic structure. For example, the set of secret keys and public keys are the set of torsion points and the set of isomorphism classes of supersingular elliptic curves, respectively. So, we cannot transfer the proof of the classical Diffie–Hellman case to SIDH. This is the main reason why the random self-reducibilities of SIDH-related problems have not been proved yet.

In contrast, as for CSIDH, we have algebraic structure, i.e., we use the action of the ideal class group $\mathcal{C}\ell(\mathbb{Z}[\sqrt{-p}])$ on the set of isomorphic classes of supersingular elliptic curves $\mathcal{E}\ell\ell_p(\mathbb{Z}[\sqrt{-p}])$, where p is a prime such that $p \equiv 3 \mod 4$. Since we still have no operations in $\mathcal{E}\ell\ell_p(\mathbb{Z}[\sqrt{-p}])$, we cannot transfer the proof of DDH case to CSIDH. However, in CSIDH, the set of secret keys is $\mathcal{C}\ell(\mathbb{Z}[\sqrt{-p}])$, the ideal class "group" of an order $\mathbb{Z}[\sqrt{-p}]$. So, we have an operation in the set of secret keys and it is homomorphic to the action. This operation enables us to transfer the proof of the CDH-case to CSIDH, which is our main contribution.

Related Works. In an independent work, de Kock et al. [9] proposed exactly the same optimally-tight CSI-GDH based (thus post-quantum) AKE, while the main focus of their study is not the random self-reducibility itself.

Moreover, Brendel et al. [2] introduced an interesting notion of *split KEM*, and they suggest that the CSI-GDH problem seems to be helpful to construct a post-quantum split KEM.

2   Hard Homogeneous Spaces and CSIDH

Commutative Supersingular Isogeny-based Diffie–Hellman, CSIDH, is one of the candidates for post-quantum cryptosystems, proposed by Castryck et al. in 2018 [5]. CSIDH realizes Hard Homogeneous Space (HHS), formulated by Couveignes [8], using elliptic curves and isogenies. We briefly introduce CSIDH here. More detailed description of CSIDH is in Appendix B.

2.1   Hard Homogeneous Space

First, we give an informal definition of HHS. See [8] for more details.

Definition 1 (Homogeneous Space). *Let $\mathbb{G}$ be a finite abelian group and X be a finite set. $(\mathbb{G}, X)$ is called homogeneous space if $\mathbb{G}$ acts on X simply transitively, i.e., for every $x \in X$, a map $\mathbb{G} \to X$ such that $g \mapsto gx$ is bijective.*

Here, we consider two problems in a homogeneous space. The first one is the vectorization, which is to invert the group action, i.e., given $x_1, x_2 \in X$, find $g \in \mathbb{G}$ such that $x_2 = gx_1$. The second one is the parallelization, which is, given x_1, x_2, x_3, compute gx_3, where $g \in \mathbb{G}$ is the unique element in $\mathbb{G}$ which enjoys $x_2 = gx_1$. We say $(\mathbb{G}, X)$ is a hard homogeneous space (HHS) if these two problems are hard.

2.2   CSIDH

Let p be a large prime such that $p \equiv 3 \mod 4$. It is a well known fact that $\mathcal{C}\ell(\mathbb{Z}[\sqrt{-p}])$, the ideal class group of an order $\mathbb{Z}[\sqrt{-p}]$ acts simply transitively on $\mathcal{E}\ell\ell(\mathbb{Z}[\sqrt{-p}])$, the set of $\mathbb{F}_p$-isomorphic classes of supersingular elliptic curves whose $\mathbb{F}_p$-endomorphism ring is isomorphic to $\mathbb{Z}[\sqrt{-p}]$. It is believed that this action is hard to invert even for quantum computers, and thus we can regard $(\mathcal{C}\ell(\mathbb{Z}[\sqrt{-p}]), \mathcal{E}\ell\ell_p(\mathbb{Z}[\sqrt{-p}]))$ as a quantum-resistant HHS.

Remark 2. A recent study [4] proposes a protocol similar to CSIDH and slightly faster than CSIDH. This protocol is called CSURF, CSIDH on the surface. Since CSURF is also an instantiation of HHS, our result is also applicable to CSURF.

2.3   Key Exchanges Based on HHS

We can construct a Diffie–Hellman type key exchange protocol from HHS as shown in [8]. For a HHS $(\mathbb{G}, X)$, the key exchange between Alice and Bob proceeds as follows:

1. Alice and Bob share a public parameter $x_0 \in X$ beforehand.
2. Alice chooses a random element $a \in \mathbb{G}$ and sends ax_0 to Bob as a public key. Bob sends bx_0 to Alice in the same way.
3. On receiving Bob's public key bx_0, Alice computes $K_A = a(bx_0) = abx_0$. Bob computes $K_B = b(ax_0) = bax_0$ in the same way.

It is easy to see that $K_A = K_B$ because $\mathbb{G}$ is an abelian group. CSIDH is a Diffie–Hellman-like key exchange of this form. As mentioned above, CSIDH is conjectured to be a quantum-resistant key exchange protocol. For more details, see Appendix B and [5].

Note that this kind of key exchange is vulnerable to the man-in-the-middle attack. Let Charlie be an attacker who conducts the man-in-the-middle attack. Charlie intercepts Bob's message bx_0 and sends $C = cx_0$ to Alice instead, where $c \in \mathbb{G}$ is chosen by Charlie. Then, receiving altered message C, Alice computes her shared key $aC = (ac)x_0$. Here, Charlie can compute this key with $cA = (ca)x_0$. So, we cannot use this kind of key exchange in insecure channels.

3   Random Self-reducibility of Isogeny-Based Problems

In this section, we discuss the random self-reducibility of isogeny-based problems. First, we review the classical Diffie–Hellman-based problems in Sect. 3.1 to see what kinds of techniques are used to prove the random self-reducibility. Then, we move on to isogeny-based problems such as SIDH-based ones (Sect. 3.2) and CSIDH-based ones (Sect. 3.3). Our main contribution, the random self-reducibility of the gap problem of CSIDH, is in Sect. 3.3.

We first define the random self-reducibility:

Definition 3 (Random Self-Reducibility). *Let f be a function $f : X \to Y$. For a problem P, which is to evaluate $f(x)$ for randomly chosen $x \in X$, we say P is k-random self-reducible when we can generate $x_1, \ldots, x_k$ in polynomial time such that (1) $x_1, \ldots, x_k$ independently follow the same distribution which x follows and (2) given any one of $(1, f(x_1)), \ldots, (k, f(x_k))$, we can compute $f(x)$ efficiently with high probability. Moreover, if the solver of P whose instance is x' is given the access to the oracle $O_{x'}$, then P is k-random self-reducible if we can simulate O_y perfectly for all generated instances y with or without the help of O_x.*

If the problem P is k-random self-reducible for arbitrary positive integer k, we say P is random self-reducible.

3.1   The Classical Diffie–Hellman-Related Problems

Firstly, we will review Diffie–Hellman based problems in this subsection.

Decisional Problems of the Classical Diffie–Hellman. Now, we will describe the random self-reducibility of the DDH problem. We start with the definition of the DDH problem.

Problem 4 (Decisional Diffie–Hellman (DDH) Problem). Let p be a large prime, $\mathbb{G}$ be a cyclic group of prime order p, and g be a generator of $\mathbb{G}$. Furthermore, a random bit $b \in \{0, 1\}$ is taken uniformly. For uniformly sampled $x, y, z \in \mathbb{Z}/p\mathbb{Z}$, $(X, Y, Z) = (g^x, g^y, g^{xy})$ if $b = 1$ and $(X, Y, Z) = (g^x, g^y, g^z)$ otherwise. Here, the problem is to guess b, given $p, \mathbb{G}, g, (X, Y, Z)$.

Here, for an adversary $\mathcal{A}$ whose guess is b', the advantage is defined as

$$\mathrm{Adv}_{\mathrm{DDH}}^{\mathcal{A}}(\lambda) = \left| \Pr[b' = b] - \frac{1}{2} \right|.$$

Note that even when $b = 0$, Z happens to be equal to g^{xy}. However, since this event happens only with negligible probability, we can ignore such a case to avoid pathologies. In other words, we assume that $Z = g^{xy}$ if and only if $b = 1$ throughout this paper.

It is a well-known fact that the DDH problem is random self-reducible.

Proposition 5. *The DDH problem is random self-reducible.*

Proof. For an instance (X, Y, Z) of the DDH problem, we generate independent exponents $u_i, v_i, w_i \in \mathbb{Z}/p\mathbb{Z}$ for all $i = 1, 2, \ldots$. Then, we generate instances of the DDH problem as

$$(X_i, Y_i, Z_i) = (X^{w_i} g^{u_i}, Y g^{v_i}, Z^{w_i} X^{v_i w_i} Y^{u_i} g^{u_i v_i}). \tag{1}$$

Finally, given j-th answer (j, b_j), we answer b_j to the original problem.

Here, we check that (X_i, Y_i, Z_i) are distributed uniformity and independently, and that $b = b_j$, where b denotes the correct answer of the original problem. Let $(X, Y, Z) = (g^x, g^y, g^z)$ and $(X_i, Y_i, Z_i) = (g^{x_i}, g^{y_i}, g^{z_i})$. Then, we have $x_i = w_i x + u_i, y_i = y + v_i, z_i = x_i y_i + w_i (z - xy)$. Since u_i and v_i are independent and uniform, we can easily check that X_i and Y_i are uniform and independent. When $b = 1$, $z = xy$, thus $z_j = x_j y_j$, which means that $b_j = 1$. Otherwise, $z - xy \neq 0$, so $u_j, v_j, w_j(z - xy)$ distribute independently and uniformly.[2] Then, X_j, Y_j, Z_j are uniform and independent, so $b_j = 0$. $\qquad\square$

Note that we use two actions of $\mathbb{Z}/p\mathbb{Z}$ on $\mathbb{G}$ in this proof. One is the "additive action", which maps $(x, h) \in \mathbb{Z}/p\mathbb{Z} \times \mathbb{G}$ to $h \cdot g^x$. This action is additive in that we regard $\mathbb{Z}/p\mathbb{Z}$ as an additive group. The other is the "multiplicative action", which maps $(x, h) \in \mathbb{Z}/p\mathbb{Z} \times \mathbb{G}$ to h^x. We regard $(\mathbb{Z}/p\mathbb{Z})^\times$ as a multiplicative group in this case. Both actions are necessary to maintain independency of generated instances. Here, we remark that the additive action cannot be used to construct HHS because we can compute $h \cdot g^{x+y}$ from $g, h, h \cdot g^x, h \cdot g^y$ easily, which means that the parallelization is easy. However, the multiplicative action is considered to form HHS, and the classical Diffie–Hellman utilizes this action to achieve a secure key exchange.

Computational Problems of the Classical Diffie–Hellman. Here, we discuss the random self-reducibility of the CDH problem. As above, we start with the definition of problem.

Problem 6 (Computational Diffie–Hellman (CDH) Problem). Let p be a large prime, $\mathbb{G}$ be a cyclic group of prime order, and g be a generator of $\mathbb{G}$. For

---

[2] As mentioned above, we assume that $z = xy$ if and only if $b = 1$ to avoid pathology.

uniformly sampled $x, y \in \mathbb{Z}/p\mathbb{Z}$, the CDH problem is to compute g^{xy} given p, g, and (g^x, g^y).

Here, for an adversary $\mathcal{A}$ whose output is Z, the advantage is defined as $\mathrm{Adv}_{\mathrm{CDH}}^{\mathcal{A}}(\lambda) = \Pr[Z = g^{xy}]$.

Note that we can prove the random self-reducibility of the CDH problem in the same way as the DDH problem.

Proposition 7. *The CDH problem is random self-reducible.*

Proof. Given a CDH instance $(X, Y) = (g^x, g^y)$, we generate instances of the CDH problem $(X_i, Y_i) = (X^{u_i}, Y^{v_i})$, where u_i and v_i are chosen independently and uniformly. Then, given j-th answer Z_j, we answer $Z_j^{(v_j u_j)^{-1}}$ to the original problem. Here, independency and uniformness are checked in a similar way to Proposition 5. As for correctness, let $(X_i, Y_i) = (g^{x_i}, g^{y_i})$. Then we have $x_i y_i = x u_i y v_i$, so $Z_j^{(v_j u_j)^{-1}} = (g^{x u_j y v_j})^{(v_j u_j)^{-1}} = g^{xy}$. □

On the contrary to the DDH-case, we use only multiplicative action in this proof. This difference enables us to prove the random self-reducibility of the computational problem of CSIDH as discussed later in Sect. 3.3.

Gap Problems of the Classical Diffie–Hellman. Now, we see the GDH problem, the gap Diffie–Hellman problem. The GDH problem is defined as follows:

Problem 8 (Gap Diffie–Hellman (GDH) problem). Notation as in Problem 6. The GDH problem is to compute g^{xy}, given access to the *decision oracle* DDH, which solves the DDH problem, i.e., $\mathsf{DDH}(g^a, g^b, g^c) = \mathsf{true}$ if and only if $c = ab$.

Here, for an adversary $\mathcal{A}$ whose output is Z, the advantage is defined as $\mathrm{Adv}_{\mathrm{GDH}}^{\mathcal{A}}(\lambda) = \Pr[Z = g^{xy}]$.

Note that if the DDH problem is solved, we have to consider the CDH problem as the GDH problem. In other words, the GDH problem is the CDH problem where the DDH problem is solved. Note that the GDH problem is also random self-reducible. The proof is almost identical to the proof of Proposition 7. The only difference is that we have to simulate the decision oracles but we have only to pass the query to the original oracle.

3.2 SIDH-Related Problems

Secondly, we discuss the problems related to SIDH, Supersingular Isogeny Diffie–Hellman. SIDH is one of the most major isogeny-based cryptosystems proposed in [16], so we cannot ignore SIDH if we try to construct an isogeny-based AKE. However, our conclusion is that it is difficult to construct a SIDH-based AKE with small security loss since there is no suitable problem assumed to be hard.

Decisional and Computational Problems of SIDH. SIDH is a key exchange protocol which can be modeled as random walks on the isogeny graphs. Thus, SIDH has a very pour algebraic structure. See [16] for the detailed protocol. Note that neither the decisional nor the computational problem of SIDH has been proved to be random self-reducible. Poor algebraic structure is one of the reason for this. Thus we cannot transfer the proof of the classical Diffie–Hellman case naively.

Since SIDH is subject to the man-in-the-middle attack, a lof of SIDH-based AKEs have been proposed [12, 19, 23]. Most of them are based on the difficulty of decisional problems, and have relatively large security loss, which indicates that the decisional problem of SIDH is not random self-reducible. To the best of our knowledge, no AKE based on the computational problem of SIDH has been proposed.

As a summary, it seems difficult to construct an AKE with small security loss from the decisional or the computational problem of SIDH.

Gap Problems of SIDH. Gap problems are very helpful to prove the security of AKE in the random oracle model, because we can capture the hash query of an adversary and check if the query contains the correct answer of the problem with the decision oracle. However, as for SIDH, it is hopeless at present as discussed below.

As pointed out in [14], gap assumptions related to SIDH sometimes do not hold, that is, there are cases that we can solve the gap problems efficiently, see [15]. So, it is risky to use a gap type problem for a security proof of SIDH-based schemes. As a result, the security of almost all existing SIDH-based AKE schemes [14, 19, 23] are proved without using gap problems contrary to the utility of gap problems in security proofs for the classical Diffie–Hellman setting. This is one of the biggest obstruction to construct SIDH-based AKE schemes.

On the other hand, there is an attempt to overcome this obstruction. The gap problem which proved to be easy has a restriction on the degree of isogenies, which is essential condition for Galbraith-Vercauteren attack [15] work. So, in order to avoid this attack, removing the condition on the degree of isogenies, Fujioka et al. propose a new gap problem, the degree-insensitive SI-GDH problem, see [12] §4. They use such a gap problem in the security proof of their AKE protocol. However, in [10], it is conjectured with an evidence that public keys in the degree-insensitive version are uniformly distributed. This conjecture shows that the degree-insensitive SI-GDH problem no longer makes sense. Here, we note that this does not mean that the AKE scheme in [12] is broken and only that its security is not supported by the computational assumption used in [12], for more detail, see [10].

Summary. As discussed above, since the mechanism of SIDH is modeled as random walks in the isogeny graph, formulating a well-defined gap problem in SIDH setting is difficult. So, in order to construct AKE from SIDH, we have to employ decisional problems. However, the random self-reducibility of SIDH-related decisional problems has not been proved yet. In conclusion, constructing tightly secure AKE schemes in SIDH setting is seems to be difficult currently.

3.3 CSIDH-Related Problems

Finally, we move on to CSIDH. Though the decisional problem of CSIDH is still likely not to be random self-reducible, we show that the computational and the gap problems are random self-reducible.

Decisional Problem of CSIDH. The decisional problem of CSIDH, CSI-DDH problem, seems not to be random self-reducible. In this subsection, we will compare the CSI-DDH problem with the DDH problem and discuss what prevents the random self-reducibility of the CSI-DDH problem.

First, we define the CSI-DDH problem.

Problem 9 (Commutative Supersingular Isogeny Decisional Diffie–Hellman (CSI-DDH) Problem). Let p be a large prime such that $p \equiv 3 \mod 4$ and E be a supersingular elliptic curve in $\mathcal{E}\ell\ell(\mathbb{Z}[\sqrt{-p}])$. Then, given $(E_1, E_2, E') = ([\mathfrak{r}]E, [\mathfrak{y}]E, E')$ for uniformly chosen $[\mathfrak{r}], [\mathfrak{y}] \in Cl(\mathbb{Z}[\sqrt{-p}])$, guess whether $E' \simeq [\mathfrak{r}\mathfrak{y}]E$ or not.

Here, for an adversary $\mathcal{A}$, the advantage is defined as $\mathrm{Adv}_{\mathrm{CSI\text{-}DDH}}^{\mathcal{A}}(\lambda) = \left| \Pr[\text{guess is correct}] - \frac{1}{2} \right|$.

One may imagine that the CSI-DDH problem is also random self-reducible since CSIDH and the classical Diffie–Hellman have very similar structures. However, to the best of our knowledge, we have not succeeded in proving so. For example, a recent study [11] leaves it as an open problem.

Here, we discuss the difference between the classical Diffie–Hellman and CSIDH, i.e., why we cannot prove the random self-reducibility of the CSI-DDH problem in the same way for the DDH problem. As for the classical Diffie–Hellman case, we use both additive action and multiplicative action to prove the random self-reducibility of the decisional problem. In CSIDH case, we have only one action, so it seems difficult to transfer the proof to the CSIDH settings. However, this lack of operation is inevitable to achieve quantum-resistance because of Shor's algorithm [22].

Remark 10. A recent study of the DDH assumption [6] shows that we can solve the DDH problem of HHS from ideal-class group action under certain circumstances. That is, if $p \equiv 1 \mod 4$, we can solve the CSI-DDH problem efficiently. This is not the case here because we restrict $p \equiv 3 \mod 4$ and in this case we cannot use this result directly. However, some modification may be required in the future even when $p \equiv 3 \mod 4$.

Computational Problems of CSIDH. Now, we will show that the computational problem of CSIDH is random self-reducible.

First, we define the CSI-CDH problem. It is defined in an analogous way to the classical Diffie–Hellman.

Problem 11 (Commutative Supersingular Isogeny-Computational Diffie-Hellman (CSI-CDH) Problem). Let p be a large prime such that $p \equiv 3 \mod 4$ and E be a supersingular elliptic curve in $\mathcal{E}\ell\ell(\mathbb{Z}[\sqrt{-p}])$. Then, given $(E_1, E_2) = ([\mathfrak{r}]E, [\mathfrak{y}]E)$ for uniformly chosen $[\mathfrak{r}], [\mathfrak{y}] \in Cl(\mathbb{Z}[\sqrt{-p}])$, compute $[\mathfrak{ry}]E$.

Here, for an adversary $\mathcal{A}$ whose output is E', the advantage is defined as $\mathrm{Adv}_{\mathrm{CSI\text{-}CDH}}^{\mathcal{A}}(\lambda) = \Pr[E' = [\mathfrak{ry}]E]$.

Here, we probe the random self-reducibility of the CSI-CDH problem.

Theorem 12. *The CSI-CDH problem is random self-reducible.*

Proof. For an instance (E_A, E_B) of the CSI-CDH problem, we generate random ideal classes $[\rho_i], [\eta_i] \in \mathcal{C}\ell(\mathcal{O})$. Then, we generate instances as $([\rho_i]E_A, [\eta_i]E_B)$. Then, given (j, E_j) for some j, we answer $[\eta_j]^{-1}[\rho_j]^{-1}E_j$ to the original problem. Since the action is simply transitive, the independency and uniformity of these generated elliptic curves are assured. Here, since $E_j = [\rho_i][\eta_i]E$, $[\eta_j]^{-1}[\rho_j]^{-1}E_j = [\mathfrak{ry}]E$, we complete the proof.   □

In this proof, we use almost the same technique as in Theorem 7, the CDH case, while it is impossible in the decisional case. This is mainly because the proof of the random self-reducibility of the CDH problem uses only one operation, whereas we use two operations in the decisional case. In the CDH problem case, as discussed above, we use only the multiplicative action to prove the random self-reducibility. Since only the multiplicative action forms HHS in the classical DH case, the action in CSIDH corresponds to the multiplicative action of the classical DH, so it is plausible that we can transfer the proof to CSIDH settings in this case.

Gap Problems of CSIDH. As a corollary of Theorem 12, we can prove that the gap problem of CSIDH, CSI-GDH problem, is also random self-reducible. We start with the definition of the problem:

Problem 13 (Commutative Supersingular Isogeny Gap Diffie–Hellman problem (CSI-GDH problem)). Notation as in Problem 11. The CSI-GDH problem is to compute $[\mathfrak{ry}]E$, given access to the decision oracle CSI-DDH, which solves the corresponding CSI-DDH problem, i.e., given $([\mathfrak{a}]E, [\mathfrak{b}]E, E')$, CSI-DDH returns true if and only if $E' = [\mathfrak{ab}]E$.

Here, for an adversary $\mathcal{A}$ whose output is E', the advantage is defined as $\mathrm{Adv}_{\mathrm{CSI\text{-}GDH}}^{\mathcal{A}}(\lambda) = \Pr[E' = [\mathfrak{ry}]E]$.

Here, we can prove that the CSI-GDH problem is also random self-reducible:

Corollary 14. *The CSI-GDH problem is random self-reducible.*

Proof. The only difference from the CSI-CDH case (Proposition 7) is that there exists a decision oracle, but we don't need special treatment since we have only to pass every query to the original oracle.   □

Moreover, we will consider the CSI-stDH problem, which is a CSIDH-version of strong DH problem used in [7]. This problem is also random self-reducible, and we will construct an AKE which is secure under the difficulty of this problem in the following section. The proof of the random self-reducibility of the CSI-stDH problem is in Appendix C.

Problem 15 (strong Diffie–Hellman (stDH) Problem). Notation as in Problem 6. The stDH problem is to compute g^{xy}, given access to the stDH oracle $\mathsf{stDH}_x(\cdot, \cdot)$, which solves the corresponding decisional problem, i.e., $\mathsf{stDH}_x(g^b, g^c) = \mathsf{true}$ if and only if $c = xb$.

Here, for an adversary $\mathcal{A}$ whose output is Z, the advantage is defined as $\mathrm{Adv}_{\mathrm{stDH}}^{\mathcal{A}}(\lambda) = \Pr[Z = g^{xy}]$.

Problem 16 (Commutative Supersingular Isogeny strong Diffie–Hellman (CSI-stDH) Problem). Let p be a large prime such that $p \equiv 3 \mod 4$ and E be a supersingular elliptic curve in $\mathcal{E}\ell\ell(\mathbb{Z}[\sqrt{-p}])$. Then, given $(E_1, E_2) = ([\mathfrak{x}]E, [\mathfrak{y}]E)$ for uniformly chosen $[\mathfrak{x}], [\mathfrak{y}] \in Cl(\mathbb{Z}[\sqrt{-p}])$, compute $[\mathfrak{x}\mathfrak{y}]E$. Here, the solver can query to the CSI-strong DH oracle $\mathsf{CSI\text{-}stDH}_{\mathfrak{x}}(\cdot, \cdot)$, such that $\mathsf{CSI\text{-}stDH}_{\mathfrak{x}}(E', E'') = \mathsf{true}$ if and only if $E'' = [\mathfrak{x}]E'$.

Here, for an adversary $\mathcal{A}$ whose output is E', the advantage is defined as $\mathrm{Adv}_{\mathrm{CSI\text{-}stDH}}^{\mathcal{A}}(\lambda) = \Pr[E' = [\mathfrak{x}\mathfrak{y}]E]$.

Summary. We have proved that the computational problem and the gap problem of CSIDH are random self-reducible. This is in stark contrast between computational problems and decisional problems. So, our conclusion is: If we hope for (somewhat) tight reduction in quantum-resistant HHS, we should use computational assumptions rather than decisional assumptions.

4 Protocol Π_{CSIDH}

In this section, we propose an isogeny-based AKE Π_{CSIDH}, which is a variation of the Protocol Π in [7]. The security loss is $O(\mu)$, where μ stands for the number of users. In the security proof, we use the random self-reducibility of the CSI-stDH problem implicitly. Since Π_{CSIDH} uses CSIDH instead of the classical Diffie–Hellman, it is expected to be quantum-resistant.

4.1 AKE Security Model

In this subsection, we discuss the security model briefly.

The security of an AKE is often defined through a game between a challenger $\mathcal{C}$ and an adversary $\mathcal{A}$. First, $\mathcal{C}$ shows public information of AKE to $\mathcal{A}$. Second,

$\mathcal{A}$ executes some attacks defined in the model. The main difference among AKE security models is which attacks are allowed to the attacker. We regard every single execution of a user as an oracle, and every attack is written in the form of a query to an oracle. Finally, $\mathcal{A}$ chooses some oracles to get real session keys or random session keys. This procedure is often called test-query, and $\mathcal{A}$ wins the game if $\mathcal{A}$ can decide correctly whether the given keys are real keys or random keys.

In this paper, we use the same model as [7]. We call this model "CCGJJ model". CCGJJ model does not allow the adversary to reveal the internal state of an oracle, compared to the CK model [3], which is one of the most popular models. In other words, we achieved an optimally tight AKE at the expense of security. The CCGJJ model is described in Appendix A in a formal way. We note here that the adversary in CCGJJ model is allowed to reveal the static key of user i if i is not the intended peer of the tested oracles, where intended peer of an oracle is the user the oracle "wants" to communicate.

4.2 Construction

Protocol Π_{CSIDH} is defined in Fig. 2. In this protocol, we use both static keys (users' key) and ephemeral keys (oracles' key) to establish a shared secret key. Note that we choose an element from $\mathcal{C}\ell(\mathcal{O})$ in the same way as in CSIDH, see Appendix B for more details.

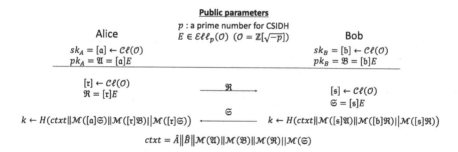

Fig. 2. Proposed AKE based on CSIDH (Protocol Π_{CSIDH}). $\mathcal{M}$ denotes the Montgomery coefficient.

4.3 Security

Here, we state the security theorem of Π_{CSIDH}. We denote the advantage of an adversary $\mathcal{A}$ against a problem P whose parameters are param as $\text{Adv}^{\mathcal{A}}_{P,\text{param}}(\lambda)$, where λ is the security parameter. We often omit some parameters when we don't have much attention to the parameters.

Theorem 17. *Let $\mathcal{A}$ be an adversary against Protocol Π_{CSIDH} in CCGJJ model under the random oracle model and assume we use $[-m, m]^n$ as a secret key* *space of CSIDH for positive integers m, n. Then, there are adversaries $\mathcal{B}_1, \mathcal{B}_2, \mathcal{B}_3$ against the CSI-stDH problem such that*

$$\mathrm{Adv}^{\mathcal{A}}_{\Pi_{\mathrm{CSIDH}}}(\lambda) \leq \mu \cdot \mathrm{Adv}^{\mathcal{B}_1}_{\mathrm{CSI\text{-}stDH}}(\lambda) + \mathrm{Adv}^{\mathcal{B}_2}_{\mathrm{CSI\text{-}stDH}}(\lambda)$$

$$+ \mu \cdot \mathrm{Adv}^{\mathcal{B}_3}_{\mathrm{CSI\text{-}stDH}}(\lambda) + \frac{\mu l^2}{(2m+1)^n},$$

where μ and l are the number of users and the maximum number of sessions per user, respectively. Moreover, the adversaries $\mathcal{B}_1, \mathcal{B}_2, \mathcal{B}_3$ all run in essentially the same time as $\mathcal{A}$ and make essentially the same number of queries to the hash oracle H.

Here, we explain the proof technique and discuss the relation to the random self-reducibility of the CSI-stDH problem. The detailed proof is in Appendix A.2.

The Security Proof of Cohn-Gordon et al. In [7], the authors construct a simulator $\mathcal{S}$ that reduce the AKE's security to the hardness of the stDH problem. We briefly review the technique they used. $\mathcal{S}$ tries to solve the stDH problem, so it gets $(X = g^x, Y = g^y)$ as an instance of the stDH problem. Then, $\mathcal{S}$ chooses one user i uniformly and sets i's static public key to $X = g^x$. Moreover, for every oracle whose intended peer is i, $\mathcal{S}$ sets its ephemeral public key to $Y \cdot g^\rho$. Here, ρ's are chosen independently and uniformly for every oracle, so embedded keys are also uniform and independent. Note that $\mathcal{S}$ cannot compute the corresponding static secret key or ephemeral secret keys in polynomial time since $\mathcal{S}$ has to solve the DLP which contradicts the assumption that stDH problem is hard.

Now, we assume that $\mathcal{A}$ test-queries to an oracle whose intended peer is i. This assumption prevents $\mathcal{A}$ from revealing the secret key of the user i. In this case, $\mathcal{A}$ has to compute $g^{x(y+\rho)}$ and query it to the hash oracle in order to decide whether the given key is real-or-random correctly with noticeable advantage. Here, in the random oracle model, the simulator can catch the hash-queries made by the adversary, so $\mathcal{S}$ catches every queries and checks if the query contains $g^{x(y+\rho)}$ with the decision oracle. Since we can eliminate ρ by computing $X^{-\rho} \cdot g^{x(y+\rho)}$, this enables the simulator to solve the stDH problem correctly whenever $\mathcal{A}$ wins the game. Since our assumption occurs with probability at least $1/\mu$, the security loss is $O(\mu)$.

Our Proof Sketch. We apply the technique above to Π_{CSIDH}. Let $\mathcal{S}$ be the simulator for an adversary $\mathcal{A}$. Given $(\mathfrak{X} = [\mathfrak{x}]E, \mathfrak{Y} = [\mathfrak{y}]E)$, $\mathcal{S}$ chooses a user i uniformly and sets i's static public key to $\mathfrak{X}$. Then, for every oracle whose intended peer is i, $\mathcal{S}$ chooses independent $[\rho] \in \mathcal{C}\ell(\mathcal{O})$ uniformly and sets the oracle's ephemeral public key to $[\rho]\mathfrak{Y}$. We assume that $\mathcal{A}$ chooses the oracle whose intended peer is i as a test oracle as above. In this case, in order for $\mathcal{A}$ to win the game, $\mathcal{A}$ has to compute $[\mathfrak{x}\rho\mathfrak{y}]E$ and query it to the random oracle, so $\mathcal{S}$ can detect it. Since $\mathcal{S}$ knows ρ and can invert it in CSIDH-setting, $\mathcal{S}$ can compute $[\mathfrak{x}\mathfrak{y}]E$ in this case, which completes the proof.

In our proof sketch above, we make multiple independent ephemeral keys, $[\rho]\mathfrak{Y}$. This is a similar rerandomization to the proof of Theorem 7. So we can regard this proof as an application of the random self-reducibility of the CSI-CDH problem or the CSI-stDH problem, which follows from our main contribution.[3]

4.4   Efficiency Analysis

In this subsection, we compare our AKE, Π_{CSIDH}, with other existing CSIDH-based AKEs in terms of efficiency. To the best of our knowledge, [13] is the only study that proposes other CSIDH-based AKEs. In [13], two CSIDH-based AKEs, CSIDH UM and CSIDH Biclique, are proposed.

Assume that we want to construct an AKE of λ-bit security, with $\mu \simeq 2^m$ users and each user conducts sessions at most $l \simeq 2^n$ times. Then, for CSIDH UM, we should use the parameters such that the 2DDH assumption provides at least $(2\lambda + 2m + 2n - 1)$-bit security. This evaluation follows from the inequality in the security proof. In the same way, the 2GDH assumption for CSIDH Biclique (resp. the CSI-stDH problem for Π_{CSIDH}) should provide at least $(\lambda + 2\max(m,n))$-bit (resp. $(\lambda + m)$-bit) security.

As a concrete example, suppose that we want to achieve 110-bit classical security ($\lambda = 110$), with 2^{16} users ($\mu = 2^{16}$ and $m = 16$), and every user conducts sessions at most 2^{16} times ($l = 2^{16}$ and $n = 16$), as done in [7]. In this case, security levels required for the underlying hard problems of CSIDH UM, CSIDH Biclique, and Π_{CSIDH} are 283, 143, and 126-bits, respectively. Though these assumptions are different, we assume that the fastest way to solve these problems is to invert the group action, since we have not found any other way to solve these problems yet.

Comparing Π_{CSIDH} with CSIDH Biclique, Π_{CSIDH} is the faster because the number of actions and the required security level of the underlying problem are both lower. As for CSIDH UM, the required security level of the underlying problem is about twice as much as that of Π_{CSIDH}. This is mainly because, in the security proof of CSIDH UM, the square of the advantage of the adversary against CSIDH UM is bounded from above by the advantage of the reduction. As a consequence, we have to use much larger parameters for CSIDH UM. Actually, we can use the parameter set CSIDH-512 [5] for the Π_{CSIDH}, since CSIDH-512 offers 128-bit classical security. On the other hand, for CSIDH UM, we should use the parameter set CSIDH-1024 (or larger parameters), since CSIDH-1024 offers 256-bit classical security. As [1] shows, the evaluation of the group action with parameters CSIDH-1024 takes more than 6 times as much as that of the CSIDH-512. So, although CSIDH UM has relatively fewer number of actions than Π_{CSIDH}, Π_{CSIDH} is the faster because every evaluation of the group action takes much less time. The comparison above is summarized in Table 2.

As a conclusion, our AKE is the fastest CSIDH-based AKE when we consider concrete security.

---

[3] Similarly, the proof of Cohn-Gordon et al. can be considered as an application of the random self-reducibility of the stDH problem.

Table 2. Efficiency analysis when we aim for 110-bit secure AKE. We assume that there exist 2^{16} users and each users executes the session at most 2^{16} times. "Assumption" shows the problems assumed to be hard. We use either CSIDH-512 or CSIDH-1024 [5] for the parameters. "Number of actions" shows how many times every user evaluates the group actions in a session. We estimated the expected clock cycles using the result of Bernstein et al. [1].

Protocol	Assumption	Parameters
CSIDH UM	2DDH	CSIDH-1024
CSIDH Biclique	2GDH	CSIDH-512
Π_{CSIDH}	CSI-stDH	CSIDH-512

Protocol	Number of actions	Expected clock cycles
CSIDH UM	3	$719,084,288 \times 3 = 2,157,252,864$
CSIDH Biclique	5	$119,995,936 \times 5 = 599,979,680$
Π_{CSIDH}	4	$119,995,936 \times 4 = 479,983,744$

5 Conclusion

In this paper, we proved that the computational problem and the gap problem of CSIDH are random self-reducible and concluded that we should use computational or gap problems to construct a CSIDH-based protocol with small security loss. Moreover, we proposed an AKE from CSIDH as an application, following the construction of Cohn-Gordon et al. [7]. This AKE is proved to be the fastest CSIDH-based AKE when we aim for a certain level of security.

Now, we have some future works. First, we have to analyse the difficulty of the CSI-GDH problem. Though the CSI-GDH problem is regarded as hard even for quantum computers at present, since few works have been done on this gap problem, we have to study this problem more. Second, we need to study more about how secure the CCGJJ model is, particularly compared to CK or CK$^+$ model. CCGJJ model is weaker than the CK model in that the states of the oracles are never leaked to the adversary. This difference seems very large, so we have to analyze the security of CCGJJ model more. Finally, if we can construct a CSIDH-based AKE in CK or CK$^+$ model with (optimally) tight proof, it is a very large contribution because one of the solution to the above future works is to construct a CSIDH-based AKE in CK model with (optimally) tight proof. Note that it seems impossible to use random self-reducibility since if we embed the instance of a problem to multiple oracles in the CK model, the adversary may try to reveal one of such oracle's secret key.

A   Authenticated Key Exchange

In this section, we give a detailed proof of Theorem 17.

A.1   CCGJJ Security Model

First, we will introduce the security model, which we call CCGJJ model in this paper. This model was introduced by [7]. The most important difference between CCGJJ model and CK model [3] is that the adversary cannot reveal an oracle's internal state, including an ephemeral secret key. In both models, we define a game between a challenger and an adversary, and if the advantage of an arbitrary efficient adversary is negligible, the protocol is regarded to be secure.

Execution Environment. Here, we describe the mathematical model of the execution environment. We assume that there exist μ users and each user $i \in \{1, \cdots, \mu\}$ has long-term public key pk_i and long-term secret key sk_i. We assume that each user i executes the protocol at most l times and each execution is regarded as an oracle. User i's s-th oracle is denoted as π_i^s. π_i^s uses not only user's static key but also its ephemeral key in the execution. Note that a static key is a user's key, so if two oracles belong to the same user, then these two oracles use the same static key, where the ephemeral keys are different with high probability. Each invocation of the protocol is called a session, and the shared secret is called a session key.

Each oracle π_i^s has an intended peer, denoted as Pid_i^s. Also, the session key of π_i^s is denoted as k_i^s, where $k_i^s = \emptyset$ if π_i^s has not computed the session key yet. The oracles send messages each other, and $\mathrm{sent}_i^s/\mathrm{recv}_i^s$ are the messages sent/received by π_i^s. Moreover, each oracle π_i^s has a role, $\mathrm{role}_i^s \in \{\emptyset, \mathrm{init}, \mathrm{resp}\}$. Here, the role of an oracle is either an initiator (denoted as init) or a responder (denoted as resp). An initiator is an oracle which sends a message first, and the responder oracle follows. In Fig. 2, Alice's oracle is the initiator and Bob's one is the responder. Note that a responder oracle computes its session key first in the session, and the initiator follows.

To describe partnering between oracles, we define two notions:

Definition 18 (Origin oracle). π_j^t is an origin oracle of π_i^s if both oracles have completed its execution and the messages sent by π_j^t are equal to the messages received by π_i^s, i.e., $\mathrm{sent}_j^t = \mathrm{recv}_i^s$.

Definition 19 (Partner oracles). π_i^s and π_j^t are called partners if (1) π_j^t is an origin oracle of π_i^s and vice versa, (2) both oracles believe the other as an intended peer, i.e., $\mathrm{Pid}_i^s = j$ and $\mathrm{Pid}_j^t = i$, and (3) their roles are distinct, i.e., $\mathrm{role}_i^s \neq \mathrm{role}_j^t$.

Attacker's Model. Since each execution is regarded as an oracle, what attacker can do are described as queries. In CCGJJ model, attacker can issue four queries, Send, RevLTK, RegisterLTK, and RevSessKey.

Send represents the ability of the adversary to control the network, i.e., Send query allows the adversary to send arbitrary message to arbitrary oracle, or even starts an oracle. RevLTK and RevSessKey stand for *Reveal Long-Term Key* and *Reveal Session Key*. The adversary can reveal arbitrary oracle's long-term key or session key. Here, the user whose oracle's long-term key is revealed with this query is said to be *corrupted*. RegisterLTK allows the adversary to add a new user. Any oracle of users added by this query is corrupted by definition.

Moreover, the adversary can issue special queries, Test.

Definition 20 (Test query). *Assume $b \in \{0,1\}$ is determined beforehand. If an adversary queries a Test query to π_i^s, π_i^s returns k_b, where k_0 is a random key and k_1 is its session key. This query is denoted as* Test(i,s).

Here, we note that all oracles use the same bit b. Now, we define a state of an oracle, *fresh*.

Definition 21 (Freshness). *We say π_i^s is fresh if following conditions hold: (1)* RevSessKey(i,s) *has not been queried, (2) when π_j^t is the partner oracle of π_i^s, neither* Test(j,t) *nor* RevSessKey(j,t) *has been issued, and (3)* Pid$_i^s$ *was not corrupted when π_i^s completed its execution if π_i^s has an origin oracle, and not corrupted at all otherwise.*

The session key of a fresh oracle is not revealed by queries (it is fresh in this sense). So, if all tested oracles are fresh and the adversary can guess b correctly, we can conclude that the adversary can break the AKE's security. The following definition of the AKE security game describe this formally. We say that an AKE is secure if all efficient adversary have negligible advantages.

Definition 22 (AKE security game). *Let $\mathcal{C}$ be a challenger and $\mathcal{A}$ be an adversary. The security game proceeds as follows:*

1. *$\mathcal{C}$ chooses μ static keys (sk_i, pk_i) $(i = 1, 2, \cdots, \mu)$ and $b \in \{0,1\}$ uniformly at random, and initializes all oracles.*
2. *$\mathcal{C}$ runs $\mathcal{A}$ with inputs $pk_1, \cdots, pk_\mu$. The model allows $\mathcal{A}$ to make some attacks on oracles as queries to an oracle, including* Test *queries. Here, $\mathcal{A}$ must keep tested oracles fresh. Otherwise, the game aborts and b' is set to be a random bit.[4]*
3. *$\mathcal{A}$ outputs b', a guess of b.*

The advantage of an adversary is

$$\text{Adv}_{\text{prot}}^{\mathcal{A}}(\lambda) = \left| \Pr[b' = b] - \frac{1}{2} \right|,$$

where λ denotes a security parameter.

---

[4] In this case, the advantage of the adversary is zero.

A.2 Detailed Security Proof of Π_{CSIDH}

In this subsection, we give a proof of Theorem 17. First, we classify the oracles into 5 types in the same way as [7].

Type I Initiator oracles whose response message is sent by a responder which has the same *ctxt* and whose intended peer is honest, i.e., not corrupted when the message is received.

Type II Other initiators whose intended peer is honest until the initiator completes the execution.

Type III Responder oracles whose initial message is sent by a initiator which has the same *ctxt* up to the responder message and whose intended peer is honest when the message is received.

Type IV Other responders whose intended peer is honest until the responder completes the execution.

Type V Oracles that are not Type I, II, III, or IV. In other words, oracles whose intended peer is corrupted.

Note that Type I, II, III, and IV oracles are fresh, whereas Type V oracles are not fresh. So we have only to consider first four types of oracles when we make a security proof, because we don't need to care the case when non-fresh oracles are tested.

Again, the security theorem is as follows:

Theorem 17. *Let $\mathcal{A}$ be an adversary against Protocol Π_{CSIDH} in CCGJJ model under the random oracle model and assume we use $[-m, m]^n$ as a secret key space of CSIDH for positive integers m, n. Then, there are adversaries $\mathcal{B}_1, \mathcal{B}_2, \mathcal{B}_3$ against the CSI-stDH problem such that*

$$
\mathrm{Adv}_{\Pi_{\mathrm{CSIDH}}}^{\mathcal{A}}(\lambda) \leq \mu \cdot \mathrm{Adv}_{\mathrm{CSI\text{-}stDH}}^{\mathcal{B}_1}(\lambda) + \mathrm{Adv}_{\mathrm{CSI\text{-}stDH}}^{\mathcal{B}_2}(\lambda)
$$
$$
+ \mu \cdot \mathrm{Adv}_{\mathrm{CSI\text{-}stDH}}^{\mathcal{B}_3}(\lambda) + \frac{\mu l^2}{(2m + 1)^n},
$$

where μ and l are the number of users and the maximum number of sessions per user, respectively. Moreover, the adversaries $\mathcal{B}_1, \mathcal{B}_2, \mathcal{B}_3$ all run in essentially the same time as $\mathcal{A}$ and make essentially the same number of queries to the hash oracle H.

In this Appendix, we give a proof of this theorem.

Proof. We prove this theorem by changing the game little by little. This technique is called "game-hopping" technique. Let S_j $(j = 0, 1, \cdots, 5)$ be events that the adversary wins in Game j.

Game 0. Game 0 is the original security game.

Game 1. In Game 1, we abort if two initiators or responders have the same ctxt. Since the size of our key space is $(2m + 1)^n$, we have

$$| \Pr[S_0] - \Pr[S_1]| \leq \frac{\mu l^2}{(2m + 1)^n} \tag{2}$$

Game 2. In Game 2, the oracles change the way they choose their session keys. Intuitively, they try to choose their session key uniformly at random, not using the hash function.

For example, let π_j^t be a Type IV oracle with $sk_j = [\mathfrak{b}]$ and $pk_j = \mathfrak{B}$. Also, let π_j^t's ephemeral secret key and ephemeral public key be $[\mathfrak{s}]$ and $\mathfrak{S}$. Moreover, for $i = \mathrm{Pid}_j^t$, let $i's$ long-term public key and ephemeral public key be $\mathfrak{A}$ and $\mathfrak{R}$, respectively.

Then, π_j^t has to query

$$x = \hat{i} || \hat{j} || \mathcal{M}(\mathfrak{A}) || \mathcal{M}(\mathfrak{B}) || \mathcal{M}(\mathfrak{R}) || \mathcal{M}(\mathfrak{S}) || \mathcal{M}([\mathfrak{s}]\mathfrak{A}) || \mathcal{M}([\mathfrak{b}]\mathfrak{R}) || \mathcal{M}([\mathfrak{s}]\mathfrak{R})$$

to the hash oracle in Game 1. If x has not been queried or "registered" to the random oracle, then π_j^t takes its session key k uniformly at random, and "register" (x, k). If (x, k') is registered to the random oracle, then π_j^t sets its session key to k'. In the beginning of the game, no queries are registered.

Other type of the oracles choose their session key in similar ways, so we omit the description. For further details, see [7].

Random oracle model assures that no difference is observable by $\mathcal{A}$, so we have

$$\Pr[S_1] = \Pr[S_2]. \tag{3}$$

Game 3. In this game, Type IV oracles choose their session keys uniformly at random and do not modify the hash oracle unless whose intended peer is corrupted.

Let π_j^t be a type IV responder and we use the same notation as in **Game 2**. Then, π_j^t must have queried

$$x = \hat{i} || \hat{j} || \mathcal{M}(\mathfrak{A}) || \mathcal{M}(\mathfrak{B}) || \mathcal{M}(\mathfrak{R}) || \mathcal{M}(\mathfrak{S}) || \mathcal{M}([\mathfrak{s}]\mathfrak{A}) || \mathcal{M}([\mathfrak{b}]\mathfrak{R}) || \mathcal{M}([\mathfrak{s}]\mathfrak{R}) \tag{4}$$

in Game 2. If queries of the form (4) do not happen before user i is corrupted, Game 2 and Game 3 are identical. So when we define the event F_i as the event that such queries are made, we have

$$| \Pr[S_2] - \Pr[S_3]| \leq \sum_i \Pr[F_i].$$

In order to make our proof simple, we define event G_i as the event that queries of the form

$$\hat{i} || \hat{j} || \mathcal{M}(\mathfrak{A}) || \mathcal{M}(\mathfrak{B}) || \mathcal{M}(\mathfrak{R}) || \mathcal{M}(\mathfrak{S}) || \mathcal{M}(\mathfrak{W}) || \star || \star, \mathfrak{W} = [\mathfrak{as}]E \tag{5}$$

are made before user i is corrupted. The symbol $\star$ means an arbitrary element. Since $\Pr[F_i] \leq \Pr[G_i]$ holds, we have

$$|\Pr[S_2] - \Pr[S_3]| \leq \sum_i \Pr[G_i]. \tag{6}$$

We can bound the righthand side by the advantage of a CSI-stDH adversary.

CSI-stDH Adversary $\mathcal{B}_1$. The reduction $\mathcal{B}_1$ is an algorithm whose inputs are two elliptic curves $(E_1, E_2) = ([\mathfrak{r}]E, [\mathfrak{y}]E) \in \mathcal{E}ll(\mathcal{O})^2$, and output is an elliptic curve E_3. The advantage of $\mathcal{B}_1$ is $\Pr[E_3 = [\mathfrak{r}\mathfrak{y}]E]$.

When $\mathcal{B}_1$ is given a tuple $(E_1, E_2) \in \mathcal{E}ll(\mathcal{O})^2$, it chooses a user i uniformly at random, and sets its static public key to E_1. Then, for every Type IV responder, $\mathcal{B}_1$ sets its ephemeral public key to $[\rho]E_2$, where each $[\rho] \in \mathcal{C}l(\mathcal{O})$ is sampled in the same way as key generation for every oracle. Here, $[\rho]$ is chosen independently for every Type IV responders.

Suppose that G_i happens in Game 2. Then, a query of the form (5) is made to the random oracle before user i is corrupted. The simulator can detect this query by querying CSI-stDH$_{\mathfrak{r}}(\mathfrak{S}, \mathfrak{W})$. If the answer is true, $\mathcal{B}_1$ outputs $[\rho]^{-1}\mathfrak{W}$, which means whenever G_i happens, the simulator can answer the CSI-stDH problem correctly. So we have

$$\Pr[G_i] \leq \mathrm{Adv}_{\mathrm{CSI\text{-}stDH}}^{\mathcal{B}_1}(\lambda). \tag{7}$$

From (6),(7), it is obvious that

$$|\Pr[S_2] - \Pr[S_3]| \leq \mu \cdot \mathrm{Adv}_{\mathrm{CSI\text{-}stDH}}^{\mathcal{B}_1}(\lambda). \tag{8}$$

We note here that other hash queries in which the identity i is included can be detected using CSI-stDH oracle.

For Game 4 and 5, the proof is similar to [7], so we just give an intuitive proof.

Game 4. In Game 4, all type III responders choose their session key at random, and do not modify the hash oracle.

Assume that the adversary $\mathcal{B}_2$ is given a CSI-stDH instance (E_1, E_2). Then, for all type I or II oracles, $\mathcal{B}_2$ generates random elements $[\rho_1] \in \mathcal{C}l(\mathcal{O})$ independently, and sets their ephemeral public keys to $[\rho_1]E_1$. Similarly, Type III oracles have ephemeral public keys $[\rho_2]E_2$. If the adversary against Game 3 does not make any hash query corresponding to Type III oracles, the Game 4 is identical to Game 3, whereas if such query is made, $\mathcal{B}_2$ can solve the strong CSIDH problem. Here, we have

$$|\Pr[S_3] - \Pr[S_4]| \leq \mathrm{Adv}_{\mathrm{CSI\text{-}stDH}}^{\mathcal{B}_2}(\lambda). \tag{9}$$

Game 5. In Game 5, all type II initiator oracles choose their session key at random and do not modify the hash oracle unless their intended peer is corrupted. The proof is identical to that of Game 3, so we have

$$|\Pr[S_4] - \Pr[S_5]| \leq \mu \cdot \mathrm{Adv}^{\mathcal{B}_3}_{\mathrm{CSI\text{-}stDH}}(\lambda) \tag{10}$$

for an adversary $\mathcal{B}_3$ against strong CSIDH problem.

Since all honest oracles choose their session keys uniformly at random in Game 5, the advantage of an arbitrary adversary against Game 5 is strictly 0. Then, we have

$$\Pr[S_5] = \frac{1}{2}. \tag{11}$$

Combining (2), (3), (8), (9), (10), and (11), we have

$$\mathrm{Adv}^{\mathcal{A}}_{\Pi_{\mathrm{CSIDH}}}(\lambda) = |\Pr[S_0] - 1/2|$$

$$\leq \left(\sum_{i=0}^{4} |\Pr[S_i] - \Pr[S_{i+1}]| \right) + |\Pr[S_5] - 1/2|$$

$$\leq \frac{\mu l^2}{(2m+1)^n} + \mu \cdot \mathrm{Adv}^{\mathcal{B}_1}_{\mathrm{CSI\text{-}stDH}}(\lambda)$$

$$+ \mathrm{Adv}^{\mathcal{B}_2}_{\mathrm{CSI\text{-}stDH}}(\lambda) + \mu \cdot \mathrm{Adv}^{\mathcal{B}_3}_{\mathrm{CSI\text{-}stDH}}(\lambda).$$

Here, we complete the proof.   □

B   CSIDH

In this section, we introduce the detailed protocol of CSIDH.

B.1   CSIDH as an Instantiation of HHS

In CSIDH, HHS is realized with the ideal class group of an imaginary quadratic field and supersingular elliptic curves. In this subsection, we see how the ideal class group $\mathcal{C}l(\mathcal{O})$ for an order $\mathcal{O}$ acts on $\mathcal{E}ll_p(\mathcal{O})$, the set of $\mathbb{F}_p$-isomorphic classes of supersingular elliptic curves whose $\mathbb{F}_p$-endomorphism ring is isomorphic to $\mathcal{O}$.

Ideal Class Group. Let K be an imaginary quadratic field and $\mathcal{O} \subset K$ be an order, a subring which is a free $\mathbb{Z}$-module of rank 2. Then, a fractional ideal of $\mathcal{O}$ is an $\mathcal{O}$-submodule of K which can be written in the form of $\alpha\mathfrak{a}$, where $\alpha \in K^\times$ and $\mathfrak{a}$ is an ideal of $\mathcal{O}$. Note that a multiplication of fractional ideals is induced by the multiplication of ideals naturally. We say a fractional ideal $\mathfrak{a}$ is invertible when there exists a fractional ideal $\mathfrak{b}$ such that $\mathfrak{a}\mathfrak{b} = \mathcal{O}$.

The set of all invertible fractional ideals $I(\mathcal{O})$ forms an abelian group under the above multiplication, and the set of all principle ideals $P(\mathcal{O})$ is a normal subgroup of $I(\mathcal{O})$. So we can define a quotient group $\mathcal{C}l(\mathcal{O}) = I(\mathcal{O})/P(\mathcal{O})$, which is called the ideal class group of $\mathcal{O}$. We denote the class containing $\mathfrak{a} \in I(\mathcal{O})$ by $[\mathfrak{a}]$. For more details, see [20].

The Action on Supersingular Elliptic Curves. For an order $\mathcal{O}$ in an imaginary quadratic field K, we define $\mathcal{E}\ell\ell_p(\mathcal{O})$ as a set of isomorphism classes of elliptic curves E over $\mathbb{F}_p$ such that $\mathrm{End}_{\mathbb{F}_p}(E) \simeq \mathcal{O}$. Here, $\mathrm{End}_{\mathbb{F}_p}(E)$ is the ring of $\mathbb{F}_p$-endomorphisms of E.

Now, we define a group action of $Cl(\mathcal{O})$ on $\mathcal{E}\ell\ell_p(\mathcal{O})$. Fix $[\mathfrak{a}] \in Cl(\mathcal{O})$ and $E \in \mathcal{E}\ell\ell_p(\mathcal{O})$, then there uniquely exist nonnegative integer r and $[\mathfrak{a}_s] \in Cl(\mathcal{O})$ such that $[\mathfrak{a}] = [(\pi\mathcal{O})]^r[\mathfrak{a}_s]$ and $\mathfrak{a}_s \not\subseteq \pi\mathcal{O}$, where π denotes the Frobenius map. For such $[\mathfrak{a}_s]$, we take an isogeny ψ from E with $\ker \psi = \bigcap_{\alpha \in \mathfrak{a}_s} \ker \alpha$. Then, for $[\mathfrak{a}]$, we take an isogeny $\pi^r \psi$, and whose codomain is denoted as $[\mathfrak{a}]E$. We can easily show that this correspondence enjoys the conditions to be a group action. A Hard Homogeneous Space can be constructed by this action.

B.2   Detailed Description of CSIDH

Let $\ell_1 \ldots \ell_n$ be small distinct odd primes such that $p = 4\ell_1 \cdots \ell_n - 1$ is a prime for some n. We can efficiently compute the class group action of $\mathfrak{l}_i = (\ell_i, \pi - 1)$ and $\mathfrak{l}_i^{-1} = (\ell_i, \pi + 1)$, since we have only to find a ℓ_i-torsion point.

Moreover, it is assumed heuristically that the map which maps $(e_1, \ldots, e_n) \in [-m, m]^n$ to $\mathfrak{l}_1^{e_1} \mathfrak{l}_2^{e_2} \cdots \mathfrak{l}_n^{e_n} \in Cl(\mathbb{Z}[\sqrt{-p}])$ is almost bijective, when m enjoys $(2m + 1)^n \geq \#Cl(\mathbb{Z}[\sqrt{-p}])$. So we can choose $e_1, \ldots, e_n$ instead of $[\mathfrak{a}]$, and its action can be computed efficiently. In this case, the size of the key space is approximately $(2m + 1)^n$.

Here, we describe how the protocol proceeds between Alice and Bob. Fix $E_0 \in \mathcal{E}\ell\ell_p(\mathbb{Z}[\sqrt{-p}])$ as a public parameter. First, Alice chooses $e_i \in [-m, m]$ for $i = 1, 2, \ldots, n$ uniformly at random, and computes $E_A = [\mathfrak{a}]E_0$, where $[\mathfrak{a}] = [\mathfrak{l}_1^{e_1} \mathfrak{l}_2^{e_2} \cdots \mathfrak{l}_n^{e_n}]$. Then, Alice sends E_A to Bob. Bob also computes $E_B = [\mathfrak{b}]E_0$, and sends it to Alice. Finally, Alice computes $[\mathfrak{a}]E_B$, and Bob computes $[\mathfrak{b}]E_A$. The shared secret is $\mathcal{M}([\mathfrak{a}]E_B) = \mathcal{M}([\mathfrak{b}]E_A)$, where $\mathcal{M}$ denotes the Montgomery coefficient.

C   Random Self-reducibility of the CSI-stDH Problem

In this section, we prove the random self-reducibility of the CSI-stDH problem. Here, we use another definition of the random self-reducibility. First, we define the CSI-stMDH problem, the multi-instance version of the CSI-stDH problem.

Problem 18 (Commutative Supersingular Isogeny strong Multi Diffie–Hellman (CSI-stMDH) Problem). Assume that a large prime p which enjoys $p \equiv 3 \mod 4$ and an elliptic $E \in \mathcal{E}\ell\ell_p(\mathcal{O})$ for $\mathcal{O} = \mathbb{Z}[\sqrt{-p}]$ are given. Then, given $(\mathfrak{X} = [\mathfrak{x}]E; (\mathfrak{Y}_i = [\mathfrak{y}_i]E)_{i \in [S]})$, the CSI-stMDH problem with parameter S is to compute $[\mathfrak{x}\mathfrak{y}_j]E$ for the index j chosen by the solver. Here, the solver is given accesses to the decision oracle $\mathsf{CSI\text{-}stDH}_{\mathfrak{x}}(\cdot, \cdot)$.

For an adversary $\mathcal{A}$ whose output is E', the advantage of $\mathcal{A}$ is defined as $\mathrm{Adv}_{\mathrm{CSI\text{-}stMDH}}^{\mathcal{A}}(\lambda) = \Pr[E' = [\mathfrak{x}\mathfrak{y}]E]$.

In this subsection, we say that the CSI-stDH is random self-reducible if we can reduce the CSI-stDH problem to the CSI-stMDH problem tightly. The only difference from the Definition 3 is that we fix the first curve $\mathfrak{X}$. Though we can prove the random self-reducibility of the CSI-stDH problem in a similar way following the Definition 3, we use this definition here so that we can see the analogy with the security proof of Π_{CSIDH} easily. Actually, $\mathfrak{X}$ corresponds to the user i's long-term public key in the security proof in Sect. A, and $\mathfrak{Y}_i$'s correspond to the ephemeral public keys of the oracles whose intended peer is i.

Here, our goal is to prove the random self-reducibility of CSI-stDH problem, i.e., the existence of tight reduction from the CSI-stDH problem to the CSI-stMDH problem:

Corollary 19 (Random Self-Reducibility of the CSI-stDH Problem).
For arbitrary adversary $\mathcal{A}$ against the CSI-stMDH problem with parameter S, there is an adversary $\mathcal{B}$ against the CSI-stDH problem such that

$$\text{Adv}_{\text{CSI-stMDH}}^{\mathcal{A},S}(\lambda) \leq \text{Adv}_{\text{CSI-stDH}}^{\mathcal{B},S}(\lambda), \text{ and } \text{Time}(\mathcal{A}) \simeq \text{Time}(\mathcal{B})$$

hold.

Proof. For an instance $(\mathfrak{X}, \mathfrak{Y}) = ([\mathfrak{x}]E, [\mathfrak{y}]E)$ of the CSI-stDH problem, $\mathcal{B}$ generates random ideal classes $[\eta_i] \in C\ell(\mathcal{O})$ for $i \in [S]$. Then, $\mathcal{B}$ generates a CSI-stMDH instance $(\mathfrak{X}; (\eta_i \mathfrak{Y})_{i \in [S]})$ and inputs this to $\mathcal{A}$. If $\mathcal{A}$ outputs $\mathfrak{z}_j$ for $j \in [S]$, $\mathcal{A}$ outputs $[\eta_j]^{-1} \mathfrak{z}_j$. For CSI-stDH$_{\mathfrak{x}}$ query made by $\mathcal{A}$, $\mathcal{B}$ queries it to its own CSI-stDH$_{\mathfrak{x}}$ oracle. Here, if $\mathcal{A}$ succeeds, $\mathcal{B}$ answers the CSI-stMDH problem correctly, which completes the proof. □

Remark 20. If we use the Definition 3 for the definition of the random self-reducibility, we also rerandomize the first curve $\mathfrak{X}$ as $\mathfrak{X}_i = [\xi_i]\mathfrak{X}$ for randomly chosen $[\xi_i] \in C\ell(\mathcal{O})$. Here, to prove the random self-reducibility, we should answer to the decision queries CSI-stDH$_{\xi_i \mathfrak{x}}(E_1, E_2)$ for every i. However, since

$$E_2 = [\xi_i \mathfrak{x}]E_1 \Leftrightarrow [\xi_i^{-1}]E_2 = [\mathfrak{x}]E_1,$$

we have CSI-stDH$_{\xi_i \mathfrak{x}}(E_1, E_2) = $ CSI-stDH$_{\mathfrak{x}}(E_1, [\xi_i^{-1}]E_2)$, thus we can simulate the oracles perfectly.

References

1. Bernstein, D.J., De Feo, L., Leroux, A., Smith, B.: Faster computation of isogenies of large prime degree. Cryptology ePrint Archive, Report 2020/341 (2020). https://eprint.iacr.org/2020/341
2. Brendel, J., Fischlin, M., Günther, F., Janson, C., Stebila, D.: Towards post-quantum security for signal's X3DH handshake. In: Selected Areas in Cryptography (SAC) (2020, to appear)
3. Canetti, R., Krawczyk, H.: Analysis of key-exchange protocols and their use for building secure channels. In: Pfitzmann, B. (ed.) EUROCRYPT 2001. LNCS, vol. 2045, pp. 453–474. Springer, Heidelberg (2001). https://doi.org/10.1007/3-540-44987-6_28

4. Castryck, W., Decru, T.: CSIDH on the surface. Cryptology ePrint Archive, Report 2019/1404 (2019). https://eprint.iacr.org/2019/1404

5. Castryck, W., Lange, T., Martindale, C., Panny, L., Renes, J.: CSIDH: an efficient post-quantum commutative group action. In: Peyrin, T., Galbraith, S. (eds.) ASI-ACRYPT 2018. LNCS, vol. 11274, pp. 395–427. Springer, Cham (2018). https://doi.org/10.1007/978-3-030-03332-3_15

6. Castryck, W., Sotáková, J., Vercauteren, F.: Breaking the decisional Diffie-Hellman problem for class group actions using genus theory. Cryptology ePrint Archive, Report 2020/151 (2020). https://eprint.iacr.org/2020/151

7. Cohn-Gordon, K., Cremers, C., Gjøsteen, K., Jacobsen, H., Jager, T.: Highly efficient key exchange protocols with optimal tightness. In: Boldyreva, A., Micciancio, D. (eds.) CRYPTO 2019. LNCS, vol. 11694, pp. 767–797. Springer, Cham (2019). https://doi.org/10.1007/978-3-030-26954-8_25

8. Couveignes, J.-M.: Hard Homogeneous Spaces. Cryptology ePrint Archive, Report 2006/291 (2006). https://eprint.iacr.org/2006/291

9. de Kock, B., Gjøsteen, K., Veroni, M.: Practical isogeny-based key-exchange with optimal tightness. In: Selected Areas in Cryptography (SAC) 2020 (2020, to appear) .

10. Dobson, S., Galbraith, S.D.: On the degree-insensitive SI-GDH problem and assumption. Cryptology ePrint Archive, Report 2019/929 (2019). https://eprint.iacr.org/2019/929

11. El Kaafarani, A., Katsumata, S., Pintore, F.: Lossy CSI-FiSh: efficient signature scheme with tight reduction to decisional CSIDH-512. In: Kiayias, A., Kohlweiss, M., Wallden, P., Zikas, V. (eds.) PKC 2020. LNCS, vol. 12111, pp. 157–186. Springer, Cham (2020). https://doi.org/10.1007/978-3-030-45388-6_6

12. Fujioka, A., Takashima, K., Terada, S., Yoneyama, K.: Supersingular isogeny Diffie-Hellman authenticated key exchange. In: ICISC 2018, pp. 177–195 (2018)

13. Fujioka, A., Takashima, K., Yoneyama, K.: One-round authenticated group key exchange from isogenies. In: Steinfeld, R., Yuen, T.H. (eds.) ProvSec 2019. LNCS, vol. 11821, pp. 330–338. Springer, Cham (2019). https://doi.org/10.1007/978-3-030-31919-9_20

14. Galbraith, S.D.: Authenticated key exchange for SIDH. Cryptology ePrint Archive, Report 2018/266 (2018). https://eprint.iacr.org/2018/266

15. Galbraith, S.D., Vercauteren, F.: Computational problems in supersingular elliptic curve isogenies. Quantum Inf. Process. **17**(10), 265 (2018)

16. Jao, D., De Feo, L.: Towards quantum-resistant cryptosystems from supersingular elliptic curve isogenies. In: Yang, B.-Y. (ed.) PQCrypto 2011. LNCS, vol. 7071, pp. 19–34. Springer, Heidelberg (2011). https://doi.org/10.1007/978-3-642-25405-5_2

17. Krawczyk, H.: HMQV: a high-performance secure Diffie-Hellman protocol. In: Shoup, V. (ed.) CRYPTO 2005. LNCS, vol. 3621, pp. 546–566. Springer, Heidelberg (2005). https://doi.org/10.1007/11535218_33

18. LaMacchia, B., Lauter, K., Mityagin, A.: Stronger security of authenticated key exchange. In: Susilo, W., Liu, J.K., Mu, Y. (eds.) ProvSec 2007. LNCS, vol. 4784, pp. 1–16. Springer, Heidelberg (2007). https://doi.org/10.1007/978-3-540-75670-5_1

19. Longa, P.: A Note on Post-Quantum Authenticated Key Exchange from Supersingular Isogenies. Cryptology ePrint Archive, Report 2018/267 (2018). https://eprint.iacr.org/2018/267

20. Neukirch, J.: Algebraic Number Theory, vol. 322. Springer, Heidelberg (2013)

21. Okamoto, T., Pointcheval, D.: The Gap-Problems: A New Class of Problems for the Security of Cryptographic Schemes. An Efficient Authenticated Key Exchange from Random Self-Reducibility on CSIDH. In: Public Key Cryptography 2001, pp. 104–118. Springer, Heidelberg (2001)
22. Shor, P.W.: Polynomial-time algorithms for prime factorization and discrete logarithms on a quantum computer. SIAM J. Comput. **26**(5), 1484–1509 (1997)
23. Xu, X., Xue, H., Wang, K., Au, M.H., Tian, S.: Strongly secure authenticated key exchange from supersingular isogenies. In: Galbraith, S.D., Moriai, S. (eds.) ASIACRYPT 2019. LNCS, vol. 11921, pp. 278–308. Springer, Cham (2019). https://doi.org/10.1007/978-3-030-34578-5_11

Constructions and Designs

A Sub-linear Lattice-Based Submatrix Commitment Scheme

Huang Lin[(✉)]

Mercury's Wing and Suterusu Project, Beijing, China
huanglinepfl@gmail.com

Abstract. Subvector commitment is a recently proposed cryptographic primitive that provides the underlying cryptographic tool to design many interesting security systems such as succinct non-interactive arguments of knowledge (SNARK), verifiable database, dynamic accumulators, etc. In this paper, we present a generalization of subvector commitment, *a public-coin-setup lattice-based submatrix commitment*, which allows a commitment of a message matrix to be opened on multiple entries of the matrix simultaneously. It exploits a conceptual similarity between Single-Instruction Multiple-Data (SIMD) in homomorphic encryption and submatrix commitment, and develops a novel position binding technique based on the Chinese Remainder Theorem. We show that the position binding property can be reduced to module-based short integer solution (SIS) problem, a standard assumption that is believed to be post-quantum secure. We also show that the commitment and opening size of our commitment scheme are both sublinear, i.e., proportional to the square root of the message size. As far as we know, this is the first public-coin-setup and post-quantum secure subvector commitment scheme.

Keywords: Subvector commitment · Homomorphic encryption · Lattice-based cryptography · Position binding

1 Introduction

Digital commitment is a crucial cryptographic primitive that serves as a fundamental building block for various cryptographic protocols such as zero-knowledge proof, accumulators, etc. Recently, we have seen multiple generalizations of commitment schemes such as vector commitment [1], subvector commitment [2], and functional commitment [3].

A vector commitment scheme allows a prover to commit to a message vector, and open the commitment to the $i-$th component of the vector, while functional commitment allows a prover to commit to a vector v and open the commitment to a function-evaluation tuple $(f, f(v))$. A subvector commitment scheme [2], which is a generalization of the vector commitment, commits to a message vector and allows the opening of multiple entries of the message vector simultaneously. A subvector commitment can be derived from a functional commitment scheme by

© Springer Nature Switzerland AG 2021
D. Hong (Ed.): ICISC 2020, LNCS 12593, pp. 87–98, 2021.
https://doi.org/10.1007/978-3-030-68890-5_5

restricting the function f to a linear map that maps the original message vector to its subvector.

Subvector commitment was proposed as the foundation for the succinct non-interactive arguments of knowledge (SNARK) scheme [2], which can be built upon probabilistically checkable proof (PCP) and subvector commitment schemes. The prover first commits to the PCP string by running a subvector commitment scheme. After receiving the indices of the entries to be inspected from the verifier, the prover runs the opening algorithm for the subvector commitment scheme to open these entries. SNARK scheme has recently found its real-world applications in building anonymous cryptocurrency such as Zcash [4]. There exists a vast literature on SNARK schemes with trusted setup. However, the trusted setup is a bottleneck because the trapdoor information of the setup phase could be exploited by an adversary to weaken the soundness property of SNARK schemes [5,6], which implies that an attacker could exploit the weakness of the trusted setup phase to generate an unlimited amount of cryptocurrency [7]. Consequently, an efficient SNARK scheme without the need of a trusted setup, i.e., a public-coin-setup SNARK, has long been considered as the holy grail in cryptography research. A public-coin-setup SNARK scheme can be derived from a public-coin-setup subvector commitment by applying the "CS proof" framework [8]. As a matter of fact, the SNARK scheme proposed in [2] was the first public-coin-setup SNARK scheme with constant argument size.

The subvector commitment scheme guarantees the position binding property [2,9] in the sense that the subvector opening is bond to its positions in the committed vector. In contrast, vector commitment scheme only accommodates the opening of one entry [9], and a straightforward subvector commitment scheme derived from a vector commitment scheme would require running multiple instances of the vector commitment schemes [9].

The first public-coin-setup subvector commitment scheme proposed in [2] has reduced its position binding property to a non-standard assumption in class groups of unknown order, i.e., adaptive root assumption over class groups of imaginary quadratic orders. It is well known that this assumption cannot resist the quantum attack [9]. Whether it is possible to design an efficient subvector commitment scheme based on plausibly post-quantum-secure assumptions remains an interesting open problem [9].

As an answer to this open problem, we propose a lattice-based submatrix commitment scheme, in which the position binding property is reduced to module-based shortest integer solution (SIS) assumption, a well-studied assumption that is believed to be resistant to quantum attacks. As a matter of fact, it is the underlying security assumption coming from one of the NIST post-quantum digital signature competition candidates, i.e., Dilithium scheme [10]. Obviously, a subvector commitment scheme can be derived from our proposed submatrix commitment scheme by simply rearranging a message vector with N entries into a $\sqrt{N} \times \sqrt{N}$ matrix. The commitment size and opening size of our scheme are both proportional to the square root of the message size.

Our proposed scheme exploits a conceptual similarity between the submatrix scheme and the SIMD (Single-Instruction Multiple-Data) technique in homomorphic encryption [11] based on the Chinese Remainder Theorem (CRT). We develop a novel technique by exploiting the composite-order feature and CRT representation to ensure the representation in the homomorphic encryption compatible with the position binding property for our submatrix commitment scheme. In fact, this technique is of independent interest for the future design of lattice-based cryptographic techniques.

The remainder of this paper is organized as follows. In Sect. 2, we introduce the definitions of related concepts and problem setting for our lattice-based commitment construction. Afterwards, in Sect. 2.3, we present preliminaries for our submatrix commitment. Sect. 3 presents the construction of our submatrix commitment scheme and its security proof. We then provide the complexity analysis of our scheme and its comparison with several related schemes in Sect. 4. We conclude the paper in Sect. 5.

2   Preliminaries

For $a, b \in \mathbb{N}$, we use $[a, b]$ to denote the set $\{a, a + 1, \cdots, b - 1, b\}$. Let $s \leftarrow S$ denote uniformly random sampling an element s from the set S. Given a matrix

$$\mathbf{w} = \begin{pmatrix} w_{0,0} & w_{0,1} & \cdots & w_{0,N-1} \\ w_{1,0} & w_{1,1} & \cdots & w_{1,N-1} \\ \vdots & \vdots & \ddots & \vdots \\ w_{h-1,0} & w_{h-1,1} & \cdots & w_{h-1,N-1} \end{pmatrix}$$

that is also denoted as $\left\{\langle w_{i,0}, w_{i,1}, \ldots, w_{i,N-1}\rangle_{i \in [0, h-1]}\right\}$, we define a submatrix $\mathbf{w_{I,J}} = \left\{\langle w_{i,j} | j \in \mathbf{J}\rangle_{i \in \mathbf{I}}\right\}$ as an ordered subset of the entries of the matrix indexed by $\mathbf{I} \subseteq [0, h - 1]$ and $\mathbf{J} \subseteq [0, N - 1]$.

2.1   System Setting

Let $\mathcal{R}$ be the cyclotomic ring $\mathcal{R} = \mathbb{Z}[X]/\langle X^N + 1\rangle$, where N is a power of 2 [12,13]. Let q be a positive integer and define $\mathcal{R}_q = \mathbb{Z}_q[X]/\langle X^N + 1\rangle$. Here $\mathbb{Z}_q$ denotes the integers modulo q. For $f(X) = \sum_i f_i X^i \in \mathcal{R}$, the norms of f are defined as $l_1 : \|f\|_1 = \sum_i |f_i|$, $l_2 : \|f\|_2 = \left(\sum_i |f_i|^2\right)^{1/2}$, $l_\infty : \|f\|_\infty = \max_i |f_i|$. In our system, q is a product of two primes p_1 and p_2. For a positive integer β, we write S_β to be the set of all elements in $\mathcal{R}_{p_2}$ with l_∞-norm at most β.

We denote by $\lambda \in \mathbb{N}$ the security parameter and by $\text{poly}(\lambda)$ the sets of polynomials in λ.

2.2   Module-SIS Assumption (MSIS)

The position binding property in our submatrix commitment scheme can be reduced to the hardness of Module-SIS, which is the underlying assumption used in the post-quantum secure signature scheme Dilithium [10], and the recently proposed lattice-based commitment scheme [13] and ring signature scheme [12]. It is a variant of the well-known SIS [14] but is defined on modules.

The homogeneous Module-SIS problem [10] (or the search Knapsack Problem as in [13]) is defined as follows.

Definition 1 (MSIS$_{n,k,\beta}$). *Given $\mathbf{A} \leftarrow \mathcal{R}_q^{n \times (k-n)}$, find a short vector $\boldsymbol{r} \in \mathcal{R}^k$ such that $(\mathbf{I}_n, \mathbf{A}) \cdot \boldsymbol{r} = 0$ and $0 < \|\boldsymbol{r}\|_2 \le \beta$. For an algorithm $\mathcal{A}$, we define $\mathrm{Adv}_{n,k,\beta}^{msis}(\mathcal{A})$ as*

$$\Pr \left[b = 1 \; \middle| \; \begin{array}{l} \mathbf{A} \leftarrow \mathcal{R}_q^{n \times (k-n)}; \\ \boldsymbol{r} \leftarrow \mathcal{A}(\mathbf{A}); \\ b := \left(\boldsymbol{r} \in \mathcal{R}^k\right) \wedge \left((\mathbf{I}_n, \mathbf{A}) \cdot \boldsymbol{r} = 0\right) \wedge \\ (0 < \|\boldsymbol{r}\|_2 \le \beta) \end{array} \right],$$

where $\wedge$ indicates the conjunctive operation. We say an algorithm $\mathcal{A}$ has at least an advantage ϵ in solving the Module-SIS$_{n,k,\beta}$ problem if $Adv_{n,k,\beta}^{msis}(\mathcal{A}) \ge \epsilon$.

2.3   Definitions Related to the Submatrix Commitment Scheme

Since a submatrix commitment scheme is a generalization to subvector commitment schemes [2,9], the definitions related to the submatrix commitment scheme also follow those of subvector commitment schemes [2,9].

Definition 2 (Submatrix commitment syntax). *A submatrix commitment scheme consists of five algorithms:*
Setup$(1^\lambda, h, N)$: *Given security parameter λ, the dimension of a matrix h and N, outputs the public parameters PP.*
Com$(\mathbf{w})$: *Given a matrix $\mathbf{w}$, outputs a commitment $\mathbf{C}$ and an auxiliary message aux.*
Open$(\mathbf{I}, \mathbf{J}, \mathbf{w}_{\mathbf{I},\mathbf{J}}, aux, \mathbf{C})$: *Given two order index sets $\mathbf{I}$, $\mathbf{J}$ and the auxiliary message aux, outputs an opening $\Lambda_{\mathbf{I},\mathbf{J}}$ that proves $\mathbf{w}_{\mathbf{I},\mathbf{J}}$ is the submatrix of the message committed under $\mathbf{C}$.*
Verify$(\mathbf{C}, \mathbf{I}, \mathbf{J}, \mathbf{w}_{\mathbf{I},\mathbf{J}}, \Lambda_{\mathbf{I},\mathbf{J}})$: *Given inputs commitment $\mathbf{C}$, two order index sets $\mathbf{I}$ and $\mathbf{J}$, the submatrix of the message $\mathbf{w}_{\mathbf{I},\mathbf{J}}$ and opening $\Lambda_{\mathbf{I},\mathbf{J}}$, outputs 1 (accept) or 0 (reject).*

Definition 3 (Correctness). *A submatrix commitment scheme is correct if for all message $\mathbf{w}$ in the message space and index sets $\mathbf{I} \subseteq [0, h-1]$ and $\mathbf{J} \subseteq [0, N-1]$, the probability*

$$\Pr \left[b = 1 \; \middle| \; \begin{array}{l} PP \leftarrow \textbf{\textit{Setup}}(1^\lambda, h, N) \\ \mathbf{C}, aux \leftarrow \textbf{\textit{Com}}(\mathbf{w}) \\ \Lambda_{\mathbf{I},\mathbf{J}} \leftarrow \textbf{\textit{Open}}(\mathbf{I}, \mathbf{J}, \mathbf{w}_{\mathbf{I},\mathbf{J}}, aux, \mathbf{C}) \\ b := \textbf{\textit{Verify}}(\mathbf{C}, \mathbf{I}, \mathbf{J}, \mathbf{w}_{\mathbf{I},\mathbf{J}}, \Lambda_{\mathbf{I},\mathbf{J}}) \end{array} \right]$$

is equal to 1.

Definition 4 (Position Binding). *A submatrix commitment scheme is position binding if for any adversary* $\mathcal{A}$*, there exists a negligible function* $negl(\lambda)$ *such that:*

$$
\Pr\left[b = 1 \,\middle|\,
\begin{array}{l}
PP \leftarrow \textbf{Setup}(1^\lambda, h, N) \\
\begin{pmatrix} \mathbf{C}, \mathbf{I}, \mathbf{J}, \mathbf{w_{I,J}}, \mathit{\Lambda_{I,J}} \\ \mathbf{I}', \mathbf{J}', \mathbf{w'_{I',J'}}, \mathit{\Lambda'_{I',J'}} \end{pmatrix} \leftarrow \mathcal{A}\,(PP) \\
b := (\, \textbf{Verify}(\mathbf{C}, \mathbf{I}, \mathbf{J}, \mathbf{w_{I,J}}, \mathit{\Lambda_{I,J}})) \wedge \\
(\, \textbf{Verify}(\mathbf{C}, \mathbf{I}', \mathbf{J}', \mathbf{w'_{I',J'}}, \mathit{\Lambda'_{I',J'}})) \wedge \\
\begin{pmatrix} \exists i \in \mathbf{I} \cap \mathbf{I}' \wedge j \in \mathbf{J} \cap \mathbf{J}' \\ s.t.\ \mathbf{w}_{i,j} \neq \mathbf{w}'_{i,j} \end{pmatrix}
\end{array}
\right]
$$

is smaller than or equal to $negl(\lambda)$.

It is important to note that the hiding property is not taken into account in the original definition of a subvector commitment [2,9] because this property is not vital in the construction of its major applications, i.e., the SNARK scheme, but one can achieve the hiding property by combining the subvector (or submatrix) commitment scheme with a general commitment scheme in a rather straightforward manner as specified in [2].

3 Submatrix Commitment Scheme

In our scheme, the modulus q is a product of two primes p_1 and p_2. The committed message of our scheme is an element in $\mathbb{Z}_{p_1}$ with p_1 being a small prime. In practice, one could set p_1 to 2. The cyclotomic polynomial $\Phi(X)$ is chosen to split into linear terms modulo $p_1 \cdot p_2$, i.e.,

$$
\Phi(X) = \left[\prod_{i=0}^{N-1} (X - \varsigma_i) \right]_{p_1 p_2}
$$

where ς_i is the i-th root of $\Phi(X)$. We will use $[\cdot]_p$ to denote $\mod p$ operation from here on. From the Chinese Remainder Theorem (CRT), one can define an isomorphism:

$$
\mathbb{Z}_q[X] / \langle \Phi(X) \rangle \mapsto (\mathbb{Z}_q[X] / \langle X - \varsigma_0 \rangle, \ldots, \mathbb{Z}_q[X] / \langle X - \varsigma_{N-1} \rangle)
$$

One could further define another isomorphism based on CRT:

$$
\mathbb{Z}_{p_1 p_2}[X] / \langle \Phi(X) \rangle \mapsto (\mathbb{Z}_{p_1}[X] / \langle \Phi(X) \rangle, \mathbb{Z}_{p_2}[X] / \langle \Phi(X) \rangle)
$$

We do not distinguish these two isomorphisms for the ease of exposition in the subsequent development. To invoke which isomorphism in the concrete steps of the scheme is determined by the outputs of the isomorphism used in the context.

Similar to the setting in [11], an element $\mathbf{m} \in \mathcal{R}_{p_1 p_2}$ can be represented either in its polynomial form $\mathbf{m}(X)$, or its evaluation representation, which is a vector of ring elements, i.e., $\mathrm{CRT}(\mathbf{m}(X)) = \langle m_0, m_1, \ldots, m_{N-1} \rangle$, where $m_j = [\mathbf{m}(X)]_{(X - \varsigma_j, p_1 p_2)} = \mathbf{m}(X) \bmod (X - \varsigma_j, p_1 p_2) = \mathbf{m}(\varsigma_j) \bmod (p_1 p_2)$.

Our scheme employs the CRT isomorphism to represent the subset map, which could be formed either as its evaluation representation, i.e., a binary vector, such as $\mathbf{J} = \langle 0, 1, 1, \cdots, 0 \rangle$[1] or its own coefficient representation, $\mathbf{J}(X) = \mathrm{CRT}^{-1}(\mathbf{J})$, where CRT^{-1} denotes the inverse CRT map. The component-wise multiplication between a vector $\mathbf{m}$ and $\mathbf{J}$, i.e.,

$$\mathbf{m} \otimes \mathbf{J} = \langle m_0 * \mathbf{J}_0, m_1 * \mathbf{J}_1, \ldots, m_{N-1} * \mathbf{J}_{N-1} \rangle$$

corresponds to the polynomial multiplication $\mathbf{m}(X) \times \mathbf{J}(X)$ under the CRT isomorphism. Here the component-wise multiplication gives us the subvector of $\mathbf{m}$ defined by $\mathbf{J}$.

Our matrix commitment scheme is based on a recently proposed commitment scheme from structured lattice [13], and we develop several techniques to prove the position binding property. Our proposed scheme compresses the commitment size by committing to the hashed value of the column vectors of the message matrix. The hashing guarantees that as long as there is any difference at any entry of the committed message column vector, the ring elements in $\mathcal{R}_q$ representing the message will be different. However, the hashing technique can only guarantee the position binding property in the message column.

To guarantee the position binding property in the message row vector, we need to make sure that the randomness used in $\mathcal{R}_{p_1}$ is equal to zero such that the difference of the message row vector in $\mathcal{R}_{p_1}$ can be translated into the difference in the randomness vector of $\mathcal{R}_{p_2}$ when the commitment in $\mathcal{R}_{p_1 \cdot p_2}$ is assumed to be identical in the proof. We define the component of $\mathbf{J}(X)$ in $\mathcal{R}_{p_2}$ as 1 so that we can reduce the position binding property to $MSIS$ problem in $\mathcal{R}_{p_2}$.

Now, we are ready to present the formal description of our submatrix commitment scheme below. t

- **Setup**$(1^\lambda, h, N)$: Choose $q = p_1 \cdot p_2$ and a cyclotomic polynomial $\Phi(X) = X^N + 1$, where p_1 and p_2 are two primes, and N is a power of 2. p_1 is chosen such that $\Phi(X)$ can be decomposed into N linear factors in $\mathbb{Z}_{p_1}$, i.e.,

$$\Phi(X) = \prod_{i=0}^{N-1} (X - \varsigma_i) \bmod p_1$$

Sample two random matrices $\mathbf{A}_1 \in \mathcal{R}_q^{n \times k}$ and $\mathbf{A}_2 \in \mathcal{R}_q^{\ell \times k}$ formed as in the KeyGen algorithm of [13], i.e.,

$$\mathbf{A}_1 = (\mathbf{I}_n, \mathbf{A}_1') \text{ with } \mathbf{A}_1' \leftarrow \mathcal{R}_q^{n \times (k-n)} \text{ and}$$
$$\mathbf{A}_2 = (0^{\ell \times n}, \mathbf{I}_\ell, \mathbf{A}_2') \text{ with } \mathbf{A}_2' \leftarrow \mathcal{R}_q^{\ell \times (k-n-\ell)},$$

---

[1] Technically, $\mathbf{J}$ is a subset of $[0, N-1]$. The evaluation representation of $\mathbf{J}$ is defined in such way that $\mathbf{J}_i$ is equal to 1 whenever $i \in [0, N-1]$ belongs to $\mathbf{J}$ and $\mathbf{J}_i = 0$ otherwise.

where $\mathbf{I}_n$ and $\mathbf{I}_\ell$ are both identity matrices. Here ℓ, k and n are chosen in the same way as in [13] to ensure Module SIS problem is hard. In our performance analysis, we assume the message matrix is a square matrix and hence we have $h = N$ equal to the square root of the message matrix size. Note since the message in our **Com** algorithm is really the hash of the column vectors of the message matrix, one could simply set ℓ as $poly(\lambda)$ to guarantee the collision-resistance property of the hash function. Select a collision-resistant cryptographic functions $H' : (\mathbb{Z}_{p_1})^h \rightarrow (\mathbb{Z}_{p_1})^\ell$ and an extensible function $H : \{0,1\}^\lambda \rightarrow S_\beta^k$.

- **Com** $\left(\left\{ \langle w_{i,0}, w_{i,1}, \ldots, w_{i,N-1} \rangle_{i \in [0,h-1]} \right\} \right)$:

Let $\mathbf{w}_i = \langle w_{i,0}, w_{i,1}, \ldots, w_{i,N-1} \rangle$, where $w_{i,j} \in \mathbb{Z}_{p_1}$.

Generate $\mathbf{w}(X) \in (\mathcal{R}_q)^\ell$ in the following way:

1. $\forall m \in [0, \ell - 1], j \in [0, N-1]$, compute $[\mathbf{w}_m(X)]_{(X-\varsigma_j, p_1)}$ as[2]

$$H'_m \left(w_{0,j} || w_{1,j} || \cdots || w_{h-1,j} \right);$$

2. $\forall m \in [0, \ell - 1]$, Set $[\mathbf{w}_m(X)]_{p_2} = 0$;
3. Compute $\mathbf{w}_m(X)$ as

$$\mathrm{CRT}^{-1} \left([\mathbf{w}_m(X)]_{p_1}, [\mathbf{w}_m(X)]_{p_2} \right);$$

where

$$[\mathbf{w}_m(X)]_{p_1} = \mathrm{CRT}^{-1} \left([\mathbf{w}_m(X)]_{(X-\varsigma_0, p_1)}, \cdots, [\mathbf{w}_m(X)]_{(X-\varsigma_{N-1}, p_1)} \right)$$

4. Set $\mathbf{w}(X) = \langle \mathbf{w}_0(X), \ldots, \mathbf{w}_{\ell-1}(X) \rangle$.

Generate $\mathbf{r}(X)$ as follows:

1. Sample random $\rho \leftarrow \{0,1\}^\lambda$;
2. Generate $[\mathbf{r}(X)]_{p_2} = H(\rho)$;
3. Set $[\mathbf{r}(X)]_{p_1} = 0$;
4. Compute $\mathbf{r}(X) = \mathrm{CRT}^{-1} \left([\mathbf{r}(X)]_{p_1}, [\mathbf{r}(X)]_{p_2} \right)$.

Generate the commitment as follows:

$$\mathbf{C} = \mathrm{Com}(\mathbf{w}(X), \mathbf{r}(X)) = \begin{pmatrix} A_1 \\ A_2 \end{pmatrix} \mathbf{r}(X) + \begin{pmatrix} 0 \\ \mathbf{w}(X) \end{pmatrix};$$

Let aux be a vector consisting of the original message $\mathbf{w}$ and the randomness ρ.

---

[2] $H'_m(\cdot)$ denote the m-th bit of the hash output.

– **Open** $(\mathbf{I}, \mathbf{J}, \mathbf{w_{I,J}}, aux, \mathbf{C})$:

Generate the opening as

$$\varLambda_{\mathbf{I,J}} = (\varLambda_1, \varLambda_2) = \left(\langle w_{i,j} | i \in [0, h-1], j \in \mathbf{J} \rangle, \rho \right).$$

– **Verify** $(\mathbf{C}, \mathbf{I}, \mathbf{J}, \mathbf{w'_{I,J}}, \varLambda_{\mathbf{I,J}})$:

For $i \in \mathbf{I}$ and $j \in \mathbf{J}$, check whether $\mathbf{w}'_{i,j} = \mathbf{w}_{i,j}$, where $\mathbf{w}_{i,j}$ belongs to $\varLambda_1$.

Recover $\mathbf{w}(X)\mathbf{J}(X)$ as follows:

1. $\forall j \in \mathbf{J}, m \in [0, \ell - 1]$, compute

$$[\mathbf{w}_m(X)\mathbf{J}(X)]_{(X-\varsigma_j, p_1)} = H'_m \left(w_{0,j}||w_{1,j}|| \cdots ||w_{h-1,j}\right).$$

2. $\forall j \in [0, N-1]/\mathbf{J}$, set $[\mathbf{w}_m(X)\mathbf{J}(X)]_{(X-\varsigma_j, p_1)} = 0$.
3. Compute

$$[\mathbf{w}_m(X)\mathbf{J}(X)]_{p_1} = \text{CRT}^{-1}\left([\mathbf{w}_m(X)\mathbf{J}(X)]_{(X-\varsigma_j, p_1)}, j \in [0, N-1]\right).$$

4. Set $[\mathbf{w}_m(X)\mathbf{J}(X)]_{p_2} = 0$.
5. Compute

$$\mathbf{w}_m(X)\mathbf{J}(X) = \text{CRT}^{-1}\left([\mathbf{w}_m(X)\mathbf{J}(X)]_{p_1}, [\mathbf{w}_m(X)\mathbf{J}(X)]_{p_2}\right).$$

6. Set $\mathbf{w}(X)\mathbf{J}(X) = \langle \mathbf{w}_m(X)\mathbf{J}(X) | m \in [0, \ell - 1] \rangle$.

Recover $\mathbf{r}(X)$ as $\text{CRT}^{-1}(0, H(\rho))$.
Compute $\mathbf{J}(X)$ as follows:

1. $\forall j \in [0, N-1]$, set $[\mathbf{J}(X)]_{(X-\varsigma_j, p_1)} = \mathbf{J}_j$
2. Set $[\mathbf{J}(X)]_{p_2} = 1$.
3. Compute $\mathbf{J}(X) = \text{CRT}^{-1}\left([\mathbf{J}(X)]_{p_1}, [\mathbf{J}(X)]_{p_2}\right)$.

Check whether the following equation holds

$$\mathbf{J}(X)\mathbf{C} = \begin{pmatrix} \mathbf{A}_1 \\ \mathbf{A}_2 \end{pmatrix} \mathbf{J}(X)\mathbf{r}(X) + \begin{pmatrix} 0 \\ \mathbf{w}(X)\mathbf{J}(X) \end{pmatrix}.$$

Theorem 1. *The proposed submatrix commitment scheme is correct in the sense of Definition 3.*

Proof. When the **Com** and **Open** algorithms are run according to the definition, the opening information contains $\{w_{i,j} | i \in [0, h-1], j \in \mathbf{J}\}$ and the randomness

ρ. When the **Verify** algorithm invokes the CRT^{-1} algorithm, $\mathbf{w}_m(X)\mathbf{J}(X)$ can be recovered since $[\mathbf{w}_m(X)\mathbf{J}(X)]_{(X-\varsigma_j,p_1)}$ can be recovered as it amounts to the hashing of the message vectors $\{w_{i,j} | i \in [0, h-1], j \in \mathbf{J}\}$. The correctness of the recovered $\mathbf{r}(X)$ can be attributed to the fact that it is recovered exactly as how it is generated in the **Com** algorithm. The verification of the final equation in **Verify** will pass because all the components are recovered exactly as in the **Com** algorithm.

Theorem 2. *Assuming $MSIS_{n,k,\beta}$ problem is hard over $\mathcal{R}_{p_2}$ and the collision-resistance property of the underlying hash functions, the proposed submatrix commitment scheme is position binding (in the sense of Definition 4) with non-negligible probability.*

Proof. Suppose $\mathcal{A}$ is the algorithm that an adversary uses to break the position binding property, we want to show that we could construct an algorithm $\mathcal{C}$ to solve the $MSIS$ problem as follows, resulting in the desired contradiction.

Assume that $\mathcal{C}$ generates $\mathbf{A}_1, \mathbf{A}_2$ as in practice, which means both $\mathbf{A}_1$ and $\mathbf{A}_2$ would have uniformly random components in $\mathcal{R}_{p_2}$. This satisfies the requirement of input matrix of the $MSIS_{n,k,\beta}$ problem. Now $\mathcal{A}$ will output $\mathbf{C}, \mathbf{I}, \mathbf{J}, \mathbf{w_{I,J}}$, $\Lambda_{\mathbf{I,J}}, \mathbf{I}', \mathbf{J}', \mathbf{w}'_{\mathbf{I,J}}, \Lambda_{\mathbf{I}',\mathbf{J}'}$ such that **Verify**$(\mathbf{C}, \mathbf{I}, \mathbf{J}, \mathbf{w_{I,J}}, \Lambda_{\mathbf{I,J}})=$**Verify**$(\mathbf{C}, \mathbf{I}', \mathbf{J}', \mathbf{w}'_{\mathbf{I,J}}, \Lambda_{\mathbf{I}',\mathbf{J}'}) = 1$ and $\exists i \in \mathbf{I} \cap \mathbf{I}' \wedge j \in \mathbf{J} \cap \mathbf{J}'$ such that $w_{i,j} \neq w'_{i,j}$.

According to the final verification equation, we have

$$\mathbf{J}(X)\mathbf{C} = \begin{pmatrix} \mathbf{A}_1 \\ \mathbf{A}_2 \end{pmatrix} \mathbf{J}(X)\mathbf{r}(X) + \begin{pmatrix} 0 \\ \mathbf{w}(X)\mathbf{J}(X) \end{pmatrix}$$

$$\mathbf{J}'(X)\mathbf{C} = \begin{pmatrix} \mathbf{A}_1 \\ \mathbf{A}_2 \end{pmatrix} \mathbf{J}'(X)\mathbf{r}'(X) + \begin{pmatrix} 0 \\ \mathbf{w}(X)\mathbf{J}'(X) \end{pmatrix}$$

and hence we have

$$\mathbf{J}'(X)\mathbf{J}(X)\mathbf{C} = \mathbf{J}'(X)\mathbf{J}(X) \left(\begin{pmatrix} \mathbf{A}_1 \\ \mathbf{A}_2 \end{pmatrix} \mathbf{r}(X) + \begin{pmatrix} 0 \\ \mathbf{w}(X) \end{pmatrix} \right)$$

$$\mathbf{J}'(X)\mathbf{J}(X)\mathbf{C} = \mathbf{J}'(X)\mathbf{J}(X) \left(\begin{pmatrix} \mathbf{A}_1 \\ \mathbf{A}_2 \end{pmatrix} \mathbf{r}'(X) + \begin{pmatrix} 0 \\ \mathbf{w}'(X) \end{pmatrix} \right)$$

By substracting these two equations, we have

$$\begin{pmatrix} \mathbf{A}_1 \\ \mathbf{A}_2 \end{pmatrix} \mathbf{J}'(X)\mathbf{J}(X)(\mathbf{r}(X) - \mathbf{r}'(X)) +$$

$$\begin{pmatrix} 0 \\ \mathbf{J}'(X)\mathbf{J}(X)(\mathbf{w}(X) - \mathbf{w}'(X)) \end{pmatrix} = \mathbf{0} \qquad (1)$$

We first prove that $\mathbf{J}'(X)\mathbf{J}(X)(\mathbf{w}(X) - \mathbf{w}'(X)) \neq \mathbf{0} \bmod q$ by contradiction. If $\mathbf{J}'(X)\mathbf{J}(X)(\mathbf{w}(X) - \mathbf{w}'(X)) = \mathbf{0} \bmod q$, we have $\mathbf{J}'(X)\mathbf{J}(X)(\mathbf{w}(X) - \mathbf{w}'(X))$ $= \mathbf{0} \bmod p_1$. Then, $\forall j \in [0, N-1]$, we have

$$[\mathbf{J}'(X)\mathbf{J}(X)\mathbf{w}(X)]_{(X-\varsigma_j,p_1)} = [\mathbf{J}'(X)\mathbf{J}(X)\mathbf{w}'(X)]_{(X-\varsigma_j,p_1)}$$

Thus,
$$[\mathbf{w}_m(\varsigma_j)]_{p_1} = [\mathbf{w}'_m(\varsigma_j)]_{p_1}, \forall j \in \mathbf{J} \cap \mathbf{J}', m \in [0, \ell - 1]$$

According to the verification, $\forall j \in \mathbf{J} \cap \mathbf{J}', m \in [0, \ell - 1]$, we further have

$$H'_m (w_{0,j} || w_{1,j} || \cdots || w_{h-1,j}) = H'_m \left(w'_{0,j} || w'_{1,j} || \cdots || w'_{h-1,j}\right)$$

However, since $\exists i \in \mathbf{I} \cap \mathbf{I}' \wedge j \in \mathbf{J} \cap \mathbf{J}'$ such that $w_{i,j} \neq w'_{i,j}$ and finding a collision for a collision-resistant hash function happens only with negligible probability, hence we have a contradiction except with negligible probability. Consequently, from Eq. (1), we have

$$\mathbf{A}_2 \mathbf{J}'(X) \mathbf{J}(X) \left(\mathbf{r}(X) - \mathbf{r}'(X)\right) \neq \mathbf{0} \bmod q$$

and hence

$$\mathbf{J}'(X) \mathbf{J}(X) \left(\mathbf{r}(X) - \mathbf{r}'(X)\right) \neq \mathbf{0} \bmod q$$

Since we have $[\mathbf{r}(X)]_{p_1} = [\mathbf{r}'(X)]_{p_1} = 0$, we can deduce

$$\mathbf{J}'(X) \mathbf{J}(X) \left(\mathbf{r}(X) - \mathbf{r}'(X)\right) \neq \mathbf{0} \bmod p_2$$

As $[\mathbf{J}(X)]_{p_2} = [\mathbf{J}'(X)]_{p_2} = 1$, we have

$$\left(\mathbf{r}(X) - \mathbf{r}'(X)\right) \neq \mathbf{0} \bmod p_2$$

and

$$\mathbf{A}_1 \left(\mathbf{r}(X) - \mathbf{r}'(X)\right) = \mathbf{0} \bmod p_2$$

from the upper part of Eq. (1), therefore we have found a solution to $MSIS$ problem with respect to $\mathbf{A}_1$ in $\mathcal{R}_{p_2}$ since both $\mathbf{r}(X)$ and $\mathbf{r}'(X)$ belong to S_β^k. This completes the proof.

4   Performance Analysis

The complexity comparison between our proposed subvector commitment scheme and the existing public-coin-setup schemes can be found in Table 1. Our proposed scheme is the first to achieve post-quantum security. The computational cost of three algorithms in our proposed scheme is mainly determined by the amounts of required polynomial multiplications, and CRT/inverse CRT operations. All these operations can be optimized through number-theoretic transformation, the details of which could be found in [15]. An optimized polynomial multiplication takes $\lambda^2 N \log(N)$ operations, where N is the degree of the polynomial, which is set to be $\sqrt{M}$ in our scheme. The message size $M = h \times N$ is assumed to be a square, which can always be satisfied through padding (appending 0's entries to the matrix).

In the following table, $W = \sqrt{M} \log\left(\sqrt{M}\right)$. $U = |w_{i,j}|i \in I, j \in J|$, i.e., the size of the opening set. We use CRH to denote collision-resistant hash and Root to denote the strong or adaptive root assumption. pp is an abbreviation for public parameters.

Table 1. Comparison of setup-free subvector commitment schemes.

Scheme	Merkle tree	[2]	This work
$|pp|$	1	λM	λ
$|\mathbf{C}|$	λ	λ^2	$\lambda^2 \sqrt{M}$
$|\Lambda|$	$\lambda U \log M$	λ^2	$\lambda U \sqrt{M}$
Com	λM	$\lambda^2 M$	$\lambda^2 W$
Open	$\lambda U \log M$	$\lambda^2(M - q^2)$	$\lambda^2(W + 1)$
Verify	$\lambda U \log M$	$\lambda^2 U$	$\lambda^2 W$
Assumption	CRH	Root	MSIS
PQ secure?	YES	NO	YES

5 Conclusion

This work presents a sub-linear lattice-based submatrix commitment scheme under $MSIS$ assumption, which is the first post-quantum secure subvector commitment scheme with sublinear commitment and opening size.

Acknowledgments. We thank Jonathan Bootle from IBM Research in Zurich for insightful discussions on subvector commitment and its application in SNARK.

References

1. Catalano, D., Fiore, D.: Vector commitments and their applications. In: Kurosawa, K., Hanaoka, G. (eds.) Public-Key Cryptography – PKC 2013. PKC 2013. Lecture Notes in Computer Science, vol. 7778. Springer, Heidelberg (2013). https://doi.org/10.1007/978-3-642-36362-7_5
2. Lai, R.W., Malavolta, G.: Succinct arguments from subvector commitments and linear map commitments (2018). https://eprint.iacr.org/2018/705
3. Libert, B., Ramanna, S., Yung, M.: Functional commitment schemes: from polynomial commitments to pairing-based accumulators from simple assumptions. In: 43rd International Colloquium on Automata, Languages and Programming (ICALP 2016) (2016)
4. Sasson, E.B., et al.: Zerocash: decentralized anonymous payments from bitcoin. In: IEEE Symposium on Security and Privacy, vol. 2014, pp. 459–474. IEEE (2014)
5. Kim, J., Lee, J., Oh, H.: Simulation-extractable zk-SNARK with a single verification. Cryptology ePrint Archive, Report 2019/586 (2019). https://eprint.iacr.org/2019/586
6. Miers, I., et al.: Decentralized anonymous payments. Ph.D. dissertation, Johns Hopkins University (2017)
7. Bowe, S., Gabizon, A., Green, M.D.: A multi-party protocol for constructing the public parameters of the Pinocchio zk-SNARK. In: Zohar, A. et al. (eds.) Financial Cryptography and Data Security. FC 2018. Lecture Notes in Computer Science, vol. 10958. Springer, Heidelberg (2019). https://doi.org/10.1007/978-3-662-58820-8_5

8. Micali, S.: Cs proofs. In: Proceedings 35th Annual Symposium on Foundations of Computer Science, pp. 436–453. IEEE (1994)
9. Boneh, D., Bünz, B., Fisch, B.: Batching techniques for accumulators with applications to IOPs and stateless blockchains. Cryptology ePrint Archive, Report 2018/1188, Tech. Rep. (2018)
10. Ducas, L., Lepoint, T., Lyubashevsky, V., Schwabe, P., Seiler, G., Stehle, D.: Crystals - dilithium: digital signatures from module lattices. Cryptology ePrint Archive, Report 2017/633 (2017). http://eprint.iacr.org/2017/633
11. Gentry, C.,,Halevi, S., Smart, N.P.: Homomorphic evaluation of the AES circuit. In: Safavi-Naini, R., Canetti, R. (eds.) Advances in Cryptology – CRYPTO 2012. CRYPTO 2012. Lecture Notes in Computer Science, vol. 7417. Springer, Heidelberg (2012). https://doi.org/10.1007/978-3-642-32009-5_49
12. Baum, C., Lin, H., Oechsner, S.: Towards practical lattice-based one-time linkable ring signatures. In: Naccache, D., et al. (eds.) Information and Communications Security. ICICS 2018. Lecture Notes in Computer Science, vol. 11149. Springer, Cham (2018). https://doi.org/10.1007/978-3-030-01950-1_18
13. Baum, C., Damgard, I., Lyubashevsky, V., Oechsner, S., Peikert, C.: More efficient commitments from structured lattice assumptions. Cryptology ePrint Archive, Report 2016/997 (2016). https://eprint.iacr.org/2016/997
14. Ajtai, M.: Generating hard instances of lattice problems (extended abstract). In: STOC, pp. 99–108 (1996)
15. Zucca, V.: Towards efficient arithmetic for Ring-LWE based homomorphic encryption. Ph.D. dissertation (2018)

PIPO: A Lightweight Block Cipher with Efficient Higher-Order Masking Software Implementations

Hangi Kim[1], Yongjin Jeon[1], Giyoon Kim[1], Jongsung Kim[1,2]([⊠]), Bo-Yeon Sim[1], Dong-Guk Han[1,2], Hwajeong Seo[3], Seonggyeom Kim[4], Seokhie Hong[4], Jaechul Sung[5], and Deukjo Hong[6]

[1] Department of Financial Information Security, Kookmin University, Seoul, Republic of Korea
jskim@kookmin.ac.kr
[2] Department of Information Security, Cryptology, and Mathematics, Kookmin University, Seoul, Republic of Korea
[3] Division of IT Convergence Engineering, Hansung University, Seoul, Republic of Korea
[4] School of Cyber Security, Korea University, Seoul, Republic of Korea
[5] Department of Mathematics, University of Seoul, Seoul, Republic of Korea
[6] Department of Information Technology and Engineering, Jeonbuk National University, Jeonju, Republic of Korea

Abstract. In this paper, we introduce a new lightweight 64-bit block cipher PIPO (PIPO stands for "Plug-In" and "Plug-Out", representing its use in side-channel protected and unprotected environments, respectively.) supporting a 128 or 256-bit key. It is a byte-oriented and bitsliced cipher that offers excellent performance in 8-bit AVR software implementations. In particular, PIPO allows for efficient higher-order masking implementations, since it uses a minimal number of nonlinear operations. Our implementations demonstrate that PIPO outperforms existing block ciphers (for the same block and key lengths) in both side-channel protected and unprotected environments, on an 8-bit AVR. Furthermore, PIPO records competitive round-based hardware implementations.

For the nonlinear layer of PIPO, we have developed a new lightweight 8-bit S-box that provides an efficient bitsliced implementation including only 11 nonlinear bitwise operations. Furthermore, its differential and linear branch numbers are both 3. This characteristic enables PIPO to thwart differential and linear attacks with fewer rounds. The security of PIPO has been scrutinized with regards to state-of-the-art cryptanalysis.

Keywords: Lightweight S-boxes · Differential and linear branch numbers · PIPO · Higher-order masking

This work was supported by Institute for Information & communications Technology Promotion (IITP) grant funded by the Korea government (MSIT) (No. 2017-0-00520, Development of SCR-Friendly Symmetric Key Cryptosystem and Its Application Modes).

D. Hong (Ed.): ICISC 2020, LNCS 12593, pp. 99–122, 2021.
https://doi.org/10.1007/978-3-030-68890-5_6

1   Introduction

Most devices in IoT environments are miniature and resource-constrained. Therefore, lightweight cryptography, which is an active area of research, is essential. Some lightweight block ciphers such as PRESENT [22] and CLEFIA [44] have been standardized by ISO/IEC. Additionally, a lightweight cryptography standardization project is ongoing at NIST. A variety of lightweight block ciphers have been introduced in the literature, many of which are hardware-friendly systems [9,22,24,33,43]. Others focus on software performance [10,28,31] or both hardware and software performance [8,12,14,32,49].

In 1996, Paul Kocher first proposed side-channel attacks, which extract secret information by analyzing side-channel information [37]. Since secure designs for mathematical cryptanalysis cannot guarantee security against side-channel attacks, various countermeasures have been studied. With side-channel attacks becoming more advanced and the associated equipment cost decreasing [47], the application of side-channel countermeasures to ciphers has become critical. Recently, various studies have been actively conducted on efficient implementations of side-channel countermeasures, especially on efficient masked implementations. To minimize performance overhead, the focus has been on reducing the number of nonlinear operations. Several lightweight block ciphers, whose design goal is a low nonlinear operation count, have been proposed [2,28,31]. These are intended for use in side-channel protected environments.

However, most existing lightweight block ciphers are unsuitable for at least one type of hardware, software, or side-channel protected implementations in low resource environments. Consequently, it is challenging to design general-purpose lightweight block ciphers capable of operating in any resource-constrained environment.

In this paper, we propose a new lightweight versatile block cipher PIPO. During the PIPO design process, the focus was on minimizing the number of nonlinear operations because this is the most significant factor for efficient higher-order masking implementations. To construct an 8-bit S-box with a small number of nonlinear operations, we devised a unbalanced-Bridge structure that accepts one 3-bit and two 5-bit S-boxes and produces 8-bit S-boxes. This structure allows us to construct a new 8-bit S-box that offers good cryptographic properties and an efficient bitsliced implementation including only 11 nonlinear bitwise operations. We also present theorems applied to the unbalanced-Bridge structure, which show the conditions that the both differential and linear branch numbers of the S-boxes constructed through the structure are greater than 2. We investigated the linear layer with the highest security against differential and linear cryptanalyses when combined with the new S-box, through which we could reduce the number of rounds of PIPO. Consequently, PIPO achieves fast higher-order masking implementations, and its execution time increases less with the number of shares (*i.e.*, the masking order) compared with other lightweight 64-bit block ciphers with 128-bit keys. Additionally, PIPO records excellent performance on 8-bit microcontrollers and competitive round-based hardware implementations.

The following notation and definitions are used throughout this paper.

DDT	Difference Distribution Table of an n-bit S-box whose $(\Delta\alpha, \Delta\beta)$ entry is $\#\{x \in \mathbb{F}_2^n \| S(x) \oplus S(x \oplus \Delta\alpha) = \Delta\beta\}$, where $\Delta\alpha, \Delta\beta \in \mathbb{F}_2^n$
LAT	Linear Approximation Table of an n-bit S-box whose $(\lambda_\alpha, \lambda_\beta)$ entry is $\#\{x \in \mathbb{F}_2^n \| \lambda_\alpha \bullet x = \lambda_\beta \bullet S(x)\} - 2^{n-1}$, where $\lambda_\alpha, \lambda_\beta \in \mathbb{F}_2^n$, and the symbol $\bullet$ denotes the canonical inner product in $\mathbb{F}_2^n$
Differential uniformity	$\max\limits_{\Delta\alpha \neq 0, \Delta\beta} \#\{x \in \mathbb{F}_2^n \| S(x) \oplus S(x \oplus \Delta\alpha) = \Delta\beta\}$
Non-linearity	$2^{n-1} - 2^{-1} \times \max\limits_{\lambda_\alpha, \lambda_\beta \neq 0} \|\Phi(\lambda_\alpha, \lambda_\beta)\|$, where $\Phi(\lambda_\alpha, \lambda_\beta)$ $= \sum\limits_{x \in \mathbb{F}_2^n} (-1)^{\lambda_\beta \bullet S(x) \oplus \lambda_\alpha \bullet x}$
DBN	Differential Branch Number of an S-box defined as $\min\limits_{a, b \neq a} (wt(a \oplus b) + wt(S(a) \oplus S(b)))$
LBN	Linear Branch Number of an S-box defined as $\min\limits_{a, b, \Phi(a, b) \neq 0} (wt(a) + wt(b))$

2 Specification of PIPO

The PIPO block cipher accepts a 64-bit plaintext and either a 128 or 256-bit key, generating a 64-bit ciphertext. It performs 13 rounds for a 128-bit key and 17 rounds for a 256-bit key. Each round is composed of a nonlinear layer denoted as the S-layer, a linear layer denoted as the R-layer, and round key and constant XOR additions. The overall structure of PIPO is depicted on the left side of Fig. 1. Here, RK_0 is a whitening key and $RK_1, RK_2, \cdots, RK_r$ are round keys, where $r = 13$ (128-bit key) or 17 (256-bit key). The i-th round constant c_i is i (the round counter) which is XORed with RK_i. During the enciphering process, the intermediate state is regarded as an 8×8 array of bits, as shown on the right side of Fig. 1, where $X[i]$ represents the i-th row byte for $i = 0 \sim 7$. The S-layer executes eight identical 8-bit S-boxes (denoted as S_8) in parallel. The S_8 is applied to each column of the 8×8 array of bits, where the uppermost bit is the least significant. The S_8 is shown in Table 7 of Appendix C.1. The R-layer rotates the bits in each row by a given offset (Fig. 2).

The key schedule of PIPO is very simple. We first split a 128-bit master key K into two 64-bit subkeys K_0 and K_1, i.e., $K = K_1 \| K_0$. The whitening and round keys are then defined as $RK_i = K_{i \bmod 2}$, where $i = 0, 1, 2, \cdots, 13$. Similarly, a 256-bit master key K is divided into four 64-bit subkeys K_0, K_1, K_2, and K_3, i.e., $K = K_3 \| K_2 \| K_1 \| K_0$. In this case, $RK_i = K_{i \bmod 4}$ where $i = 0, 1, 2, \cdots, 17$. Some test vectors for PIPO are provided in Appendix A. Note that resistance to related-key attacks was not considered when designing the PIPO cipher.

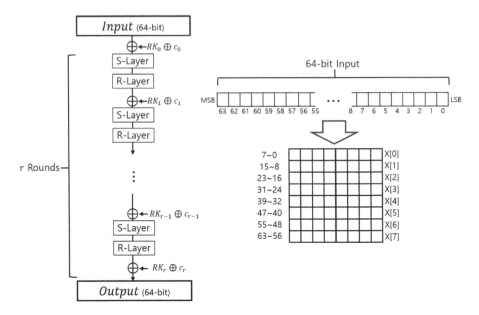

Fig. 1. Overall structure (left) and intermediate state (right) of PIPO

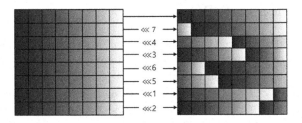

Fig. 2. R-layer

3   Design Rationale of PIPO

3.1   S-Layer

Overall Structure. We focused on the following three criteria when designing our 8-bit S-box, S_8.

1. It should offer an efficient bitsliced implementation including 11 or fewer nonlinear operations.
2. Its differential and linear branch numbers (DBN and LBN) should both be greater than 2.
3. Its differential uniformity should be 16 or less, and its non-linearity should be 96 or more.

Criterion 1 minimizes the number of nonlinear operations required by PIPO, which allows for efficient higher-order masking implementations. Criteria 2 and 3 ensure the cryptographic strengths of the S_8 against differential cryptanalysis (DC) and linear cryptanalysis (LC). Any inferior criteria will lead to the implementation of more rounds to achieve acceptable security against these attacks, eventually resulting in a weak proposal. The thresholds of the criteria were selected based on the properties of the existing lightweight 8-bit S-boxes (refer to Table 1).

The Bridge structure was first proposed in [36], and revisited in [15]. In order to construct an S_8 satisfying all the aforementioned three criteria, we employed the unbalanced-Bridge structure depicted in Fig. 3, where S_i^j represents the j-th and i-bit S-box in the structure. This structure has the following three characteristics. First, it uses 3-bit and 5-bit S-boxes instead of 4-bit S-boxes. We observe that 8-bit S-box constructions using three 4-bit S-boxes would have difficulty satisfying criterion 1, even though they conform to criteria 2 and 3. Second, all eight output bits are generated from at least two smaller S-boxes (to meet criterion 3). Finally, at least one non-bijective smaller S-box can be adopted to increase the number of possible combinations of smaller S-boxes.

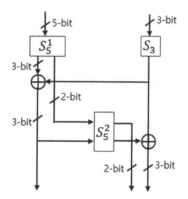

Fig. 3. The unbalanced-Bridge structure

The notation used in this section is introduced below.

$$\rho_c : \mathbb{F}_2^5 \to \mathbb{F}_2^5, \quad \rho_c(x||y) = y||x, \quad \text{for } x \in \mathbb{F}_2^3, \, y \in \mathbb{F}_2^2,$$
$$\tau_n : \mathbb{F}_2^5 \to \mathbb{F}_2^n, \quad \tau_n(x||y) = x, \quad \text{for } x \in \mathbb{F}_2^n, \, y \in \mathbb{F}_2^{5-n}, \, n \in \{1, 2, 3, 4\},$$
$$\tau_n' : \mathbb{F}_2^5 \to \mathbb{F}_2^n, \quad \tau_n'(x||y) = y, \quad \text{for } x \in \mathbb{F}_2^{5-n}, \, y \in \mathbb{F}_2^n, \, n \in \{1, 2, 3, 4\},$$
$$\mathfrak{F}_A^1 : \mathbb{F}_2^3 \to \mathbb{F}_2^5, \quad \mathfrak{F}_A^1(X) = (S_5^1)^{-1}(X||A) \text{ for } A \in \mathbb{F}_2^2,$$
$$\mathfrak{F}_A^2 : \mathbb{F}_2^3 \to \mathbb{F}_2^5, \quad \mathfrak{F}_A^2(X) = S_5^2(X||A) \text{ for } A \in \mathbb{F}_2^2.$$

Now we can define an 8-bit S-box constructed by the unbalanced-Bridge. Let $S_8(X_L||X_R) = C_L(X_L, X_R)|| C_R(X_L, X_R)$, where X_L and X_R represent the input variables of the S_8 which are in $\mathbb{F}_2^5$ and $\mathbb{F}_2^3$, respectively. Then $C_L(X_L, X_R) = \tau_3(S_5^1(X_L)) \oplus S_3(X_R)$ and $C_R(X_L, X_R) = \rho_c(S_5^2(S_5^1(X_L) \oplus (S_3(X_R)||0^{(2)})))\oplus (0^{(2)}||S_3(X_R))$ with $C_L : \mathbb{F}_2^5 \times \mathbb{F}_2^3 \to \mathbb{F}_2^3$ and $C_R : \mathbb{F}_2^5 \times \mathbb{F}_2^3 \to \mathbb{F}_2^5$.

Proposition 1 shows the conditions for the 8-bit S-box constructed by the unbalanced-Bridge to be bijective.

Proposition 1. *The 8-bit S-box constructed using the unbalanced-Bridge is bijective if and only if the following three conditions are all satisfied:*

  i) S_3 *is bijective.*
  ii) S_5^1 *is bijective.*
  iii) For all $y \in \mathbb{F}_2^3$, $f_y(x) = \tau_2'(S_5^2(y||x))$ *is a bijective function with* $f_y : \mathbb{F}_2^2 \to \mathbb{F}_2^2$.

Proof. Refer to Appendix B.1.

Construction of 8-Bit S-Boxes with DBN> 2 and LBN> 2 and Our S_8 Selection. We present here how to construct 8-bit S-boxes with DBN> 2 and LBN> 2. Our framework is to eliminate all the input and output differences (masks) where the sum of their Hamming weights is 2. During this elimination process, we can obtain some conditions of smaller S-boxes. Theorems 1 and 2 present the necessary and sufficient conditions of smaller S-boxes so that the 8-bit S-boxes constructed by Fig. 3 have both differential and linear branch numbers greater than 2.

Theorem 1. *The DBN of bijective 8-bit S-boxes constructed using the unbalanced-Bridge is greater than 2 if and only if conditions i), ii), and iii) are all satisfied ($\Delta\alpha$ and $\Delta\beta$ below represent arbitrary differences where $wt(\Delta\alpha) = wt(\Delta\beta) = 1$):*

  i) For each $\Delta\alpha, \Delta\beta \in \mathbb{F}_2^3$, *at least one of the entry* $(\Delta\alpha, \Delta\beta)$ *in DDT of* S_3 *and the entry* $(\Delta\beta||0^{(2)}, \Delta\beta||0^{(2)})$ *in DDT of* S_5^2 *is 0,*
  ii) For each $\Delta\alpha, \Delta\beta \in \mathbb{F}_2^2$, *for each* $A, B(\neq A) \in \mathbb{F}_2^2$, *at least one of* $\mathfrak{F}_A^1(X) \oplus \mathfrak{F}_B^1(X) = \Delta\alpha$ *and* $\mathfrak{F}_A^2(X) \oplus \mathfrak{F}_B^2(X) = \Delta\beta$ *has no solution* X, *where* $X \in \mathbb{F}_2^2$,
  iii) For each $\Delta\alpha \in \mathbb{F}_2^3$ *and* $\Delta\beta \in \mathbb{F}_2^5$, *for each* $A, B \in \mathbb{F}_2^2$, *at least one of* $\mathfrak{F}_A^1(X) \oplus \mathfrak{F}_B^1(X \oplus \Delta\alpha) = \Delta\beta$ *and* $\mathfrak{F}_A^2(X) \oplus \mathfrak{F}_B^2(X \oplus \Delta\alpha) = \Delta 0$ *has no solution* X, *where* $X \in \mathbb{F}_2^3$.

Proof. Refer to Appendix B.2.

The following theorem concerning the LBN can be similarly obtained.

Theorem 2. *The LBN of bijective 8-bit S-boxes constructed using the unbalanced-Bridge is greater than 2 if and only if conditions i), ii), and iii) are all satisfied (λ_α and λ_β below represent arbitrary masks where $wt(\lambda_\alpha) = wt(\lambda_\beta) = 1$):*

  i) For each $\lambda_\alpha, \lambda_\beta \in \mathbb{F}_2^3$, *at least one of the entry* $(\lambda_\alpha, \lambda_\beta)$ *in LAT of* S_3 *and the entry* $(0, \lambda_\beta||0^{(2)})$ *in LAT of* S_5^2 *is 0,*
  ii) For each $\lambda_\alpha \in \mathbb{F}_2^5$ *and* $\lambda_\beta \in \mathbb{F}_2^3$, $\sum_{A \in \mathbb{F}_2^2} X \cdot Y = 0$ *where* X *is the entry* $(\lambda_\beta, \lambda_\alpha)$ *in LAT of* $\mathfrak{F}_A^1$ *and* Y *is the entry* $(\lambda_\beta, \lambda_\beta||0^{(2)})$ *in LAT of* $\mathfrak{F}_A^2$,

iii) For each $\lambda_\alpha, \lambda_\beta \in \mathbb{F}_2^5$ satisfying $\tau_3(\lambda_\beta) = 0$, $\sum_{A \in \mathbb{F}_2^2} X \cdot Y = 0$ where X is the entry $(0, \lambda_\alpha)$ in LAT of $\mathfrak{F}_A^1$ and Y is the entry $(0, \lambda_\beta)$ in LAT of $\mathfrak{F}_A^2$.

Proof. Refer to Appendix B.3.

Our S_8 search process is outlined as follows. First, we generated 3-bit and 5-bit S-box sets; for 3-bit S-boxes we ran an exhaustive search with AND, OR, XOR, and NOT instructions while restricting the number of nonlinear (resp. linear) operations to 3 (resp. 4), and for 5-bit S-boxes we ran an exhaustive search with AND, OR, and XOR instruction while restricting the number of nonlinear (resp. linear) operations to 4 (resp.7) with a differential uniformity of 8 or less. Second, we classified two 5-bit S-boxes and one 3-bit S-box that satisfy the conditions of Proposition 1, Theorems 1 and 2. During this process, the search space for the S_8 could be significantly reduced because the early abort technique was used to select S_3, S_1^5, and S_2^5. Third, we randomly chose the combination of S_3, S_5^1, and S_5^2 to verify whether the corresponding 8-bit S-box satisfies criterion 3, and no fixed point. During the search, we found more than 8,000 candidates for the S_8. We selected the one that leads to the best resistance to differential and linear attacks when combined with the linear layer of PIPO (refer to section 3.2). The final selected input/output values of the S_8 are presented in Table 7; its bitsliced implementation is given in Appendix C.2.

Table 1 compares the cryptographic properties and operations with those of other 8-bit S-boxes built from smaller 3 S-boxes.

Table 1. Comparison of bitslice 8-bit S-boxes with respect to cryptographic properties and number of operations

Blockcipher	PIPO	FLY	Fantomas	Robin	LILLIPUT
Differential uniformity	16	16	16	16	8
DBN	3	3	2	2	2
Non-linearity	96	96	96	96	96
LBN	3	3	2	2	2
Algebraic degree	5	5	5	6	6
#(Fixed points)	0	2	0	16	1
#(Nonlinear operations)	11	12	11	12	12
#(Linear operations)	23	24	27	24	27
Construction method	*U-Bridge	Lai-Massey	*U-MISTY	MISTY	Feistel
Reference	This paper	[35]	[31]	[31]	[1]

*'U-' represents 'Unbalanced-'.

**Nonlinear (resp. linear) operations represent AND, OR (resp. XOR, NOT).

3.2   R-Layer

To ensure efficient hardware and software implementations, we chose the R-layer to be a bit permutation which only uses bit-rotations in bytes. Its bitsliced implementation is given in Listing 1.1. During the design of the R-layer, the following criteria were considered.

1. The number of rounds to achieve full diffusion – through which any input bit can affect the entire output bits – should be minimized.
2. Combining the R-layer with the S-layer should enable the cipher to have the best resistance to DC and LC (among all bit permutations satisfying the first criterion).

To meet the first criterion, we adopted a bit permutation that enables PIPO to achieve full diffusion in two rounds by using rotation offsets $0 \sim 7$ for all rows. The second criterion was taken into account when deciding which rotation to use for which row. We applied all $5,040(=7!)$ R-layers (except for all rotation equivalences) to the S-layer and selected one with the lowest probabilities of 6 and 7-round best differential and linear trails. Our analysis found that the selected combination of the S and R layers provides superior resistance to DC and LC than any other combinations even when other S-boxes among the aforementioned candidates were chosen. Note that most combinations of S and R layers candidates could not provide best 7-round differential and linear trails with less than probability 2^{-64}.

Listing 1.1. Bitsliced implementation of R-layer (in C code)

```
//Input: (MSB) X[7], X[6], X[5], X[4], X[3], X[2], X[1], X[0] (LSB)
X[1] = ((X[1] << 7)) | ((X[1] >> 1));
X[2] = ((X[2] << 4)) | ((X[2] >> 4));
X[3] = ((X[3] << 3)) | ((X[3] >> 5));
X[4] = ((X[4] << 6)) | ((X[4] >> 2));
X[5] = ((X[5] << 5)) | ((X[5] >> 3));
X[6] = ((X[6] << 1)) | ((X[6] >> 7));
X[7] = ((X[7] << 2)) | ((X[7] >> 6));
//Output: (MSB) X[7], X[6], X[5], X[4], X[3], X[2], X[1], X[0] (LSB)
```

4 Security Evaluation of PIPO

Table 2 shows the maximum numbers of rounds of characteristics and key recovery attacks that we found for each attack [3,18–20,40,42,46]. In addition to the cryptanalysis shown in Table 2, we conducted algebraic attack [23], integral attack [48], statistical saturation attack [25], invariant subspace attack [38,39], nonlinear invariant attack [45] and slide attack [21], but they were not applied more effectively than DC or LC.

Table 2. The numbers of rounds of the best characteristics for each cryptanalysis

Key length	Cryptanalysis	Best characteristic	Key recovery attack
128-bit	Differential	6-round	9-round
	Linear	6-round	9-round
	Impossible differential	4-round	6-round
	Boomerang/Rectangle	6-round	8-round
	Meet-in-the-Middle	6-round	6-round
256-bit	Differential	6-round	11-round
	Linear	6-round	11-round
	Impossible differential	4-round	8-round
	Boomerang/Rectangle	6-round	10-round
	Meet-in-the-Middle	10-round	10-round

One of the major design considerations for PIPO is to adopt a compact number of rounds (not enough rounds to guarantee security that is (too) high) based on thorough security analyses. We discovered that the best attacks applied to PIPO are DC and LC. An exhaustive search (based on the branch and bound technique [41]) for the DC and LC distinguishers was performed, in which the best reaches 6 rounds. Our analyses could recover the key of up to 9 and 11 rounds of PIPO-64/128 and PIPO-64/256, respectively.

5 Performance Evaluation of Higher-Order Masking Implementations of **PIPO**

Bitsliced implementations, initially proposed by Biham [17], are known to be efficient when applying Boolean masking, since secure S-box computations can be carried out in parallel [29–31, 34]. Thus, we used an S-box that can be efficiently implemented in this way, and only involves 11 nonlinear bitwise operations. The number of nonlinear operations is very important for Boolean masking schemes, since they have a quadratic complexity, *i.e.*, $O(d^2)$, compared with the linear complexity, *i.e.*, $O(d)$, for other operations.

We constructed PIPO using higher-order masked S-layer and R-layer. There are several variations of ISW-AND [6,7,16], however, in this paper, we apply original ISW-AND. Since logical OR of two inputs a and b satisfies $a \vee b = (a \wedge b) \oplus a \oplus b$, thus, ISW-OR can be calculated by replacing logical AND with ISW-AND.

Table 3. Comparison of required ROM (bytes) for round constant, number of nonlinear bitwise operations, and permutation layers of round functions

Block cipher	Table size	#(nonlinear bitwise operations)	Permutation
PIPO-64/128	0	1,144	7 bit-rotations in bytes
PRIDE-64/128	80	1,280	MixColumns*
SIMON-64/128	62	1,408	3 bit-rotations in 32-bit words
RoadRunneR-64/128	0	1,536	24 bit-rotations in bytes
RECTANGLE-64/128	25	1,600	3 bit-rotations in 16-bit words
CRAFT-64/128	64	1,984	MixColumns*, PermuteNibbles
PRESENT-64/128	0	1,984	Bit permutation
SKINNY-64/128	62	2,304	ShiftRows, MixColumns*

* : multiply with binary matrix

We compare our proposed PIPO with 64-bit block ciphers with 128-bit keys as shown in Table 3. All the ciphers compared were implemented using bitslice techniques, and only round constants were precomputed. There is no need to precompute round constants of PIPO, RoadRunneR, and PRESENT, because they are the i or $NR - i$ for $i = 0, 1, \cdots, NR - 1$, where NR is the number of rounds. Therefore, the required ROM for round constants is shown in Table 3. Only CRAFT used an additional 16-byte diffusion table for generating tweakeys. The same secure logical operations of PIPO were applied to implement higher-order masking structures.

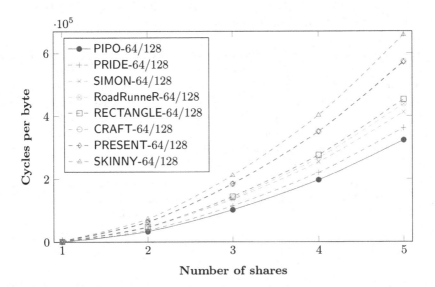

Fig. 4. Execution times of one-block encryptions according to the number of shares in an Atmel AVR XMEGA128 (1 means unprotected)

Figure 4 shows the execution times for different numbers of shares on an 8-bit AVR processor. Especially, it shows that the more nonlinear operations, the greater increase in execution time with the number of shares, refer to Table 3. PIPO has the smallest number of nonlinear operations.

6 Performance Evaluation of Software and Hardware Implementations of PIPO

6.1 Software Implementations

The PIPO block cipher consists of permutation (R-layer) and S-box (S-layer) computations. The permutation routine is performed in 8-bit rotation operations, and 22 XOR, 6 AND, 5 OR, 1 COM and 24 MOV instructions are used to compute the S-box. This uses a total of 21 general-purpose registers: six for temporal storage, one for a zero constant, eight for a plaintext, four for address pointers and two for counter variables.

The developers of SIMON and SPECK have proposed a new metric to measure overall performance on low-end devices, namely RANK [11]. This is calculated as follows:

$$RANK = (10^6/CPB)/(ROM + 2 \times RAM).$$

In this metric, higher values of RANK correspond to better performance. Table 4 compares results for several block ciphers on an 8-bit AVR platform. Here, we used Atmel Studio 6.2, and compiled all implementations with optimization level 3. The target processor was an ATmega128 running at 8 MHz [4]. PIPO requires 320 bytes of code, 31 bytes of RAM and an execution time of 197 CPB. We used the RANK metric to compare the ciphers' overall performances, finding that PIPO achieved the highest score among block ciphers with the same parameter lengths.

Table 4. Comparison of block ciphers on 8-bit AVR*

Block cipher	Code size (bytes)	RAM (bytes)	Execution time (cycles per byte)	RANK
PIPO-64/128	320	31	197	13.31
SIMON-64/128 [11]	290	24	253	11.69
RoadRunneR-64/128 [10]	196	24	477	8.59
RECTANGLE-64/128 [26]	466	204	403	2.84
PRIDE-64/128 [26]	650	47	969	1.39
SKINNY-64/128 [26]	502	187	877	1.30
PRESENT-64/128 [27]	660	280	1,349	0.61
CRAFT-64/128 [13]	894	243	1,504	0.48
PIPO-64/256	320	47	253	9.54

*The code size represents ROM, and RAM metric includes STACK.

6.2   Hardware Implementations

We implemented PIPO-64/128 and PIPO-64/256 in Verilog, and synthesized the proposed architectures using the Synopsys Design Compiler with 130 nm CMOS technology. Figure 5 shows the datapath of an area-optimized encryption-only PIPO block cipher, which performs one round per clock cycle (*i.e.*, uses a 64-bit-wide datapath). The S-layer uses the same 8-bit S-box 8 times, whereas the R-layer is implemented in wiring. For lightweight key generation, we obtain the round key from the master key, directly. This feature avoids including the key storage. Our implementations require 13 and 17 clock cycles to encrypt a 64-bit plaintext with 128-bit and 256-bit keys, respectively.

Table 5 shows the areas required by PIPO-64/128 and PIPO-64/256. Most of the areas are taken up by the S-layer, in order to compute eight 8-bit S-boxes in parallel. The flip-flops are used for storing plaintext and counter, and the other areas consist of MUX and other logical operations.

Table 6 compares the results for several different block ciphers implemented as ASICs. Compared with the other block ciphers using the same parameter lengths, PIPO needs more gates than CRAFT, Piccolo and SIMON but its cycles per block are much lower, resulting in the highest figure of merit FOM (nano

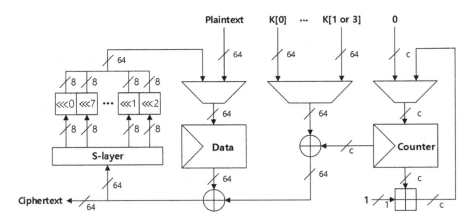

Fig. 5. Datapath of an area-optimized version of PIPO

Table 5. Area requirement of PIPO-64/128 and PIPO-64/256.

Module	PIPO-64/128		PIPO-64/256	
	GE	%	GE	%
Data and Counter States	341	24	360	22
S-layer	581	40	581	36
Add Round Key	170	12	170	11
Others	354	24	491	31
Total	1,446	100	1,602	100

Table 6. Comparison of round-based and area optimized implementations for block ciphers using 130 nm ASIC library.

Block cipher	Area [GE]	Throughput (Kbps@100KHz)	cycles /block	FOM $[\frac{bits \times 10^9}{clk \times GE^2}]$
PIPO-64/128	1,446	492	13	2,355
CRAFT-64/128 [13]	949	200	32	2,221
Piccolo-64/128 [43]	1,197	194	33	1,354
SIMON-64/128 [12]	1,417	133	48	664
RECTANGLE-64/128 [49]	2,064	246	26	578
PIPO-64/256	1,602	376	17	1,467

bits per clock cycle per GE squared [5,32]). It is obvious that the high FOM of PIPO requires less energy and battery consumption.

7 Conclusion

In this paper, we proposed a new lightweight versatile block cipher PIPO suitable for diverse resource-constrained environments. In particular, PIPO exhibits excellent performance in both side-channel protected and unprotected environments on 8-bit microcontrollers, and fast round-based hardware implementations as well. Furthermore, a thorough security analysis of PIPO was conducted.

A   Test Vectors

The following test vectors are represented in big endian representation.

– PIPO-64/128
  • Secret key: 0x6DC416DD_779428D2_7E1D20AD_2E152297
  • Plaintext: 0x098552F6_1E270026
  • Ciphertext: 0x6B6B2981_AD5D0327

– PIPO-64/256
  • Secret key:0x009A3AA4_76A96DB5_54A71206_26D15633_6DC416DD _779428D2_7E1D20AD_2E152297
  • Plaintext: 0x098552F6_1E270026
  • Ciphertext: 0x816DAE6F_B6523889

B   Proofs of Proposition and Theorems

B.1   Proof of Proposition 1

$(\Rightarrow)$

If S_3 or S_5^1 is non-bijective, there are two different inputs $X_L || X_R, X_L' || X_R'$ satisfying $(S_5^1(X_L), S_3(X_R)) = (S_5^1(X_L'), S_3(X_R'))$. Then, it is easy to see that $S_8(X_L || X_R) = S_8(X_L' || X_R')$, and thus two conditions $i)$ and $ii)$ should hold. Assume that the f_y in condition $iii)$ is non-bijective for some $y \in \mathbb{F}_2^3$. Then there should be two different inputs a, a' satisfying $f_y(a) = f_y(a')$. It induces $\tau_2'(S_5^2(y||a)) = \tau_2'(S_5^2(y||a'))$. On the other hand, we can take a pair X_R, X_R' satisfying $\tau_3(S_5^2(y||a)) \oplus S_3(X_R) = \tau_3(S_5^2(y||a')) \oplus S_3(X_R')$, and thus $C_R = C_R'$. Combining the above two equations yields $S_5^2(y||a) \oplus (S_3(X_R)||0^{(2)}) = S_5^2(y||a') \oplus (S_3(X_R')||0^{(2)})$. And, we take a pair X_L, X_L' satisfying $S_5^1(X_L) = (y \oplus S_3(X_R))||a$ and $S_5^1(X_L') = (y \oplus S_3(X_R'))||a'$. Since $a \neq a'$, we have $X_L \neq X_L'$ satisfying $S_8(X_L || X_R) = S_8(X_L' || X_R')$. Therefore, condition $iii)$ should also hold.

$(\Leftarrow)$

Assume that $X_L \neq X_L'$ and $X_R = X_R'$. If $\tau_3(S_5^1(X_L)) \neq \tau_3(S_5^1(X_L'))$, then $C_L(X_L, X_R) \neq C_L(X_L', X_R')$. Let $\tau_3(S_5^1(X_L)) = \tau_3(S_5^1(X_L'))$. It leads to $C_L(X_L, X_R) = C_L(X_L', X_R')$, and $\tau_2'(S_5^1(X_L)) \neq \tau_2'(S_5^1(X_L'))$. Because of condition $iii)$, $\tau_2(C_R(X_L, X_R)) \neq \tau_2(C_R(X_L', X_R'))$. Assume that $X_L = X_L'$ and $X_R \neq X_R'$. Since $S_3(X_R) \neq S_3(X_R')$, $C_L(X_L, X_R) \neq C_L(X_L', X_R')$. Assume that $X_L \neq X_L'$, $X_R \neq X_R'$. If $C_L(X_L, X_R) = C_L(X_L', X_R')$, either $\tau_2'(S_5^1(X_L)) \neq \tau_2'(S_5^1(X_L'))$ or $\tau_2'(S_5^1(X_L)) = \tau_2'(S_5^1(X_L'))$. The former case leads to $\tau_2(C_R(X_L, X_R)) \neq \tau_2(C_R(X_L', X_R'))$, and the latter case leads to $\tau_3'(C_R(X_L, X_R)) \neq \tau_3'(C_R(X_L', X_R'))$. Therefore, the 8-bit S-box is bijective. ∎

B.2   Proof of Theorem 1

We define the following notation for ease of expression.

$$Y = S_5^1(X_L), \ Z = S_5^1(X_L) \oplus (S_3(X_R)||0^{(2)}), \ A = \tau_2'(Y) = \tau_2'(Z), \ Y = Y'||A, \ Z = Z'||A.$$

Then, the expression of the C_L and C_R is

$$C_L(X_L, X_R) = \tau_3(Y) \oplus S_3(X_R) = \tau_3(Z),$$
$$C_R(X_L, X_R) = \rho_c(S_5^2(Y \oplus (S_3(X_R)||0^{(2)}))) \oplus S_3(X_R) = \rho_c(Z) \oplus S_3(X_R).$$

For convenience, we do not write 0 paddings on MSBs of smaller-bit data operating with larger-bit data; here, the 5-bit operand $S_3(X_R)$ represents $0^{(2)}||S_3(X_R)$.

$(0^{(5)}||\Delta a, 0^{(3)}||\Delta c)$: It happens if and only if there exists at least one (X_L, X_R) satisfying both $C_L(X_L, X_R) \oplus C_L(X_L, X_R \oplus \Delta a) = \Delta 0$ and $C_R(X_L, X_R) \oplus C_R(X_L, X_R \oplus \Delta a) = \Delta c$. The first equation is expressed as

$$\tau_3(Y) \oplus S_3(X_R) \oplus \tau_3(Y) \oplus S_3(X_R \oplus \Delta a) = S_3(X_R) \oplus S_3(X_R \oplus \Delta a) = \Delta 0.$$

Since S_3 is bijective, the $(0^{(5)}||\Delta a, 0^{(3)}||\Delta c)$ case dose not happen.

$\underline{(0^{(5)}||\Delta a, \Delta d||0^{(5)})}$: It happens if and only if there exists at least one (X_L, X_R) satisfying both $C_L(X_L, X_R) \oplus C_L(X_L, X_R \oplus \Delta a) = \Delta d$ and $C_R(X_L, X_R) \oplus C_R(X_L, X_R \oplus \Delta a) = \Delta 0$. The first equation is expressed as

$$\tau_3(Y) \oplus S_3(X_R) \oplus \tau_3(Y) \oplus S_3(X_R \oplus \Delta a) = S_3(X_R) \oplus S_3(X_R \oplus \Delta a) = \Delta d. \tag{1}$$

Similarly, the second equation $C_R(X_L, X_R) \oplus C_R(X_L, X_R \oplus \Delta a) = \Delta 0$ is expressed as

$$\rho_c(S_5^2(Y \oplus (S_3(X_R)||0^{(2)}))) \oplus S_3(X_R)$$
$$\oplus \rho_c(S_5^2(Y \oplus (S_3(X_R \oplus \Delta a)||0^{(2)}))) \oplus S_3(X_R \oplus \Delta a)$$
$$= \rho_c(S_5^2(Y \oplus (S_3(X_R)||0^{(2)}))) \oplus \rho_c(S_5^2(Y \oplus ((S_3(X_R) \oplus \Delta d)||0^{(2)}))) \oplus \Delta d = \Delta 0.$$

By applying ρ_c^{-1}, we have

$$S_5^2(Y \oplus (S_3(X_R)||0^{(2)})) \oplus S_5^2(Y \oplus ((S_3(X_R) \oplus \Delta d)||0^{(2)})) = \Delta d||0^{(2)}.$$

By applying Z, we obtain

$$S_5^2(Z) \oplus S_5^2(Z \oplus (\Delta d||0^{(2)})) = \Delta d||0^{(2)}. \tag{2}$$

Since the function $(X_L, X_R) \mapsto (Z, X_R)$ is bijective, the $(0^{(5)}||\Delta a, \Delta d||0^{(5)})$ case does not happen if and only if there is no (Z, X_R) satisfying both Eqs. (1) and (2), which is equivalent to condition i) where $\Delta\alpha = \Delta a$, $\Delta\beta = \Delta d$.

$\underline{(\Delta b||0^{(3)}, 0^{(3)}||\Delta c)}$: It happens if and only if there exists at least one (X_L, X_R) satisfying both $C_L(X_L, X_R) \oplus C_L(X_L \oplus \Delta b, X_R) = \Delta 0$ and $C_R(X_L, X_R) \oplus C_R(X_L \oplus \Delta b, X_R) = \Delta c$. The first equation is expressed as

$$\tau_3(S_5^1(X_L)) \oplus S_3(X_R) \oplus \tau_3(S_5^1(X_L \oplus \Delta b)) \oplus S_3(X_R) = \tau_3(S_5^1(X_L)) \oplus \tau_3(S_5^1(X_L \oplus \Delta b)) = \Delta 0.$$

Since S_5^1 is bijective, for a non-zero difference $\Delta\omega \in \mathbb{F}_2^2$, the above equation becomes

$$S_5^1(X_L) \oplus S_5^1(X_L \oplus \Delta b) = \Delta\omega.$$

The equation is rewritten as

$$S_5^1(X_L \oplus \Delta b) = S_5^1(X_L) \oplus \Delta\omega.$$

By applying $(S_5^1)^{-1}$, we obtain

$$X_L \oplus \Delta b = (S_5^1)^{-1}(S_5^1(X_L) \oplus \Delta\omega).$$

By using the variables Y, Y' and A, we have

$$(S_5^1)^{-1}(Y) \oplus (S_5^1)^{-1}(Y \oplus \Delta\omega) = \Delta b,$$

$$(S_5^1)^{-1}(Y'||A) \oplus (S_5^1)^{-1}(Y'||(A \oplus \Delta\omega)) = \Delta b. \tag{3}$$

And the second equation $C_R(X_L, X_R) \oplus C_R(X_L \oplus \Delta b, X_R) = \Delta c$ is expressed as

$$\rho_c(S_5^2(S_5^1(X_L) \oplus (S_3(X_R)||0^{(2)}))) \oplus S_3(X_R)$$
$$\oplus \rho_c(S_5^2(S_5^1(X_L \oplus \Delta b) \oplus (S_3(X_R)||0^{(2)}))) \oplus S_3(X_R)$$
$$= \rho_c(S_5^2(Z)) \oplus \rho_c(S_5^2(Z \oplus \Delta\omega)) = \Delta c.$$

By applying ρ_c^{-1}, we obtain

$$S_5^2(Z) \oplus S_5^2(Z \oplus \Delta\omega) = \rho_c^{-1}(\Delta c).$$

This gives the equation

$$S_5^2(Z'||A) \oplus S_5^2(Z'||(A \oplus \Delta\omega)) = \rho_c^{-1}(\Delta c). \tag{4}$$

For each A, the above Eqs. (3) and (4) are equivalent to

$$\mathfrak{F}_A^1(Y') \oplus \mathfrak{F}_{A\oplus\Delta\omega}^1(Y') = \Delta b, \tag{5}$$

$$\mathfrak{F}_A^2(Z') \oplus \mathfrak{F}_{A\oplus\Delta\omega}^2(Z') = \rho_c^{-1}(\Delta c). \tag{6}$$

Here, $\Delta\omega$ is arbitrary nonzero 2-bit difference, and thus we can define $B = A \oplus \Delta\omega$ i.e., $B \neq A$. Since the function $(X_L, X_R) \mapsto (Y', A, Z')$ is bijective, the $(\Delta b||0^{(3)}, 0^{(3)}||\Delta c)$ case does not happen if and only if there is no (Y', A, Z') satisfying both Eqs. (5) and (6) for all $B(\neq A)$, which is equivalent to condition ii) where $\Delta\alpha = \Delta b$, $\Delta\beta = \rho_c^{-1}(\Delta c)$.

$(\Delta b||0^{(3)}, \Delta d||0^{(5)})$: It happens if and only if there exists at least one (X_L, X_R) satisfying both $C_L(X_L, X_R) \oplus C_L(X_L \oplus \Delta b, X_R) = \Delta d$ and $C_R(X_L, X_R) \oplus C_R(X_L \oplus \Delta b, X_R) = \Delta 0$. The first equation is expressed as

$$\tau_3(S_5^1(X_L)) \oplus S_3(X_R) \oplus \tau_3(S_5^1(X_L \oplus \Delta b)) \oplus S_3(X_R) = \tau_3(S_5^1(X_L)) \oplus \tau_3(S_5^1(X_L \oplus \Delta b)) = \Delta d.$$

For a difference $\Delta\omega \in \mathbb{F}_2^2$, the above equation becomes

$$S_5^1(X_L) \oplus S_5^1(X_L \oplus \Delta b) = \Delta d||\Delta\omega.$$

As in Eq. (3), we obtain

$$(S_5^1)^{-1}(Y'||A) \oplus (S_5^1)^{-1}((Y' \oplus \Delta d)||(A \oplus \Delta\omega)) = \Delta b. \tag{7}$$

And the second equation is expressed as

$$\rho_c(S_5^2(S_5^1(X_L) \oplus (S_3(X_R)||0^{(2)}))) \oplus S_3(X_R)$$
$$\oplus \rho_c(S_5^2(S_5^1(X_L \oplus \Delta b) \oplus (S_3(X_R)||0^{(2)}))) \oplus S_3(X_R)$$
$$= \rho_c(S_5^2(Z)) \oplus \rho_c(S_5^2(Z \oplus (\Delta d||\Delta\omega))) = \Delta 0.$$

Clearly,

$$S_5^2(Z) \oplus S_5^2(Z \oplus (\Delta d||\Delta\omega)) = \Delta 0.$$

It becomes

$$S_5^2(Z'||A) \oplus S_5^2((Z' \oplus \Delta d)||(A \oplus \Delta\omega)) = \Delta 0. \tag{8}$$

For each A, the above Eqs. (7) and (8) are equivalent to

$$\mathfrak{F}_A^1(Y') \oplus \mathfrak{F}_{A\oplus\Delta\omega}^1(Y' \oplus \Delta d) = \Delta b, \tag{9}$$

$$\mathfrak{F}_A^2(Z') \oplus \mathfrak{F}_{A\oplus\Delta\omega}^2(Z' \oplus \Delta d) = \Delta 0. \tag{10}$$

Similarly to the case above, we define $B = A \oplus \Delta\omega$. In this time, B can be either A or not, since $\Delta\omega$ can be a zero difference. The $(\Delta b||0^{(3)}, \Delta d||0^{(5)})$ case does not happen if and only if there is no (Y', A, Z') satisfying both Eqs. (9) and (10) for all B, which is equivalent to condition $iii)$ where $\Delta\alpha = \Delta d$, $\Delta\beta = \Delta b$.   ∎

B.3   Proof of Theorem 2

We use Y, Y', Z, Z', and A defined in proof B.2.

$\underline{(0^{(5)}||\lambda_a, 0^{(3)}||\lambda_c)}$: This case is expressed as $X_R \bullet \lambda_a = C_R(X_L, X_R) \bullet \lambda_c$. It follows $X_R \bullet \lambda_a = (\rho_c(S_5^2(S_5^1(X_L) \oplus (S_3(X_R)||0^{(2)}))) \oplus S_3(X_R)) \bullet \lambda_c$. By applying the variable Z, the equation becomes $X_R \bullet \lambda_a \oplus S_3(X_R) \bullet \lambda_c = \rho_c(S_5^2(Z)) \bullet \lambda_c$. Note that the function $(X_L, X_R) \mapsto (Z, X_R)$ is bijective. Suppose $\tau_2(\lambda_c) \neq 0$. Then, the equation becomes $X_R \bullet \lambda_a = \rho_c(S_5^2(Z)) \bullet \lambda_c$. This should have zero bias because the equation $X_R \bullet \lambda_a = 0$ has zero bias, and Z and X_R are independent variables. Now, suppose $\tau_2(\lambda_c) = 0$. The equation $X_R \bullet \lambda_a \oplus S_3(X_R) \bullet \lambda_c = \rho_c(S_5^2(Z)) \bullet \lambda_c$ has zero bias if and only if at least one of the entries $(\lambda_a, \tau_3'(\lambda_c))$ in LAT of S_3 and $(0, \tau_3'(\lambda_c)||0^{(2)})$ in LAT of S_5^2 is zero. This is due to the fact that Z is independent of X_R. It is equivalent to condition $i)$

$\underline{(0^{(5)}||\lambda_a, \lambda_d||0^{(5)})}$: This case is expressed as $X_R \bullet \lambda_a = C_L(X_L, X_R) \bullet \lambda_d$. It follows $X_R \bullet \lambda_a = (\tau_3(S_5^1(X_L)) \oplus S_3(X_R)) \bullet \lambda_d$. The equation becomes $X_R \bullet \lambda_a = \tau_3(Z) \bullet \lambda_d$ by using the definition of Z. So, this case has zero bias, because $\tau_3(Z)$ is independent of X_R.

$\underline{(\lambda_b||0^{(3)}, 0^{(3)}||\lambda_c)}$: This case is expressed as $X_L \bullet \lambda_b = C_R(X_L, X_R) \bullet \lambda_c$. It follows $X_L \bullet \lambda_b = (\rho_c(S_5^2(S_5^1(X_L) \oplus (S_3(X_R)||0^{(2)}))) \oplus S_3(X_R)) \bullet \lambda_c$. We can replace the equation to

$$X_L \bullet \lambda_b \oplus S_5^1(X_L) \bullet \lambda_t$$
$$= (S_5^1(X_L) \oplus (S_3(X_R)||0^{(2)})) \bullet \lambda_t \oplus \rho_c(S_5^2(S_5^1(X_L) \oplus (S_3(X_R)||0^{(2)}))) \bullet \lambda_c,$$

where $\lambda_t = \tau_3'(\lambda_c)||0^{(2)}$ (here, $0^{(2)}$ can be replaced by $01, 10$ or $1^{(2)}$). By applying the variables of Y and Z, this becomes equivalent to the following equations

$$(S_5^1)^{-1}(Y) \bullet \lambda_b \oplus Y \bullet \lambda_t = Z \bullet \lambda_t \oplus (\rho_c(S_5^2(Z))) \bullet \lambda_c,$$

$$(S_5^1)^{-1}(Y'||A) \bullet \lambda_b \oplus (Y'||A) \bullet \lambda_t = (Z'||A) \bullet \lambda_t \oplus (\rho_c(S_5^2(Z'||A))) \bullet \lambda_c.$$

For all $A \in \mathbb{F}_2^2$, we have

$$\mathfrak{F}_A^1(Y') \bullet \lambda_b \oplus (Y'||A) \bullet \lambda_t = (Z'||A) \bullet \lambda_t \oplus (\rho_c(\mathfrak{F}_A^2(Z'))) \bullet \lambda_c.$$

Clearly,

$$\mathfrak{F}_A^1(Y') \bullet \lambda_b \oplus Y' \bullet \tau_3(\lambda_t) = Z' \bullet \tau_3(\lambda_t) \oplus (\rho_c(\mathfrak{F}_A^2(Z'))) \bullet \lambda_c.$$

A collection of (Y', Z') that satisfies the above equation is equivalent to

$$\{Y'|0 = \mathfrak{F}_A^1(Y') \bullet \lambda_b \oplus Y' \bullet \tau_3(\lambda_t)\} \times \{Z'|0 = Z' \bullet \tau_3(\lambda_t) \oplus (\rho_c(\mathfrak{F}_A^2(Z'))) \bullet \lambda_c\}$$
$$\cup \{Y'|1 = \mathfrak{F}_A^1(Y') \bullet \lambda_b \oplus Y' \bullet \tau_3(\lambda_t)\} \times \{Z'|1 = Z' \bullet \tau_3(\lambda_t) \oplus (\rho_c(\mathfrak{F}_A^2(Z'))) \bullet \lambda_c\}$$

Then the number of the above set is $(4 + a_A)(4 + b_A) + (4 - a_A)(4 - b_A) = 32 + 2a_A b_A$, where a_A and b_A are the entries of $(\tau_3(\lambda_t), \lambda_b)$ and $(\tau_3(\lambda_t), \rho_c^{-1}(\lambda_c))$ in LAT of $\mathfrak{F}_A^1$ and $\mathfrak{F}_A^2$, respectively. The above equation has zero bias if and only if

$$\sum_{A \in \mathbb{F}_2^2} (32 + 2a_A b_A) = 2(\sum_{A \in \mathbb{F}_2^2} a_A b_A) + 128 = 128$$

It leads to $\sum_{A \in \mathbb{F}_2^2} a_A b_A = 0$. Because $\tau_3(\lambda_t) = \tau_3'(\lambda_c)$, it is equivalent to condition $ii)$ (when $\tau_3'(\lambda_c) \neq 0$) and condition $iii)$ (when $\tau_3'(\lambda_c) = 0$).

$\underline{(\lambda_b||0^{(3)}, \lambda_d||0^{(5)})}$: This case is expressed as $X_L \bullet \lambda_b = C_L(X_L, X_R) \bullet \lambda_d$. It follows $X_L \bullet \lambda_b = (\tau_3(S_5^1(X_L)) \oplus S_3(X_R)) \bullet \lambda_d$. The equation becomes $X_L \bullet \lambda_b = Z' \bullet \lambda_d$ by using the definition of Z'. We note that the function $(X_L, X_R) \mapsto (X_L, Z')$ is bijective, and X_L and Z' are independent variables. So, this equation has zero bias. ∎

C   8-bit S-box of **PIPO**, S_8

C.1   Table of the S_8

Table 7 shows the S_8.

Table 7. 8-bit S-box of PIPO in hexadecimal notation: For example, $S_8(31){=}86$.

$S_8(x\|\|y)$	y															
		0	1	2	3	4	5	6	7	8	9	A	B	C	D	E	F
x	0	5E	F9	FC	00	3F	85	BA	5B	18	37	B2	C6	71	C3	74	9D
	1	A7	94	0D	E1	CA	68	53	2E	49	62	EB	97	A4	0E	2D	D0
	2	16	25	AC	48	63	D1	EA	8F	F7	40	45	B1	9E	34	1B	F2
	3	B9	86	03	7F	D8	7A	DD	3C	E0	CB	52	26	15	AF	8C	69
	4	C2	75	70	1C	33	99	B6	C7	04	3B	BE	5A	FD	5F	F8	81
	5	93	A0	29	4D	66	D4	EF	0A	E5	CE	57	A3	90	2A	09	6C
	6	22	11	88	E4	CF	6D	56	AB	7B	DC	D9	BD	82	38	07	7E
	7	B5	9A	1F	F3	44	F6	41	30	4C	67	EE	12	21	8B	A8	D5
	8	55	6E	E7	0B	28	92	A1	CC	2B	08	91	ED	D6	64	4F	A2
	9	BC	83	06	FA	5D	FF	58	39	72	C5	C0	B4	9B	31	1E	77
	A	01	3E	BB	DF	78	DA	7D	84	50	6B	E2	8E	AD	17	24	C9
	B	AE	8D	14	E8	D3	61	4A	27	47	F0	F5	19	36	9C	B3	42
	C	1D	32	B7	43	F4	46	F1	98	EC	D7	4E	AA	89	23	10	65
	D	8A	A9	20	54	6F	CD	E6	13	DB	7C	79	05	3A	80	BF	DE
	E	E9	D2	4B	2F	0C	A6	95	60	0F	2C	A5	51	6A	C8	E3	96
	F	B0	9F	1A	76	C1	73	C4	35	FE	59	5C	B8	87	3D	02	FB

C.2   Bitsliced Implementations of the S_8 and Its Inverse

Listing 1.2 is the bitsliced implementation of the S_8.[1] The bitsliced implementation of the inverse S_8 cannot be obtained by reversing the bitsliced implementation of the S_8 because the input bits of S_5^2 are not all given. The Listing 1.3 shows how to implement the inverse S_8 with the given input bits. Since the S_8 applies each column of 8×8 array of bits depicted in Fig. 1, we can implement the S-layer by replacing bit $x[i]$ with byte $X[i]$ which represents the i-th row value, where $i = 0, 1, 2, \cdots, 7$.

Listing 1.2. The bitsliced implementation of the S_8 (in C code)

```
//(MSb: x[7], LSb: x[0]) :"b" represents bit
// Input: x[7], x[6], x[5], x[4], x[3], x[2], x[1], x[0]
// S5_1
x[5] ^= (x[7] & x[6]);
x[4] ^= (x[3] & x[5]);
x[7] ^= x[4];
x[6] ^= x[3];
x[3] ^= (x[4] | x[5]);
x[5] ^= x[7];
```

---

[1] For a higher resistance against DC and LC, swapping bits is additionally conducted in the S_8 design (refer to section 3.2).

```c
x[4] ^= (x[5] & x[6]);
// S3
x[2] ^= x[1] & x[0];
x[0] ^= x[2] | x[1];
x[1] ^= x[2] | x[0];
x[2] = ~x[2];
// Extend XOR
x[7] ^= x[1]; x[3] ^= x[2]; x[4] ^= x[0];
//S5_2
t[0] = x[7]; t[1] = x[3]; t[2] = x[4];
x[6] ^= (t[0] & x[5]);
t[0] ^= x[6];
x[6] ^= (t[2] | t[1]);
t[1] ^= x[5];
x[5] ^= (x[6] | t[2]);
t[2] ^= (t[1] & t[0]);
// truncate XOR and swap
x[2] ^= t[0]; t[0] = x[1] ^ t[2]; x[1] = x[0]^t[1];
x[0] = x[7]; x[7] = t[0];
t[1] = x[3]; x[3] = x[6]; x[6] = t[1];
t[2] = x[4]; x[4] = x[5]; x[5] = t[2];
// Output: x[7], x[6], x[5], x[4], x[3], x[2], x[1], x[0]
```

Listing 1.3. The bitsliced implementation of the inverse S_8 (in C code)

```c
//(MSb: x[7], LSb: x[0]) :"b" represents bit
// Input: x[7], x[6], x[5], x[4], x[3], x[2], x[1], x[0]
t[0] = x[7]; x[7] = x[0]; x[0] = x[1]; x[1] = t[0];
t[0] = x[7]; t[1] = x[6]; t[2] = x[5];
// S52 inv
x[4] ^= (x[3] | t[2]);
x[3] ^= (t[2] | t[1]);
t[1] ^= x[4];
t[0] ^= x[3];
t[2] ^= (t[1] & t[0]);
x[3] ^= (x[4] & x[7]);
// Extended XOR
x[0] ^= t[1]; x[1] ^= t[2]; x[2] ^= t[0];
t[0] = x[3]; x[3] = x[6]; x[6] = t[0];
t[0] = x[5]; x[5] = x[4]; x[4] = t[0];
// Truncated XOR
x[7] ^= x[1]; x[3] ^= x[2]; x[4] ^= x[0];
// Inv_S5_1
x[4] ^= (x[5] & x[6]);
x[5] ^= x[7];
x[3] ^= (x[4] | x[5]);
x[6] ^= x[3];
x[7] ^= x[4];
x[4] ^= (x[3] & x[5]);
x[5] ^= (x[7] & x[6]);
```

```
// Inv_S3
x[2] = ~x[2];
x[1] ^= x[2] | x[0];
x[0] ^= x[2] | x[1];
x[2] ^= x[1] & x[0];
// Output: x[7], x[6], x[5], x[4], x[3], x[2], x[1], x[0]
```

References

1. Adomnicai, A., et al.: Lilliput-AE: a new lightweight tweakable block cipher for authenticated encryption with associated data. Submission to the NIST Lightweight Cryptography Standardization Process (2019)
2. Albrecht, M.R., Driessen, B., Kavun, E.B., Leander, G., Paar, C., Yalçın, T.: Block ciphers – focus on the linear layer (feat. PRIDE). In: Garay, J.A., Gennaro, R. (eds.) CRYPTO 2014. LNCS, vol. 8616, pp. 57–76. Springer, Heidelberg (2014). https://doi.org/10.1007/978-3-662-44371-2_4
3. Aoki, K., Sasaki, Yu.: Preimage attacks on one-block MD4, 63-step MD5 and more. In: Avanzi, R.M., Keliher, L., Sica, F. (eds.) SAC 2008. LNCS, vol. 5381, pp. 103–119. Springer, Heidelberg (2009). https://doi.org/10.1007/978-3-642-04159-4_7
4. Atmel Corporation, ATmega128(L) Datasheet. www.microchip.com/wwwproducts/en/ATmega128. Accessed 23 Apr 2019
5. Badel, S., et al.: ARMADILLO: a multi-purpose cryptographic primitive dedicated to hardware. In: Mangard, S., Standaert, F.-X. (eds.) CHES 2010. LNCS, vol. 6225, pp. 398–412. Springer, Heidelberg (2010). https://doi.org/10.1007/978-3-642-15031-9_27
6. Barthe, G., Dupressoir, F., Faust, S., Grégoire, B., Standaert, F.-X., Strub, P.-Y.: Parallel implementations of masking schemes and the bounded moment leakage model. In: Coron, J.-S., Nielsen, J.B. (eds.) EUROCRYPT 2017. LNCS, vol. 10210, pp. 535–566. Springer, Cham (2017). https://doi.org/10.1007/978-3-319-56620-7_19
7. Battistello, A., Coron, J.-S., Prouff, E., Zeitoun, R.: Horizontal side-channel attacks and countermeasures on the ISW masking scheme. In: Gierlichs, B., Poschmann, A.Y. (eds.) CHES 2016. LNCS, vol. 9813, pp. 23–39. Springer, Heidelberg (2016). https://doi.org/10.1007/978-3-662-53140-2_2
8. Banik, S., Pandey, S.K., Peyrin, T., Sasaki, Yu., Sim, S.M., Todo, Y.: GIFT: a small present. In: Fischer, W., Homma, N. (eds.) CHES 2017. LNCS, vol. 10529, pp. 321–345. Springer, Cham (2017). https://doi.org/10.1007/978-3-319-66787-4_16
9. Banik, S., et al.: Midori: a block cipher for low energy. In: Iwata, T., Cheon, J.H. (eds.) ASIACRYPT 2015. LNCS, vol. 9453, pp. 411–436. Springer, Heidelberg (2015). https://doi.org/10.1007/978-3-662-48800-3_17
10. Baysal, A., Şahin, S.: RoadRunneR: a small and fast bitslice block cipher for low cost 8-bit processors. In: Güneysu, T., Leander, G., Moradi, A. (eds.) LightSec 2015. LNCS, vol. 9542, pp. 58–76. Springer, Cham (2016). https://doi.org/10.1007/978-3-319-29078-2_4
11. Beaulieu, R., Shors, D., Smith, J., Treatman-Clark, S., Weeks, B., Wingers, L.: The SIMON and SPECK block ciphers on AVR 8-bit microcontrollers. In: Eisenbarth, T., Öztürk, E. (eds.) LightSec 2014. LNCS, vol. 8898, pp. 3–20. Springer, Cham (2015). https://doi.org/10.1007/978-3-319-16363-5_1

12. Beaulieu, R., Shors, D., Smith, J., Treatman-Clark, S., Weeks, B., Wingers, L.: The SIMON and SPECK families of lightweight block ciphers, Cryptology ePrint Archive (2013)

13. Beierle, C., Leander, G., Moradi, A., Rasoolzadeh, S.: CRAFT: lightweight tweakable block cipher with efficient protection against DFA attacks. IACR Trans. Symmetric Cryptol. **2019**(1), 5–45 (2019)

14. Beierle, C., et al.: The SKINNY family of block ciphers and its low-latency variant MANTIS. In: Robshaw, M., Katz, J. (eds.) CRYPTO 2016. LNCS, vol. 9815, pp. 123–153. Springer, Heidelberg (2016). https://doi.org/10.1007/978-3-662-53008-5_5

15. Bilgin, B., De Meyer, L., Duval, S., Levi, I., Standaert, F.X.: Low AND depth and efficient inverses: a guide on s-boxes for low-latency masking. IACR Trans. Symmetric Cryptol. **2020**(1), 144–184 (2020)

16. Belaïd, S., Benhamouda, F., Passelègue, A., Prouff, E., Thillard, A., Vergnaud, D.: Randomness complexity of private circuits for multiplication. In: Fischlin, M., Coron, J.-S. (eds.) EUROCRYPT 2016. LNCS, vol. 9666, pp. 616–648. Springer, Heidelberg (2016). https://doi.org/10.1007/978-3-662-49896-5_22

17. Biham, E.: A fast new DES implementation in software. In: Biham, E. (ed.) FSE 1997. LNCS, vol. 1267, pp. 260–272. Springer, Heidelberg (1997). https://doi.org/10.1007/BFb0052352

18. Biham, E., Biryukov, A., Shamir, A.: Cryptanalysis of skipjack reduced to 31 rounds using impossible differentials. In: Stern, J. (ed.) EUROCRYPT 1999. LNCS, vol. 1592, pp. 12–23. Springer, Heidelberg (1999). https://doi.org/10.1007/3-540-48910-X_2

19. Biham, E., Dunkelman, O., Keller, N.: The rectangle attack — rectangling the serpent. In: Pfitzmann, B. (ed.) EUROCRYPT 2001. LNCS, vol. 2045, pp. 340–357. Springer, Heidelberg (2001). https://doi.org/10.1007/3-540-44987-6_21

20. Biham, E., Shamir, A.: Differential cryptanalysis of DES-like cryptosystems. In: Menezes, A.J., Vanstone, S.A. (eds.) CRYPTO 1990. LNCS, vol. 537, pp. 2–21. Springer, Heidelberg (1991). https://doi.org/10.1007/3-540-38424-3_1

21. Biryukov, A., Wagner, D.: Advanced slide attacks. In: Preneel, B. (ed.) EUROCRYPT 2000. LNCS, vol. 1807, pp. 589–606. Springer, Heidelberg (2000). https://doi.org/10.1007/3-540-45539-6_41

22. Bogdanov, A., et al.: PRESENT: an ultra-lightweight block cipher. In: Paillier, P., Verbauwhede, I. (eds.) CHES 2007. LNCS, vol. 4727, pp. 450–466. Springer, Heidelberg (2007). https://doi.org/10.1007/978-3-540-74735-2_31

23. Boura, C., Canteaut, A., De Cannière, C.: Higher-order differential properties of KECCAK and *Luffa*. In: Joux, A. (ed.) FSE 2011. LNCS, vol. 6733, pp. 252–269. Springer, Heidelberg (2011). https://doi.org/10.1007/978-3-642-21702-9_15

24. Borghoff, J., et al.: PRINCE – a low-latency block cipher for pervasive computing applications. In: Wang, X., Sako, K. (eds.) ASIACRYPT 2012. LNCS, vol. 7658, pp. 208–225. Springer, Heidelberg (2012). https://doi.org/10.1007/978-3-642-34961-4_14

25. Collard, B., Standaert, F.-X.: A statistical saturation attack against the block cipher PRESENT. In: Fischlin, M. (ed.) CT-RSA 2009. LNCS, vol. 5473, pp. 195–210. Springer, Heidelberg (2009). https://doi.org/10.1007/978-3-642-00862-7_13

26. Dinu, D., Biryukov, A., Großschädl, J., Khovratovich, D., Corre, Y.L., Perrin, L.: FELICS-fair evaluation of lightweight cryptographic systems. In: NIST Workshop on Lightweight Cryptography (2015)

27. Engels, S., Kavun, E.B., Paar, C., Yalçin, T., Mihajloska, H.: A non-linear/linear instruction set extension for lightweight ciphers. In: IEEE 21st Symposium on Computer Arithmetic, pp. 67–75 (2013)

28. Gérard, B., Grosso, V., Naya-Plasencia, M., Standaert, F.-X.: Block ciphers that are easier to mask: how far can we go? In: Bertoni, G., Coron, J.-S. (eds.) CHES 2013. LNCS, vol. 8086, pp. 383–399. Springer, Heidelberg (2013). https://doi.org/10.1007/978-3-642-40349-1_22

29. Goudarzi, D., Journault, A., Rivain, M., Standaert, F.-X.: Secure multiplication for bitslice higher-order masking: optimisation and comparison. In: Fan, J., Gierlichs, B. (eds.) COSADE 2018. LNCS, vol. 10815, pp. 3–22. Springer, Cham (2018). https://doi.org/10.1007/978-3-319-89641-0_1

30. Goudarzi, D., Rivain, M.: How fast can higher-order masking be in software? In: Coron, J.-S., Nielsen, J.B. (eds.) EUROCRYPT 2017. LNCS, vol. 10210, pp. 567–597. Springer, Cham (2017). https://doi.org/10.1007/978-3-319-56620-7_20

31. Grosso, V., Leurent, G., Standaert, F.-X., Varıcı, K.: LS-designs: bitslice encryption for efficient masked software implementations. In: Cid, C., Rechberger, C. (eds.) FSE 2014. LNCS, vol. 8540, pp. 18–37. Springer, Heidelberg (2015). https://doi.org/10.1007/978-3-662-46706-0_2

32. Guo, J., Peyrin, T., Poschmann, A., Robshaw, M.: The LED block cipher. In: Preneel, B., Takagi, T. (eds.) CHES 2011. LNCS, vol. 6917, pp. 326–341. Springer, Heidelberg (2011). https://doi.org/10.1007/978-3-642-23951-9_22

33. Hong, D., et al.: HIGHT: a new block cipher suitable for low-resource device. In: Goubin, L., Matsui, M. (eds.) CHES 2006. LNCS, vol. 4249, pp. 46–59. Springer, Heidelberg (2006). https://doi.org/10.1007/11894063_4

34. Journault, A., Standaert, F.-X.: Very high order masking: efficient implementation and security evaluation. In: Fischer, W., Homma, N. (eds.) CHES 2017. LNCS, vol. 10529, pp. 623–643. Springer, Cham (2017). https://doi.org/10.1007/978-3-319-66787-4_30

35. Karpman, P., Grégoire, B.: The littlun s-box and the fly block cipher. In: Lightweight Cryptography Workshop (2016)

36. Kim, J., Lee, C., Sung, J., Hong, S., Lee, S., Lim, J.: Seven new block cipher structures with provable security against differential cryptanalysis. IEICE Trans. **91-A**(10), 3047–3058 (2008)

37. Kocher, P.C.: Timing attacks on implementations of Diffie-Hellman, RSA, DSS, and other systems. In: Koblitz, N. (ed.) CRYPTO 1996. LNCS, vol. 1109, pp. 104–113. Springer, Heidelberg (1996). https://doi.org/10.1007/3-540-68697-5_9

38. Leander, G., Abdelraheem, M.A., AlKhzaimi, H., Zenner, E.: A cryptanalysis of PRINTCIPHER: the invariant subspace attack. In: Rogaway, P. (ed.) CRYPTO 2011. LNCS, vol. 6841, pp. 206–221. Springer, Heidelberg (2011). https://doi.org/10.1007/978-3-642-22792-9_12

39. Leander, G., Minaud, B., Rønjom, S.: A generic approach to invariant subspace attacks: cryptanalysis of Robin, iSCREAM and Zorro. In: Oswald, E., Fischlin, M. (eds.) EUROCRYPT 2015. LNCS, vol. 9056, pp. 254–283. Springer, Heidelberg (2015). https://doi.org/10.1007/978-3-662-46800-5_11

40. Matsui, M.: Linear cryptanalysis method for DES cipher. In: Helleseth, T. (ed.) EUROCRYPT 1993. LNCS, vol. 765, pp. 386–397. Springer, Heidelberg (1994). https://doi.org/10.1007/3-540-48285-7_33

41. Matsui, M.: On correlation between the order of S-boxes and the strength of DES. In: De Santis, A. (ed.) EUROCRYPT 1994. LNCS, vol. 950, pp. 366–375. Springer, Heidelberg (1995). https://doi.org/10.1007/BFb0053451

42. Sasaki, Yu., Aoki, K.: Finding preimages in full MD5 faster than exhaustive search. In: Joux, A. (ed.) EUROCRYPT 2009. LNCS, vol. 5479, pp. 134–152. Springer, Heidelberg (2009). https://doi.org/10.1007/978-3-642-01001-9_8

43. Shibutani, K., Isobe, T., Hiwatari, H., Mitsuda, A., Akishita, T., Shirai, T.: *Piccolo*: an ultra-lightweight blockcipher. In: Preneel, B., Takagi, T. (eds.) CHES 2011. LNCS, vol. 6917, pp. 342–357. Springer, Heidelberg (2011). https://doi.org/10.1007/978-3-642-23951-9_23

44. Shirai, T., Shibutani, K., Akishita, T., Moriai, S., Iwata, T.: The 128-bit blockcipher CLEFIA (extended abstract). In: Biryukov, A. (ed.) FSE 2007. LNCS, vol. 4593, pp. 181–195. Springer, Heidelberg (2007). https://doi.org/10.1007/978-3-540-74619-5_12

45. Todo, Y., Leander, G., Sasaki, Y.: Nonlinear invariant attack - practical attack on full SCREAM, iSCREAM, and Midori64. In: Cheon, J.H., Takagi, T. (eds.) ASIACRYPT 2016. LNCS, vol. 10032, pp. 3–33. Springer, Heidelberg (2016). https://doi.org/10.1007/978-3-662-53890-6_1

46. Wagner, D.: The boomerang attack. In: Knudsen, L. (ed.) FSE 1999. LNCS, vol. 1636, pp. 156–170. Springer, Heidelberg (1999). https://doi.org/10.1007/3-540-48519-8_12

47. Worthman, E.: ChaoLogix: integrated security. Semiconductor Eng. (2015)

48. Z'aba, M.R., Raddum, H., Henricksen, M., Dawson, E.: Bit-pattern based integral attack. In: Nyberg, K. (ed.) FSE 2008. LNCS, vol. 5086, pp. 363–381. Springer, Heidelberg (2008). https://doi.org/10.1007/978-3-540-71039-4_23

49. Zhang, W., Bao, Z., Lin, D., Rijmen, V., Yang, B., Verbauwhede, I.: RECTANGLE: a bit-slice lightweight block cipher suitable for multiple platforms. Sci. China Inf. Sci. **58**(12), 1–15 (2015)

Efficient Implementations

Curve448 on 32-Bit ARM Cortex-M4

Hwajeong Seo[1][✉] and Reza Azarderakhsh[2,3]

[1] IT Department, Hansung University, Seoul, South Korea
hwajeong84@gmail.com
[2] Department of Computer and Electrical Engineering and Computer Science,
Florida Atlantic University, Boca Raton, FL, USA
razarderakhsh@fau.edu
[3] PQSecure Technologies, LLC, Boca Raton, USA

Abstract. Public key cryptography is widely used in key exchange and digital signature protocols. Public key cryptography requires expensive primitive operations, such as finite-field and group operations. These finite-field and group operations require a number of clock cycles to execute. By carefully optimizing these primitive operations, public key cryptography can be performed with reasonably fast execution timing. In this paper, we present the new implementation result of Curve448 on 32-bit ARM Cortex-M4 microcontrollers. We adopted state-of-art implementation methods, and some previous methods were re-designed to fully utilize the features of the target microcontrollers. The implementation was also performed with constant timing by utilizing the features of microcontrollers and algorithms. Finally, the scalar multiplication of Curve448 on 32-bit ARM Cortex-M4@168 MHz microcontrollers requires 6,285,904 clock cycles. To the best of our knowledge, this is the first optimized implementation of Curve448 on 32-bit ARM Cortex-M4 microcontrollers. The result is also compared with other ECC and post-quantum cryptography (PQC) implementations. The proposed ECC and the-state-of-art PQC results show the practical usage of hybrid post-quantum TLS on the target processor.

Keywords: ARM Cortex-M4 · Curve448 · Public key cryptography · Hybrid post-quantum TLS

1 Introduction

Public key cryptography is widely used in key exchange and digital signature protocols. For public key cryptography, implementation is a challenge with low-end microcontrollers, which have the disadvantages of low energy, performance, and memory. In particular, the efficiency of elliptic curve cryptography (ECC) depends on the compact implementation of finite-field arithmetic and group operation. For this reason, the optimized implementation of finite-field arithmetic and group operation should be considered. In this paper, we present the first Curve448 implementation result on 32-bit ARM Cortex-M4 microcontrollers. The motivations of this work may be summarized as follows:

© Springer Nature Switzerland AG 2021
D. Hong (Ed.): ICISC 2020, LNCS 12593, pp. 125–139, 2021.
https://doi.org/10.1007/978-3-030-68890-5_7

- Curve448 offers 224-bit security and is designed for use with the elliptic curve Diffie-Hellman (ECDH) key agreement scheme [1]. The curve was favored by the Internet Research Task Force Crypto Forum Research Group (IRTF CFRG) for inclusion in transport layer security (TLS) standards along with Curve25519. The curve is an approved elliptic curve for use by the US federal government, which is confirmed in FIPS 186-5. However, the implementation of algorithms has not been actively conducted. This work fills this gap.
- The target microcontroller, namely the 32-bit ARM Cortex-M4, is the most widely used in practice because it has relatively powerful computation abilities in terms of the arithmetic logic unit (ALU), frequency of the CPU, RAM, and ROM in comparison to legacy embedded processors, such as 8-bit AVR ATmega and 16-bit MSP430(X) microcontrollers. Furthermore, the NIST recommended this board for evaluation of post-quantum cryptography (PQC). For this reason, a number of cryptographic implementations have been recently done over 32-bit ARM Cortex-M4 microcontrollers [2–5]. However, Curve448 had not been implemented on this target microcontroller. This work evaluated Curve448 on ARM Cortex-M4 microcontrollers for the first time.

For high performance, we adopted state-of-art implementation methods and some previous methods were re-designed to fully utilize the features of the target microcontrollers. This was the first implementation of Curve448 on this target processor. The result was compared with those of other 128-bit security ECC implementations. The scalar multiplication of Curve448 on 32-bit ARM Cortex-M4@168 MHz microcontrollers requires 6,285,904 clock cycles. The result shows that Curve448 is reasonably fast enough on the target microcontroller. The result was also compared with other PQC implementations. This shows the practical usage of hybrid post-quantum TLS on the target processor is available.

1.1 Contribution

Detailed contributions are as follows:

First Implementation of Curve448 on 32-Bit ARM Cortex-M4. This paper presents the first implementation of Curve448 on 32-bit ARM Cortex-M4 processors. State-of-art techniques were applied to improve the performance. The result shows that the implementation is practically fast enough.

Secure and Efficient Implementation of Primitive Operations. All primitive operations such as finite-field arithmetic and group operation were implemented in a secure and efficient way. By using constant and regular implementation, the timing attack was efficiently prevented. Furthermore, cache attack were prevented by avoiding the pre-computed value access. All requirements for constant timing on ARM Cortex-M4 specifically are also presented for interested cryptographic researchers.

In-Depth Comparison of Pre-quantum and Post-quantum Cryptography. We compared pre-quantum and post quantum cryptography on the target processors. The performance report shows the availability of hybrid post-quantum TLS. Furthermore, we discuss the trade-off between performance and security in detail.

First Curve448 on ARM Cortex-M4 as an Open Source. The implementation will be public domain after publication. The source code will be a helpful resource for researchers.

The remainder of this paper is organized as follows. In Sect. 2, we introduce the target curve (Curve448), the target microcontroller (32-bit ARM Cortex-M4), and previous implementations. The optimized implementation techniques for Curve448 on 32-bit ARM Cortex-M4 are presented in Sect. 3. In Sects. 4, we evaluate and compare implementation results. Finally, we conclude the paper in Sect. 6.

2   Related Works

In this section, we introduce the target curve (Curve448), target microcontroller (32-bit ARM Cortex-M4), and previous implementations.

2.1   Target Curve: Curve448

Edwards curves, which were suggested in [6] provide complete addition formulas, which does not have a case (division by zero). The one proper Edwards curve for cryptography is Curve448–Goldilocks, which is faster and simpler than traditional NIST curves [1]. Curve448–Goldilocks provides high-security level (224-bit) and the related equation is as follows:

$$E : y^2 + x^2 = 1 + dx^2y^2$$

defined over the field $\mathbb{F}_{2^{448}-2^{224}-1}$ with curve parameter $d = -39081$. Curve448 satisfies the requirement of SafeCurves [7] and is one of ECC standards for TLS 1.3 [8].

2.2   Target Microcontroller: 32-Bit ARM Cortex–M4

The ARM Cortex–M4 microcontroller is a small and energy-efficient ARM processor. The microcontroller supports the ARMv7E-M instruction set, which comprises Thumb-2 instructions and additional DSP extensions. The Cortex-M4 architecture has a 3-stage pipeline with branch speculation. It includes 16 32-bit registers (R0:R15), and supports a mix of 16 and 32-bit operations corresponding to Thumb-2.

Instructions that are relevant for the proposed implementation include 32-bit arithmetic and logical instructions, such as addition (ADD) and addition with carry (ADC), as well as memory instructions that perform multiple-data loading/storing (LDM/STM).

The microcontroller is equipped with powerful single-cycle multiply and multiply-and-accumulate instructions from DSP extensions, including UMUL, UMLAL, and UMAAL. These instructions compute the product 32×32-bit $\rightarrow$ 64-bit (UMUL), plus a 64-bit accumulation with a single 64-bit value (UMLAL) or plus a 64-bit accumulation with two 32-bit values (UMAAL). The core instruction set is presented in detail in Table 1.

Table 1. Instruction set summary for ARM Cortex-M4.

Inst	Operands	Description	Operation
ADD	C, A, B	Addition without Carry	C ← A+B
ADC	C, A, B	Addition with Carry	C ← A+B+Carry
SUB	C, A, B	Subtraction without Carry	C ← A−B
MOV	C, A	Move 32-bit word between registers	C ← A
UMAAL	D, C, A, B	Multiplication with Accumulation	{D|C} ← A×B+C+D
LDM	A!, {B−C}	Loading data from memory to registers	–
STM	A!, {B−C}	Storing data from registers to memory	–

2.3   Previous Implementations

Since Curve448 was recently presented, it has become a new ECC standard as a TLS 1.3. For this reason, only few implementations of Curve448 on low-end microcontrollers are available. In [9], the first Curve448 implementations on both 8-bit AVR ATmega and 16-bit MSP430 microcontrollers were presented. These works achieved 103,228,541 and 73,477,660 clock cycles for scalar multiplication of Curve448 on 8-bit AVR ATmega and 16-bit MSP430 microcontrollers, respectively. To improve the performance, the Karatsuba algorithm is utilized for multi-precision multiplication. On the 32-bit ARM Cortex-M4 microcontroller, several studies have investigated optimized implementations of the well-known Curve25519 [10–13]. Curve25519 provides a 128-bit security level (i.e. short-term security), while Curve448 provides a 224-bit security level (i.e. long-term security). For long-term security, implementation of Curve448 should be considered rather than Curve25519. In this paper, we present an optimized implementation of Curve448 on the 32-bit ARM Cortex-M4 microcontroller for the first time.

3   Optimization Techniques for Curve448 on 32-Bit ARM Cortex-M4

ECC implementations consist of finite-field arithmetic and group operation. For finite-field arithmetic, modular addition, subtraction, multiplication, squaring, and inversion operations are required. For group operations, point addition, point doubling, and scalar multiplication operations are required.

3.1   Finite-Field Operations

Finite-Field Addition/Subtraction. The 448-bit addition and subtraction operations are performed together with modular reduction for finite-field addition and subtraction. First, addition or subtraction is performed. Then, modular reduction is performed when the addition or subtraction generates a carry bit or borrow bit as follows:

Integer Addition/Subtraction → Modular Reduction

According to the school-book approach, modular reduction is performed whenever a carry bit or borrow bit is captured. This approach generates leakage information from branch statements. For this reason, modular reduction is always performed regardless of the carry or borrow bit, which removes the relation between secret information and modular reduction. When the carry bit or borrow bit is set, the mask value is generated from it. For example, when the borrow bit is set, the value is 0xFFFFFFFF, which is used to mask the modulus. When the carry bit is set, the zero value is subtracted by the carry bit, which also generates 0xFFFFFFFF mask. Afterward, the masked modulus is added/subtracted to/from the intermediate results for modular subtraction and modular addition, respectively.

For efficient memory access, the usage of registers is also optimized further because the register access is much faster than the memory access. The 32-bit ARM Cortex-M4 microcontroller provides 14 general purpose registers. These registers cannot maintain all operands and intermediate results throughout the computation to reduce the number of memory accesses. For this reason, only some of the intermediate results are maintained in registers, while the others are stored in memory. For this purpose, 9, 2, and 3 registers are used for intermediate results, temporal storage, and memory pointers, respectively.

Finite-Field Multiplication. Multiplication is the most expensive operation of ECC implementation. The multiplication consists of integer multiplication and modular reduction. The proposed implementation performs each operation separately.

Integer Multiplication → Modular Reduction

To improve the multiplication performance on the 32-bit ARM Cortex-M4 microcontroller, the operand caching (OC) method is utilized [14]. The OC method caches many operands in registers, which reduces the number of memory accesses efficiently. In a previous work, the OC method with a width of 4 was adopted utilizing general purpose registers [4]. The order of instructions was also optimized to reduce the number of pipeline stalls.

Figure 1 illustrates strategies for implementing 448-bit multiplication on 32-bit ARM Cortex-M4 microcontroller. Let A and B be operands of length 448 bits each. Each operand is written as $A = (A[13], ..., A[1], A[0])$ and $B = (B[13], ..., B[1], B[0])$. The result $C = A \cdot B$ is represented as $C = (C[27], ...,$

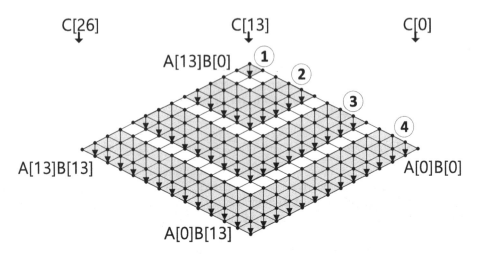

Fig. 1. 448-bit Operand Scanning multiplication at the word-level on ARM Cortex-M4 [4].

$C[1]$, $C[0]$). In the rhombus form, the lowest indices $(i, j = 0)$ of the product appear at the rightmost corner, whereas the highest indices $(i, j = 13)$ appear at the leftmost corner. A black arrow over a point indicates the processing of a partial product. The lowermost points represent the results $C[i]$ from the rightmost corner $(i = 0)$ to the leftmost corner $(i = 27)$. Computation is performed from ① to ④. Because the length of the operand caching is set to 4, the process is divided into 4 sections.

Finally, the implementation achieved 566 clock cycles for 448-bit multiplication. Because this approach achieves the best performance, we adopted our implementation. For better performance, the Karatsuba algorithm was also considered but performance improvement was not observed during the experiment due to the high efficiency of UMAAL instructions. For the modular reduction, the fast reduction method introduced in [9] was adopted. Detailed descriptions of Curve448 are given in Algorithm 1. All general purpose registers are utilized to maintain the intermediate results. In Step 1, both operands $A[2]$ and $A[3]$ are added and output the intermediate result $(\varepsilon 0 \| T)$. The intermediate result (T) is maintained in registers, and the carry bit $(\varepsilon 0 \| T)$ is stored in STACK.

In Step 2, both operands $A[0]$ and $\varepsilon 0 \| T$ are added and output intermediate result $(\varepsilon 1 \| C[0])$. The intermediate result $(\varepsilon 1 \| C[0])$ is stored in STACK, while the intermediate result (T) is maintained in registers.

In Step 3, the operand $(A[1])$ is loaded and the intermediate result $(\varepsilon 0 \| T)$ is added. Then, the operand $(A[3])$ is added to the intermediate result and output the intermediate result $(\varepsilon 2 \| C[1])$.

From Step 4 to Step 7, carry bits are added to the intermediate result. Both intermediate results $(C[0]$ and $C[1])$ are maintained in registers. Two registers are utilized to handle carry bits, while part of registers are stored in STACK.

Algorithm 1. Fast reduction Curve448 [9].

Require: 896-bit intermediate result `A` (`A[3]~A[0]` in 224-bit)
Ensure: 448-bit result `C` (`C[1]||C[0]` in 224-bit)

1: $\varepsilon 0\|$`T` $\leftarrow$ `A[2]+A[3]`

2: $\varepsilon 1\|$`C[0]` $\leftarrow$ `A[0]+`$\varepsilon 0\|$`T`
3: $\varepsilon 2\|$`C[1]` $\leftarrow$ `A[1]+A[3]+`$\varepsilon 0\|$`T`

4: $\varepsilon 3\|$`C[0]` $\leftarrow$ `C[0]+`$\varepsilon 2$
5: $\varepsilon 4\|$`C[1]` $\leftarrow$ `C[1]+(`$\varepsilon 1$`+`$\varepsilon 2$`+`$\varepsilon 3$`)`

6: $\varepsilon 5\|$`C[0]` $\leftarrow$ `C[0]+`$\varepsilon 4$
7: `C[1]` $\leftarrow$ `C[1]+(`$\varepsilon 4$`+`$\varepsilon 5$`)`

8: **return** `C`

Finite-Field Inversion. The finite-field inversion can be performed by following Fermat's Theorem. The prime of Curve448 is $p = 2^{448} - 2^{224} - 1$ and the computation of inversion is $a = z^{-1} \equiv z^{2^{448}-2^{224}-3} \bmod p$. The inversion operation can be performed with 447 modular squaring and 13 modular multiplication operations. Detailed descriptions are given in Algorithm 2.

3.2 Group Operations

The scalar multiplication of Curve448 requires a number of point addition and point doubling operations. The school-book approach to perform the scalar multiplication executes the addition operation depending on the secret value (i.e. branch statement). In order to ensure the constant execution timing for scalar multiplication, Montgomery ladder algorithm is utilized [15]. The Montgomery ladder algorithm performs point addition and point doubling in a regular pattern. The inner routine of the point addition performs addition of two points, including $P1(x1, y1, z1, e1, h1)$ in extended projective coordinates and $P2(u2, v2, w2)$ in extended affine coordinates. This point addition outputs the point $P3(x3, y3, z3, e3, h3)$ in extended projective coordinates. The detailed procedure of point addition is given in Algorithm 3.

The inner routine of the point doubling performs doubling of one point, including $P1(x1, y1, z1, e1, h1)$ in extended projective coordinates. This point doubling outputs the point $P3(x3, y3, z3, e3, h3)$ in extended projective coordinates. The detailed procedure of point doubling is given in Algorithm 4.

3.3 Side-Channel Attack Protection

The cryptography implementation may include the conditional branch depending on the secret. The proposed implementation of finite-field operation is

Algorithm 2. Fermat-based inversion for Curve448 ($p = 2^{448} - 2^{224} - 1$).

Require: Integer z satisfying $1 \leq z \leq p - 1$.
Ensure: Inverse $t_7 = z^{p-2} \bmod p = z^{-1} \bmod p$.

1: $z_3 \leftarrow z^{2^1} \cdot z$	{ cost: 1S+1M}
2: $t_0 \leftarrow z_3^{2^2} \cdot z_3$	{ cost: 2S+1M}
3: $t_1 \leftarrow t_0^{2^1} \cdot z$	{ cost: 1S+1M}
4: $t_2 \leftarrow t_1^{2^4} \cdot t_0$	{ cost: 4S+1M}
5: $t_3 \leftarrow t_2^{2^9} \cdot t_2$	{ cost: 9S+1M}
6: $t_4 \leftarrow (t_3^{2^{18}} \cdot t_3)^2 \cdot z$	{ cost: 19S+2M}
7: $t_5 \leftarrow (t_4^{2^{37}} \cdot t_4)^{2^{37}} \cdot t_4$	{ cost: 74S+2M}
8: $t_6 \leftarrow t_5^{2^{111}} \cdot t_5$	{ cost: 111S+1M}
9: $t_7 \leftarrow (t_6^{2^1} \cdot z^{2^{223}} \cdot t_6)^{2^2} \cdot z$	{ cost: 226S+3M}

10: **return** t_7

Table 2. Evaluation of finite-field operation and group operation on the 32-bit ARM Cortex-M4 microcontrollers in speed (in clock cycles).

Frequency	Finite-field Operation				Group Operation		
	Addition	Subtraction	Multiplication	Inversion	Addition	Doubling	Scalar Multiplication
24 MHz	164	161	821	363,485	6,566	6,567	6,218,135
168 MHz	181	172	838	363,626	6,686	6,674	6,285,904

performed with constant timing by replacing the conditional branch with masked operation. The mask generation is as follows:

$$mask \leftarrow 0 - (carry \ or \ borrow)$$

Furthermore, the legacy ARM Cortex-M3 has early termination issues depending on the input values [23]. Because the ARM Cortex-M4 is the successor of the ARM Cortex-M3, all arithmetic and logical operations are performed in one clock cycle. This satisfies one requirement for constant timing.

For the case of group operation, the Montgomery ladder algorithm always performs point doubling and point addition in regular fashion [24]. When the target processor equips the cache, the implementation must prevent a cache attack. The cache is activated when memory accesses happen frequently depending on a certain regular pattern of input (i.e. pre-computed result). The proposed implementation does not utilize the pre-computed result to avoid a cache attack.

With the above approaches, the implementation achieved constant timing and this is a basic requirement for cryptographic implementation (i.e. timing attack resistant). The checklist for constant timing is presented in Table 4. The proposed implementation satisfies all requirements for constant timing.

Algorithm 3. Point Addition for Curve448.

Require: Point $P1 = (x1, y1, z1, e1, h1)$ in extended projective coordinates, Point $P2 = (u2, v2, w2)$ in extended affine coordinates
Ensure: $P3 = (x3, y3, z3, e3, h3)$ in extended projective coordinates

1: $t1 \leftarrow e1 \cdot h1$
2: $e3 \leftarrow y1 - x1$
3: $h3 \leftarrow y1 + x1$
4: $x3 \leftarrow e3 \cdot v2$ $\{ A = (y1 - x1) \cdot (y2 - x2) \}$
5: $y3 \leftarrow h3 \cdot u2$ $\{ B = (y1 + x1) \cdot (y2 + x2) \}$
6: $e3 \leftarrow y3 - x3$ $\{ E = B - A \}$
7: $h3 \leftarrow y3 + x3$ $\{ H = B + A \}$
8: $x3 \leftarrow t1 \cdot w2$ $\{ C = t1 \cdot w2 \}$
9: $t1 \leftarrow z1 - x3$ $\{ F = z1 - C \}$
10: $x3 \leftarrow z1 + x3$ $\{ G = z1 + C \}$
11: $z3 \leftarrow t1 \cdot x3$ $\{ Z3 = F \cdot G \}$
12: $y3 \leftarrow x3 \cdot h3$ $\{ Y3 = G \cdot H \}$
13: $x3 \leftarrow e3 \cdot t1$ $\{ X3 = E \cdot F \}$

14: **return** $P3(x3, y3, z3, e3, h3)$

4   Evaluation

In this section, we first evaluate the proposed implementations of finite-field operation and group operation for 448-bit wise on the 32-bit ARM Cortex-M4 microcontroller. Then, a comparison of scalar multiplication on low-end processors will be presented.

A benchmark result was obtained on an STM32F4 Discovery board equipped with 32-bit ARM Cortex-M4 microcontrollers. The execution timing in clock cycles was obtained at two frequencies (24 MHz and 168 MHz). The high frequency (i.e. 168 MHz) was for the real-world application, and it showed the highest performance. The low frequency (i.e. 24 MHz) is to avoid wait cycles due to the speed of the memory controller, which ensures the correct clock cycles. All implementations of arithmetic were implemented in the ARM assembly, and the libraries were compiled with GCC with optimization flags set to -O3.

The results of finite-field operation and group operation on the 32-bit ARM Cortex-M4 microcontroller is presented in Table 2. Finite-field addition, subtraction, multiplication, and inversion operations require 164/181, 161/172, 821/838, and 363,485/363,626 clock cycles for 24 MHz/168 MHz, respectively. Clock cycles 24 MHz show better performance than 168 MHz case because the frequency does not have a wait delay. Group addition, doubling, and scalar multiplication operations require 6,566/6,686, 6,567/6,674, and 6,218,135/6,285,904 clock cycles for 24 MHz/168 MHz, respectively.

In Table 3, a comparison of scalar multiplication on 8-bit AVR, 16-bit MSP, and 32-bit ARM processors is presented. For other low-end microcontrollers, NIST P-256 shows the worst performance among 128-bit security ECCs. The

---

Algorithm 4. Point Doubling for Curve448.

Require: Point $P1 = (x1, y1, z1, e1, h1)$ in extended projective coordinates
Ensure: $P3 = (x3, y3, z3, e3, h3)$ in extended projective coordinates

1: $e3 \leftarrow x1 \cdot x1$	$\{A = x1 \cdot x1\}$
2: $h3 \leftarrow y1 \cdot y1$	$\{B = y1 \cdot y1\}$
3: $t1 \leftarrow e3 - h3$	$\{G = A - B\}$
4: $h3 \leftarrow e3 + h3$	$\{H = A + B\}$
5: $x3 \leftarrow x1 + y1$	
6: $e3 \leftarrow x3 \cdot x3$	
7: $e3 \leftarrow h3 - e3$	$\{E = H - (x1 + y1) \cdot (x1 + y1)\}$
8: $y3 \leftarrow z1 \cdot z1$	
9: $y3 \leftarrow 2 \cdot y3$	$\{C := 2 \cdot z1 \cdot z1\}$
10: $y3 \leftarrow t1 + y3$	$\{F := G + C\}$
11: $x3 \leftarrow e3 \cdot y3$	$\{X3 := E \cdot F\}$
12: $z3 \leftarrow y3 \cdot t1$	$\{Z3 := F \cdot G\}$
13: $y3 \leftarrow t1 \cdot h3$	$\{Y3 := G \cdot H\}$

14: **return** $P3(x3, y3, z3, e3, h3)$

---

fastest performance is achieved in the implementation of FourQ. Implementations of Curve25519 show middle performance. The 224-bit security ECC (i.e. Curve448) on 8-bit AVR ATmega and 16-bit MSP430 requires 103M and 73M clock cycles, respectively. The performance of Curve448 is relatively slower than that of 128-bit security ECC implementations because of its parameters. On the 32-bit ARM Cortex-M4 microcontroller, the fastest implementation of Curve25519 requires 847,048 clock cycles [13], while the FourQ requires 542,900 clock cycles [20]. The proposed implementation of Curve448 requires 6,218,135 clock cycles. Compared with other ECC implementations, the implementation of Curve448 is 86% and 91% slower than Curve25519 and FourQ because these curves are defined over small finite-fields, which ensure compact finite-field implementations on the target processor. ROM and RAM sizes are 3,828 bytes and 2,128 bytes, respectively.

4.1 Trade-Off Between Performance and Security

Performance and security have trade-off relations between them. In the implementation, we focused on security first. The recommended security level by 2030 is 128-bit (i.e. Curve25519 and FourQ) [25,26]. Even though the performance of 128-bit security ECCs (i.e. Curve25519 and FourQ) is better than that of 224-bit security ECC (i.e. Curve448), security-sensitive services should ensure high security levels. This is even more secure against quantum attacks. The quantum resources for the 224-bit security ECC are significantly more than those for 128-bit security ECCs [27].

Table 3. Comparison of scalar multiplication on 8-bit AVR ATmega, 16-bit MSP430(X), and 32-bit ARM Cortex-M4 processors in speed (in clock cycles).

Target	Implementation	128-bit security			224-bit security
		NIST P-256	Curve25519	FourQ	Curve448
8-bit AVR ATmega	Wenger et al. [16]	34,930 000	–	–	–
	Hutter and Schwabe [17]	–	22,791,580	–	–
	Nascimento et al. [18]	–	20,153,658	–	–
	Düll et al. [19]	–	13,900,397	–	–
	Liu et al. [20]	–	–	7,296,000	–
	Seo [9]	–	–	–	103,228,541
16-bit MSP430	Wenger et al. [16]	22,170 000	–	–	–
	Gouvêa and López [21]	20,476,234	–	–	–
	Seo [9]	–	–	–	73,477,660
16-bit MSP430X	Hinterwälder et al. [22]	–	6,513,011	–	–
	Düll et al. [19]	–	5,301,792	–	–
	Liu et al. [20]	–	–	4,826,100	–
32-bit ARM Cortex-M4	Groot [10]	–	1,816,351	–	–
	Santis and Sigl [11]	–	1,563,852	–	–
	Fujii and Aranha [12]	–	907,240	–	–
	Haase and Labrique [13]	–	847,048	–	–
	Liu et al. [20]	–	–	542,900	–
	This work	–	–	–	6,218,135

Table 4. Checklist for ECC implementations in constant timing.

Masked implementation	Early termination prevention	Montgomery ladder	w/o look-up table
√	√	√	√

5 Hybrid Post-Quantum TLS

During the transition from pre-quantum cryptography to post-quantum cryptography, both algorithms should be supported in real-world applications. Recently, AWS cryptography proposed supersingular isogeny key encapsulation (SIKE) based hybrid post-quantum transport layer security (TLS) algorithms[1]. Because SIKE is an alternative candidates, this algorithm should be counted for PQC[2] Classical TLS 1.2 and hybrid post-quantum TLS 1.2 are compared in detail in Table 5. The protocol performs two independent key exchanges (one classical and one post-quantum). Then, both keys are combined into a single TLS master secret. The hybrid post-quantum TLS allows network connections to be secure when one of the key exchanges (i.e. classical or post-quantum) for TLS is compromised by hackers. One of the potential scenarios is quantum computers. If a large-scale quantum computer is developed in the near future, the current

---

[1] https://aws.amazon.com/ko/blogs/security/round-2-hybrid-post-quantum-tls-benchmarks/.

[2] https://csrc.nist.gov/News/2020/pqc-third-round-candidate-announcement.

discrete logarithm problem (DLP) and integer factorization (IF)-based public key cryptography will be vulnerable. Under this difficult condition, the hybrid post-quantum TLS still keeps the connection in secret. Similarly, PQC is not completely proven to be secure against the quantum computer and quantum algorithm. When PQC has a backdoor, the legacy PKC still ensures security.

In Table 6, the performance of isogeny based post-quantum cryptography (i.e. SIKE) is presented. The execution timing for SIKEp434, SIKEp503, SIKEp610, and SIKEp751 require 184, 257, 493, and 770 million clock cycles, respectively. Implementations on the 168 MHz Cortex-M4 take 1.09, 1.53, 2.94, and 4.58 s for SIKEp434, SIKEp503, SIKEp610, and SIKEp751, respectively. The performance is not as fast as pre-quantum PKC but it is still practically fast enough for real-world applications, considering that PKC is not frequently performed. When ECC and SIKE cryptography systems are adopted for hybrid post-quantum TLS, the multiplication part can be shared. This optimizes the code size. It is also possible to adopt other PQC candidates for protocols. This is our future work.

Table 5. Comparison result between classical TLS 1.2 and hybrid post-quantum TLS 1.2 [28].

Classical TLS 1.2	Hybrid Post-Quantum TLS 1.2
premaster_secret = ECDHE_KEY	premaster_secret = ECDHE_KEY ‖ PQ_KEY
seed = "master secret"	seed = "hybrid master secret"
‖ ClientHello.random	‖ ClientHello.random
‖ ServerHello.random	‖ ServerHello.random
master_secret=HMAC(premaster_secret,seed)	master_secret=HMAC(premaster_secret,seed)

Table 6. SIKE implementations on the ARM Cortex-M4 microcontrollers.

Implementation	Timings [$cc \times 10^6$]				Timings [second]			
	KeyGen	Encaps	Decaps	Total	KeyGen	Encaps	Decaps	Total
SIKEp434 (AES-128)								
Seo et al. [4]	74	122	130	252	0.44	0.73	0.77	1.50
Seo et al. [29]	54	89	95	184	0.32	0.53	0.56	1.09
SIKEp503 (SHA-256)								
Seo et al. [4]	104	172	183	355	0.62	1.02	1.09	2.11
Seo et al. [29]	76	125	133	257	0.45	0.74	0.79	1.53
SIKEp610 (AES-192)								
Seo et al. [29]	134	246	248	493	0.80	1.46	1.47	2.94
SIKEp751 (AES-256)								
Seo et al. [4]	282	455	491	946	1.68	2.71	2.92	5.63
Seo et al. [29]	229	371	399	770	1.36	2.21	2.37	4.58

6   Conclusion

In this paper, we presented the first optimized implementation of Curve448 on the 32-bit ARM Cortex-M4 microcontroller. State-of-art implementation techniques are used to achieve the optimal performance. The proposed implementation achieved 6,218,135 clock cycles. This is practically fast enough considering that the target microcontroller supports 168 MHz operating frequency. Furthermore, the implementation is secure against timing attacks by avoiding conditional branch and cache access.

Our future work is practical implementation of a hybrid post-quantum TLS protocol for pre-quantum and post-quantum cryptography algorithms. We will investigate the secure and efficient implementation of both protocols to achieve the highest performance.

Acknowledgement. This work of Hwajeong Seo was supported by Institute for Information & communications Technology Planning & Evaluation (IITP) grant funded by the Korea government(MSIT) (<Q|Crypton>, No.2019-0-00033, Study on Quantum Security Evaluation of Cryptography based on Computational Quantum Complexity). This work of Reza Azarderakhsh was supported by ARO grant W911NF2010328.

References

1. Hamburg, M.: Ed448-Goldilocks, a new elliptic curve. IACR Cryptol. ePrint Arch. **2015**, 625 (2015)
2. Kannwischer, M.J., Rijneveld, J., Schwabe, P., Stoffelen, K.: pqm4: testing and benchmarking NIST PQC on ARM Cortex-M4 (2019)
3. Kannwischer, M.J., Rijneveld, J., Schwabe, P.: Faster multiplication in $\mathbb{Z}_{2m}[x]$ on Cortex-M4 to speed up NIST PQC candidates. In: Deng, R., Gauthier-Umana, V., Ochoa, M., Yung, M. (eds.) ACNS 2019. LNCS, vol. 11464, pp. 281–301. Springer, Cham (2019). https://doi.org/10.1007/978-3-030-21568-2_14
4. Seo, H., Jalali, A., Azarderakhsh, R.: SIKE round 2 speed record on ARM Cortex-M4. In: Mu, Y., Deng, R., Huang, X. (eds.) CANS 2019. LNCS, vol. 11829, pp. 39–60. Springer, Cham (2019). https://doi.org/10.1007/978-3-030-31578-8_3
5. Botros, L., Kannwischer, M.J., Schwabe, P.: Memory-efficient high-speed implementation of Kyber on Cortex-M4. In: Buchmann, J., Nitaj, A., Rachidi, T. (eds.) AFRICACRYPT 2019. LNCS, vol. 11627, pp. 209–228. Springer, Cham (2019). https://doi.org/10.1007/978-3-030-23696-0_11
6. Edwards, H.: A normal form for elliptic curves. Bull. Am. Math. Soc. **44**(3), 393–422 (2007)
7. Bernstein, D.J., Lange, T., et al.: SafeCurves: choosing safe curves for elliptic-curve cryptography (2013). http://safecurves.cr.yp.to
8. Rescorla, E., et al.: The transport layer security (TLS) protocol version 1.3 (2017). https://tools.ietf.org/html/draft-ietf-tls-tls13-21
9. Seo, H.: Compact implementations of Curve Ed448 on low-end IoT platforms. ETRI J. **41**(6), 863–872 (2019)
10. de Groot, W.: A performance study of X25519 on Cortex-M3 and M4, Ph. D. thesis, Eindhoven University of Technology (2015)

11. De Santis, F., Sigl, G.: Towards side-channel protected X25519 on ARM Cortex-M4 processors. In: Proceedings of Software Performance Enhancement for Encryption and Decryption, and Benchmarking, Utrecht, The Netherlands, pp. 19–21 (2016)

12. Fujii, H., Aranha, D.F.: Curve25519 for the Cortex-M4 and beyond. In: Lange, T., Dunkelman, O. (eds.) LATINCRYPT 2017. LNCS, vol. 11368, pp. 109–127. Springer, Cham (2017). https://doi.org/10.1007/978-3-030-25283-0_6

13. Haase, B., Labrique, B.: AuCPace: efficient verifier-based PAKE protocol tailored for the IIoT. IACR Trans. Cryptogr. Hardw. Embed. Syst. 1–48, 2019 (2019)

14. Hutter, M., Wenger, E.: Fast multi-precision multiplication for public-key cryptography on embedded microprocessors. In: Preneel, B., Takagi, T. (eds.) CHES 2011. LNCS, vol. 6917, pp. 459–474. Springer, Heidelberg (2011). https://doi.org/10.1007/978-3-642-23951-9_30

15. Montgomery, P.L.: Speeding the pollard and elliptic curve methods of factorization. Math. Comput. 48(177), 243–264 (1987)

16. Wenger, E., Unterluggauer, T., Werner, M.: 8/16/32 shades of elliptic curve cryptography on embedded processors. In: Paul, G., Vaudenay, S. (eds.) INDOCRYPT 2013. LNCS, vol. 8250, pp. 244–261. Springer, Cham (2013). https://doi.org/10.1007/978-3-319-03515-4_16

17. Hutter, M., Schwabe, P.: NaCl on 8-bit AVR microcontrollers. In: Youssef, A., Nitaj, A., Hassanien, A.E. (eds.) AFRICACRYPT 2013. LNCS, vol. 7918, pp. 156–172. Springer, Heidelberg (2013). https://doi.org/10.1007/978-3-642-38553-7_9

18. Nascimento, E., López, J., Dahab, R.: Efficient and secure elliptic curve cryptography for 8-bit AVR microcontrollers. In: Chakraborty, R.S., Schwabe, P., Solworth, J. (eds.) SPACE 2015. LNCS, vol. 9354, pp. 289–309. Springer, Cham (2015). https://doi.org/10.1007/978-3-319-24126-5_17

19. Düll, M., et al.: High-speed Curve25519 on 8-bit, 16-bit, and 32-bit microcontrollers. Des. Codes Cryptogr. 77(2–3), 493–514 (2015). https://doi.org/10.1007/s10623-015-0087-1

20. Liu, Z., Longa, P., Pereira, G., Reparaz, O., Seo, H.: FourQ on embedded devices with strong countermeasures against side-channel attacks. IEEE Trans. Depend. Secure Comput. 17, 536–549 (2018)

21. Gouvêa, C.P.L., López, J.: Software implementation of pairing-based cryptography on sensor networks using the MSP430 microcontroller. In: Roy, B., Sendrier, N. (eds.) INDOCRYPT 2009. LNCS, vol. 5922, pp. 248–262. Springer, Heidelberg (2009). https://doi.org/10.1007/978-3-642-10628-6_17

22. Hinterwälder, G., Moradi, A., Hutter, M., Schwabe, P., Paar, C.: Full-size high-security ECC implementation on MSP430 microcontrollers. In: Aranha, D.F., Menezes, A. (eds.) LATINCRYPT 2014. LNCS, vol. 8895, pp. 31–47. Springer, Cham (2015). https://doi.org/10.1007/978-3-319-16295-9_2

23. Franck, C., Großschädl, J., Le Corre, Y., Tago, C.L.: Energy-scalable montgomery-curve ECDH key exchange for ARM cortex-M3 microcontrollers. In: 2018 6th International Conference on Future Internet of Things and Cloud Workshops (FiCloudW), pp. 231–236. IEEE (2018)

24. Joye, M., Yen, S.-M.: The montgomery powering ladder. In: Kaliski, B.S., Koç, K., Paar, C. (eds.) CHES 2002. LNCS, vol. 2523, pp. 291–302. Springer, Heidelberg (2003). https://doi.org/10.1007/3-540-36400-5_22

25. Barker, E., Barker, W., Burr, W., Polk, W., Smid, M., et al.: Recommendation for key management: Part 1: General. National Institute of Standards and Technology, Technology Administration (2006)

26. Orman, H., Hoffman, P.: Determining strengths for public keys used for exchanging symmetric keys. Technical report, BCP 86, RFC 3766, April 2004
27. Roetteler, M., Naehrig, M., Svore, K.M., Lauter, K.: Quantum resource estimates for computing elliptic curve discrete logarithms. In: Takagi, T., Peyrin, T. (eds.) ASIACRYPT 2017. LNCS, vol. 10625, pp. 241–270. Springer, Cham (2017). https://doi.org/10.1007/978-3-319-70697-9_9
28. Campagna, M., Crockett, E.: Hybrid post-quantum key encapsulation methods (PQ KEM) for transport layer security 1.2 (TLS). Internet Engineering Task Force, Internet-Draft draft-campagna-tls-bike-sike-hybrid-01 (2019)
29. Seo, H., Anastasova, M., Jalali, A., Azarderakhsh, R.: Supersingular isogeny key encapsulation (SIKE) round 2 on ARM Cortex-M4. IACR Cryptol. ePrint Arch. **2020**, 410 (2020)

Efficient Implementation of SHA-3 Hash Function on 8-Bit AVR-Based Sensor Nodes

YoungBeom Kim, Hojin Choi, and Seog Chung Seo$^{(\boxtimes)}$ (iD)

Department of Information Security, Cryptology, and Mathematics,
Kookmin University, Seoul, South Korea
{darania,ondoli0312,scseo}@kookmin.ac.kr

Abstract. The Keccak algorithm was selected by NIST as the standard SHA-3 hash algorithm for replacing currently used SHA-2 algorithm in 2015. Despite SHA-3's improved security compared to SHA-2, its low performance in software implementation limits its wide use. In this paper, we propose an optimized SHA-3 implementation on 8-bit AVR microcontrollers (MCU) which are dominantly used for sensor devices in WSNs. Until now, there are only a few researches on optimization of SHA-3 in spite of its security importance. Furthermore, it is very challenging to optimize hash function, especially, SHA-3, on 8-bit AVR MCUs. This is because the internal state of SHA-3 is 1,600-bit which is much larger than internal state of symmetric algorithms (typically, 128-bit) like AES, ARIA, and so on. In other words, it is difficult to accommodate the whole of SHA-3's internal state on the registers of AVR MCUs, which incurs heavy memory accesses during computation. Thus, we analyzed the structure of SHA-3 algorithm and found that each lane of the internal state can be executed independently for each process in SHA-3. By using this fact, we propose an optimization method which can reduce efficiently the times of memory accesses to the internal state. With this proposed method minimizing the memory accesses, our implementation of SHA3-256 achieves around 25.0% of performance improvement when hashing 500 bytes message compared with the previous best work on 8-bit AVR MCU. To the best of our knowledge, our software is the fastest SHA-3 implementation on AVR platforms until now. In addition, the proposed optimization method can be easily extended to other embedded MCUs such as 16-bit MSP430, 32-bit RISC-V and ARM-based MCUs.

Keywords: SHA-3 · Keccak algorithm · 8-bit AVR MCUs · Embedded · Microcontroller · WSN

1 Introduction

As the recent Internet of Things (IoT) era has arrived, many applications are being implemented with Wireless Sensor Networks (WSNs) [1]. However, due

This work was supported by the National Research Foundation of Korea (NRF) grant funded by the Korea government (MSIT) (No. 2019R1F1A1058494).

D. Hong (Ed.): ICISC 2020, LNCS 12593, pp. 140–154, 2021.
https://doi.org/10.1007/978-3-030-68890-5_8

to the characteristics of wireless communication in WSNs, the transmitted data can be easily modified or forged. Hash function is cryptographic algorithm that provides data integrity and be used to prove whether the data have been modified or not. Furthermore, it is also used in other various cryptographic applications, e.g., PBKDF2, HMAC, DRBG, and digital signature algorithms such as DSA, ECDSA, and RSA-PSS. Hash functions (e.g., SHA-1 or SHA2) are the most significant cryptographic algorithms from an integrity perspective. However, several vulnerabilities in SHA-1 have been identified; thus, the US National Institute of Standards and Technology (NIST) currently does not recommend the use of SHA-1 [2–6]. Public attacks on the SHA-2 family have continued since 2008. Note that SHA-1 and SHA-2 differ significant; however, they share a similar algorithm (i.e., SHA) [7,8]. Therefore, SHA-1 and SHA-2 are vulnerable to attacks with some identical digests. In some scenarios, the security of SHA-2 is superior to that of SHA-1, except SHA-2 uses larger inputs and outputs [7,8]. In addition, several theoretical preimage attacks on SHA-2 have been found [9–11]. Prior to the discovery of preimage attacks on SHA-2, NIST held the SHA-3 competition, concerning that SHA-2 had a similar structure to SHA-1. The Keccak algorithm was the winner and selected to be the next-generation hash function standard, i.e., SHA-3 [12].

Through numerous verification, the security of SHA-3(Keccak) has been to be secure against existing attacks threatening the security of SHA-1 and SHA-2. However, in spite of higher security of SHA-3, it is currently used less in the field. One of the biggest reasons is SHA-3 has low performance in software [13–20]. In the software implementation, SHA-3 (256-bit) is three times slower than SHA-256 in the 8-bit AVR platform [12,15–17] and SHA-3 (512-bit) is two times slower than SHA-512 in a CPU environment [2,18–20]. Therefore, software optimization studies on SHA-3 applicable to various platforms are essential in the future.

In this paper, we propose optimization methods for SHA-3 software implementation. Also, we demonstrate the efficiency of the proposed optimization method, for typical low-end 8-bit AVR microcontrollers which are mainly used for sensor devices in WSNs. The proposed SHA-3 optimization method exploits the fact that each of the 25 lanes of the internal state can be operated independently. Unlike the general implementation order of the existing implementation methods, we combine θ process and ρ process into a single process to reduce memory accesses to the internal state in an efficient manner. By using this fact, π process can be executed implicitly. In addition, the proposed optimization method does not require additional operations or lookup tables. This advantage is especially efficient for limited embedded devices which used in WSNs.

Using the propose method, we present a carefully-optimized assembler implementation of SHA-3 for 8-bit AVR microcontrollers, specifically the ATmega128. When hashing a 500 bytes message, we obtained an execution time of 1073 Cycles Per Byte (CPB) using the Assembly implementation. This is 79.1% and 25.0% faster than the latest implementations proposed by Otte et al. [15] and Balasch et al. [16], respectively. In addition, our optimized implementation of SHA-3 is faster than the implementation proposed by Balasch et al., which is currently the fastest SHA-3 software implementation on the same platform. Furthermore, the proposed method can be applied to various platforms, e.g., 16-bit MSP430, 32-bit ARM-based MCUs, 32-bit RISC-V, CPUs, GPUs [16], and so on.

The remainder of this paper is organized as follows. In Sect. 2, we briefly review SHA-3 and provides an overview of the 8-bit AVR microcontroller. Section 3 analysis a related work on the 8-bit AVR microcontroller. Section 4 propose a new optimization method for SHA-3. Section 5 compares the performance. Finally, Sect. 6 concludes the paper.

2   Background

2.1   Overview of SHA-3

In 1993, NIST proposed Secure Hash Algorithm 0 (SHA-0) hash function. Later, SHA-1 and SHA-2 were proposed and standardized. However, Stevens et al. proposed the collision pair of SHA-1 [2], and, as the security of SHA-2 hash algorithm has improved [8,21,22], NIST selected the Keccak algorithm by Bertoni et al. as SHA-3 in order to substitute SHA-2 in a competition on August 5, 2015 [12]. The Keccak algorithm is based on a sponge construction structure, which differs from the structure of SHA-2; thus, the Keccak algorithm is resistant against attacks that are applicable to SHA-2 [12].

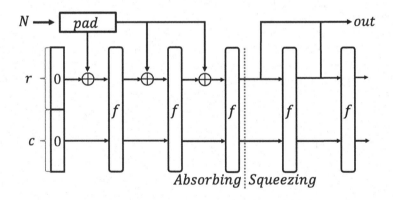

Fig. 1. Overview of sponge structure

Sponge Structure. Figure 1 shows the sponge structure of SHA-3. The operational process of SHA-3 assumes a sponge structure comprising absorbing and squeezing processes. The absorbing process compresses the message using the b-bit permutation f-function and a padding function to *pad* the message. Then, using the exclusive-OR (XOR) operation, the padded message is calculated with the output of f-function. The digest is calculated in the squeezing process. In SHA-3, b-bit permutation means the size of the state and is fixed with $b \in \{25,$

50, 100, 200, 400, 800, 1600}. Here, b comprises bitrate (r) and capacity (c), and satisfies $b = r + c$. In the squeezing process, if the required digest length is greater than r, then f-function is called to update the internal state. In this paper, the values of the SHA-3 parameters are $b = 1600$, $r = 1088$, and $c = 512$, and the digest length is 256 bits. These parameters are generally used for safety [12].

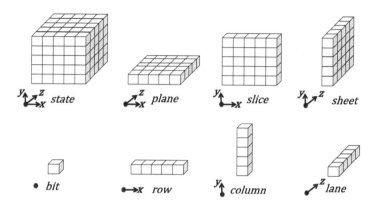

• bit

Fig. 2. State of SHA-3[12]

Table 1. the values of w and l for each b

b (bit)	25	50	100	200	400	800	1600
w (bit)	1	2	4	8	16	32	64
l	0	1	2	3	4	5	6

State of SHA-3. f-function is the primary process of SHA-3. f-function takes the state as input data. Figure 2 shows the structure of the state in SHA-3.

The *state* of SHA-3 is a three-dimensional $x \times y \times z$ matrix, where row x and column y are both fixed to five. The *state* comprises 25 lanes and the length of each lane depends on parameters b, w, and l in SHA-3. According to b, the *state* is composed of $5 \times 5 \times w$. Table 1 shows the w and l values for each value of b. As shown in Table 1, l is equal to $log_2(b/25)$, and w is equal to 2^l. In SHA-3, f-function repeats the same process by the number of rounds (denoted n_r). Using parameter l, n_r is represented by $12 + 2 \times l$.

f-function. f-function is an internal function of the sponge structure. f-function is a b-bit permutation. In SHA-3, f-function comprise five processes: θ, π, ρ, χ and ι). These five processes are used to update state, and f-function repeats the processes in times of n_r rounds.

The effect of θ process is to XOR each bit in the *state* with parties of two columns in the array. In the θ process, if this process operates for the bit (x_0, y_0, z_0), the x-coordinates of the required columns are $(x_0 + 1) \bmod 5$, $(x_0 - 1) \bmod 5$. In addition, in this process, the z-coordinate of the column with x-coordinate $(x_0 - 1) \bmod 5$ is $(z_0 - 1) \bmod w$. Algorithm 1 shows the pseudocode of the θ process. Line 2 in Algorithm 1 is referred to as the initial θ, which generates $D[x, z]$.

Algorithm 1. θ Process

Require: *state* A
Ensure: *state* A'
1: For all pairs(x, z) such that $0 \le x < 5$ and $0 \le z < w$
    $C[x, z] = A[x, 0, z] \oplus A[x, 1, z] \oplus A[x, 2, z] \oplus A[x, 3, z] \oplus A[x, 4, z]$;
2: For all pairs(x, z) such that $0 \le x < 5$ and $0 \le z < w$
    //This step is initial θ
    $D[x, z] = C[(x - 1) \bmod 5, z] \oplus C[(x + 1) \bmod 5, (z - 1) \bmod w]$;
3: For all triples(x, y, z) such that $0 \le x, y < 5$ and $0 \le z < w$
    $A'[x, y, z] = A[x, y, z] \oplus D[x, z]$;
4: **return** A'

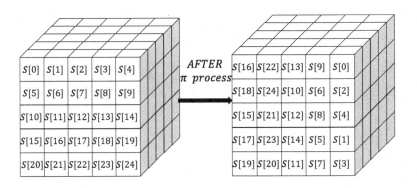

Fig. 3. Overview of π process

The effect of π process is to rearrange the positions of the lanes, as shown in Fig. 3. Here, $S[i]$ $(i \in [0, 24])$ is each lane of *state*. Note that $S[12]$ is a lane of $(x = 0, y = 0)$ in *state* [12].

The effect of ρ process is to right-rotate the bits of each lane by length (referred to as the offset), which depends on the fixed x- and y-coordinates of the lane [12]. Equivalently, for each bit in the lane, the z-coordinate is modified by adding the offset, modulo the lane size. The offsets for each lane that require the operation are listed in Table 2.

Table 2. Offsets of ρ Process

	x = 3	x = 4	x = 0	x = 1	x = 2
y = 2	153	231	3	10	171
y = 1	55	276	36	300	6
y = 0	28	91	0	1	190
y = 4	120	78	210	66	253
y = 3	21	136	105	45	15

The effect of χ process is to XOR each bit with a nonlinear function of two other bits in its row [12]. Note that the difference between χ process and the other processes (i.e., θ, π, and ρ processes) is that χ process should be operated in row form and implemented accordingly.

ι process execute an XOR operation, for the lane of $(x = 0, y = 0)$ of the *state* and constants RC [12]. Since ι process operates only one lane, in the most implementations, χ process and ι process are combined into a single process.

2.2 Overview of 8-Bit AVR MCUs

The 8-bit AVR microcontroller is an embedded device made into a single integrated circuit by adding memory and I/P to the microprocessor. And this microcontroller is currently the most used worldwide in the WSNs environment [1].

Table 3. 8-bit AVR Assembly Instructions [23,24], cc means clock cycles.

Asm	Operands	Description	Operation	cc
ADD	Rd, Rr	Add without Carry	Rd ← Rd+Rr	1
ADC	Rd, Rr	Add with Carry	Rd ← Rd+Rr+C	1
MOV	Rd, Rr	Copy Register	Rd ← Rr	1
LDI	Rd, K	Load Immediate	Rd ← K	1
LD	Rd, X	Load Indirect	Rd ← (X)	2
ST	Z, Rr	Store Indirect	(Z) ← Rr	2
EOR	Rd, Rr	Exclusive OR	Rd ← Rd⊕Rr	1
LSL	Rd	Logical Shift Left	C|Rd ← Rd<<1	1
LSR	Rd	Logical Shift Right	Rd|C ← 1>>Rd	1
ROL	Rd	Rotate Left Through Carry	C|Rd ← Rd<<1||C	1
ROR	Rd	Rotate Right Through Carry	Rd|C ← C||1>>Rd	1
BST	Rd, b	Bit store from Bit in Reg to T Flag	T ← Rd(b)	1
BLD	Rd, b	Bit load from T Flag to a Bit in Reg	Rd(b) ← T	1

The 8-bit AVR microcontroller's commands comprise operation codes and an operand. Table 3 shows the operands and clock cycles of the commands used in this paper. The 8-bit AVR-MCU consists of flash memory, SRAM, and EEP-ROM with a Harvard architecture. Our target device is ATmega128 which is widely used for sensor nodes in Wireless Sensor Networks [25]. ATmega128 has a 128 KB of flash memory, 4 KB SRAM, and 4 KB EEPROM. The device supports throughput of 16 MIPS 16 MHz and operates between 4.5-5.5 volts [26]. The AVR-MCU has 32 8-bit general-purpose resisters, which are used for various purposes, e.g., basic private operations and bit operations. Specifically, the R26-R31 registers can be combined and used as three 16-bit registers, i.e., X, Y, and Z registers. These registers (X,Y, and Z) are used as pointers to indirectly specify a 16-bit address for data memory. The Status REGister (SREG) shows the status and result after Arithmetic Logic Unit (ALU) calculations.

3   Analysis of Existing Implementations of Hash Functions on 8-Bit AVR MCUs

Since Keecak algorithm was selected as SHA-3 standard in 2012, it has been implemented in various low-end-processors. Basically, the hardware implementation of SHA-3 is faster than SHA-2 and SHA-1 [27]. In particular, ARM-v8 architectures include special commands to increase the speed of the SHA-3 algorithm [28]. However, the software implementation of SHA-3 is much slower than SHA-2 [13–17]. Currently, the existing software-implemented SHA-3 algorithm in various IoT devices, including 8-bit AVR MCUs, follows the order of the standard SHA-3 implementation listed by NIST [13–16]. In the implementations based on order of standard way [13–16], π and ρ processes were implemented in once ($\pi \sim \rho$ process), because rotate-operation can be executed while executing π process. Note that standard implementation execute f-function as follows: θ process $\rightarrow$ $\pi \sim \rho$ process $\rightarrow$ $\chi \sim \iota$ process.

We found two representative SHA-3 implementations in 8-bit AVR MCUs [15, 16]. First, we analyzed the implementation of Otte et al. in AVR-Crypto-Lib; however, this implementation is based on the C-language (not 8-bit AVR assembly) [15]. The Otte et al.'s implementation of SHA-3 (256-bit) from AVR-Crypto-Lib requires 2,570,828 clock cycles to compute the digest of a 500 bytes message, which corresponds to a hash rate of 5,142 (CPB). This incurs nearly seven times the cost of 783 (CPB) of the SHA-2 implementation proposed by Otte et al. [15]. Therefore, we analyzed the SHA-3 implementation presented by Balasch et al., which is the most popular hash function implementations with an assembly on the 8-bit AVR microcontroller [16]. The Balash et al.'s implementation of SHA-3 (256-bit) requires 716,483 (CPB) to compute the digest of a 500 bytes message, which corresponds to a hash rate of 1,432 (CPB). This is the fastest implementation among existing SHA-3 implementations.

Balash et al. implemented the efficient shift-rotation for operating $\pi \sim \rho$ process. The only rotation instructions in the 8-bit AVR microcontroller are rotations of 8-bit operands by 1 bit to the left (ROL) and right (ROR). In SHA-3, for $b = 1600$ and $w = 64$, the length of a single lane is 64 bits. to rotate 64-bit in ρ process, eight general-purpose registers are required. Typically, in an 8-bit AVR MCUs, a 1-bit left-rotation is implemented via a 1-bit logical left-shift (LSL) followed by seven 1-bit left-rotations through carry (ROL) and an addition with carry (ADC) [29]. In contrast, a 1-bit right-rotation is implemented by a 1-bit store to T in SREG ((BST)) followed by eight 1-bit right-rotations through carry (ROR) and a 1-bit load from T in SREG (BLD). Rotation of 64-bit data by n bits, where $1 < n < 8$, can be calculated by repeating the sequence of instructions to rotate a 64-bit quantity by one bit in the same direction n times. However, the execution time is not equal when executing 1-bit left-rotation and 1-bit right-rotation for 64-bit data; therefore, for efficient implementation, the n-bit rotation or $(64-n)$-bit rotation is employed as required [29]. In addition, for n-bit rotations greater than 8, the actual execution time of all n-bit shift-rotation operations can be reduced to 40 clock cycles or fewer if $x = (x \ggg n)$ operations are replaced by $x = (x \ggg n\%8)$. In this process, operation of $x = (x \ggg n/8)$ directly allocate and store in memory. When storing to memory (ρ process), the implementation of Balasch et al. combines π and ρ processes into a single process ($\pi \sim \rho$). Note that Balasch et al. implemented SHA-3 based on the order of standard implementation [12].

However, even though the implementation presented by Balasch et al. is efficient for the 8-bit AVR MCUs, SHA-3 remains much slower than SHA-2. In low-end processors, accessing memory requires longer execution time compared to arithmetic operations. In addition, the *state* in SHA-3 where $b = 1600$ requires at least 200 bytes, which is a heavier memory requirement compared to symmetric ciphers, where the *state* is only 128 bits. Therefore, in 8-bit AVR MCUs, it is important to optimize memory access to the *state* during SHA-3 execution.

Table 4. Number of times to memory accesses to the *state* of previous implementation, e.g., Balasch et al. and Otte et al. [13–17]

Standard Method	initial θ	θ process	$\pi \sim \rho$ process	$\chi \sim \iota$ process	Total Access
Load	O	O	O	O	7
Store	X	O	O	O	

Table 4 shows the number of times to memory access to *state* in the standard implementation of SHA-3 recommended by NIST. Here, $\pi \sim \rho$ process indicates that π and ρ processes are combined, and $\theta \sim \rho$ process indicates that θ process and ρ process are combined. χ and ι processes are also combined with the same logic. Note that initial θ, which creates $D[x, z]$ in Algorithm 1, is part of θ process. Table 4 shows that *state* stored in memory is accessed three times during θ process and is accessed twice when $\pi \sim \rho$ process is executed. In addition, the *state* is access twice in the $\chi \sim \iota$ process. Therefore, existing standard implementations require seven memory accesses to the *state*. The size of *state* is

200 bytes ($25 \times 64/8$), where b = 1600, w = 64, and l = 16. In addition, the θ, π, ρ, χ, and ι processes should be executed in 24 rounds ($12 + 6 \times 2$). Therefore, each execution of f-function of SHA-3 results in 168 = 24 $\times$ 7 memory accesses for 200 bytes (200 bytes are *state* stored in memory).

In the embedded device, currently used in the WSNs, the high number of memory accesses to *state* causes low performance of the algorithm; therefore, in Sect. 4, we propose a new optimization method that reduces memory access without additional computation and lookup tables.

4   Proposed SHA-3 Implementations in 8-Bit AVR MCUs

4.1   Main Idea

Here, we propose a generic SHA-3 optimization method that can be used for various platforms, e.g., low-end and high-end processors. In addition, we present the optimized techniques of SHA-3 on 8-bit AVR microcontroller. As mentioned in Sect. 3, in low-end-processors, memory access to *state* requires longer execution time compared to arithmetic operations. Therefore, it is important to schedule the use of general-purpose registers for the *state* efficiently and optimize memory access during SHA-3.

The main idea is to execute π process implicitly at minimum cost and combine the θ and ρ processes into a single process ($\theta \sim \rho$ process) to reduce accesses to the state. Figure 4 shows an overview of proposed SHA-3 method. Each lane of the *state* is operated independently in the θ process; thus, ρ process can be applied to each lane. In other words, after generating $D[x, z]$ through the initial θ, we can execute the remained θ process and ρ process. Note that initial θ is part of θ process. Twice memory accesses (Load and Store) occur for *state* executing $\theta \sim \rho$ process. π process is performed implicitly while executing $\theta \sim \rho$ process; also, changes only the position of the lane without directly affecting the value of the lane in the *state*. Therefore, our main idea implicitly executes π process, when storing the calculated *state* in the memory.

Table 5. Number of memory accesses to the *state*

Standard Method	initial θ	θ process	$\pi \sim \rho$ process	$\chi \sim \iota$ process	Total Access
Load	O	O	O	O	**7**
Store	**X**	O	O	O	
Proposed Method	initial θ	$\theta \sim \rho$ process	π process	$\chi \sim \iota$ process	Total Access
Load	O	O	**X**(Implied)	O	**5**
Store	**X**	O	**X**(Implied)	O	

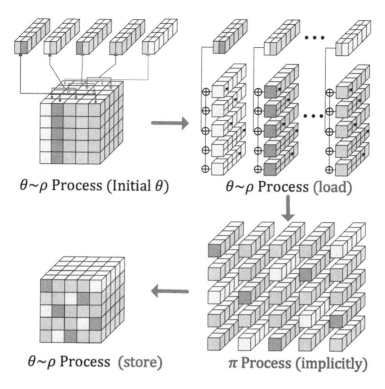

$\theta{\sim}\rho$ Process (Initial θ) $\theta{\sim}\rho$ Process (load)

$\theta{\sim}\rho$ Process (store) π Process (implicitly)

Fig. 4. Proposed main idea

Table 5 shows the number of accesses to the *state* for the proposed implementation, compared the previous implementations based on the order of standard methods. The proposed method reduces memory accesses twice compared to the standard method. Therefore, a total of five accesses to the *state* occur, and when f-function is called, 120 (24×5) memory accesses to the 200-byte *state* occur. In other words, the proposed implementation reduces the amount of memory access 48 (168 - 120) compared to the standard implementation.

Implementations of SHA-3 on the 8-bit AVR MCUs depend on the value of b. If b is less than 200 (w is less than 7), the general-purpose register can store all lanes of the state. However, if b is 400, 800, or 1600, the length of a single lane is 16-bit, 32-bit, and 64-bit, respectively; therefore, it is difficult to store all lanes of state in a general-purpose register. Therefore, an important point in implementing SHA-3 is how efficiently memory is accessed when the length of the lane is more than 8-bit. In this section, we present the SHA-3 implementation technique where $b = 1600$, which conforms to the Korean Cryptographic Module Validation Program (KCMVP) [30].

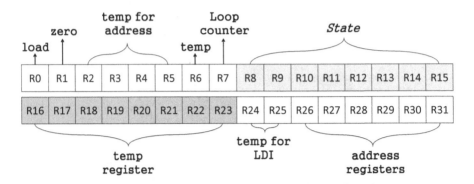

Fig. 5. Register scheduling for proposed implementation in 8-bit AVR MCUs

4.2   Proposed Implementation Technique on 8-Bit AVR MCUs

Figure 5 shows the register scheduling for the proposed method on the 8-bit AVR microcontroller. Here, eight registers are required to compute a single lane (64-bit) within the *state*. Here, registers R8-R15 are used to perform operations on the single lane. Note that registers R16-R23 and R8-R15 can also store a single lane. Registers R2-R5 are required to handle the *state* address value of in $\theta \sim \rho$ process. R26:R27 keep the address value of state of f-function. R28:R29 keep the address value of $D[x, z]$ stored initial θ. In addition, R30:R31 keep the address value of the constant data used in ι process.

Algorithm 2 shows the pseudocode for computation of the initial four lanes of the *state* in $\theta \sim \rho$ process. Here, load_state is a macro that uses registers R24 and R25 as offsets to perform an XOR operation using the $D[x, z]$ made in initial θ, and the results are stored in registers R8-R15. load_state comprises assembly instructions as 2 ADD, 2 ADC, 16 LD, and 8 EOR. First, load_state loads the value of the lane to the registers R8-R15. Then, load_state loads 8-bit of $D[x, z]$ to R0 and execute the EOR instruction for registers R8 (EOR R8, R0). In load_state macro, the sequence of loading $D[x, z]$ and executing EOR instruction is repeated eight times, because one of lane is 64-bit: thus, load_state completes the computational part of θ process for a single lane. In other words, for θ process, only the memory update to the *state* remains. load_temp is a macro with the same logic as load_state that stores the computational value in registers R16-R23. rotate_store_s is a macro that executes ρ process for the eight registers that have completed θ process stored in registers R8-R15 and updates registers R8-R15 to *state* using the offset stored in register R24. rotate_store_t is a macro of logic such as rotate_store_s that uses R16-R23.

In line 1 of Algorithm 2, load_state stores the results of θ process for $S[0]$. This value is $S'[0]$. Here, θ process has been completed; however, to execute π process implicitly, $S[4]$ must be loaded to registers R16-R23, because the lane of $S[0]$ move to $S[4]$ in π process. Line 2 (Algorithm 2) sets the offset of $S[4]$, and line 3 sets offset $D[4]$. Then, line 4 makes $S'[4]$ into the same logic as line 1, and, at line 5, ρ process is executed to make $S'[0]$ to $\bar{S}[0]$. Line 5 updates $S[4]$

Algorithm 2. AVR Assembly Codes for proposed combined $\theta \sim \rho$ process for implicitly executing π process with initial four lane, $D[i]$ $(i \in [0,4])$: $D[i,z]$ of initial θ, $S[j]$: 64-bit data of one lane of *state*, $S'[j]$: 64-bit data with θ process calculated, $S'[j]$: 64-bit data with θ and ρ process calculated, $j \in [0,24]$.

$S[4] \leftarrow \bar{S}[0]$ computation	$S[17] \leftarrow \bar{S}[14]$ computation
1: `load_state`	10: `LDI` $R17, 136$ // $S[17]$
//R8-R15 : $S'[0] \leftarrow (S[0] \oplus D[0])$	11: `LDI` $R17, 16$ // $D[2]$
2: `LDI` $R24, 32$ // $S[4]$	12: `load_temp`
3: `LDI` $R25, 32$ // $D[4]$	//R16-R23 : $S'[17] \leftarrow (S[17] \oplus D[2])$
4: `load_temp`	13: `rotate_store_s` // $S[17] \leftarrow \bar{S}[14]$
//R16-R23 : $S'[4] \leftarrow (S[4] \oplus D[4])$	
5: `rotate_store_s` // $S[4] \leftarrow \bar{S}[0]$	
$S[14] \leftarrow \bar{S}[4]$ computation	$S[15] \leftarrow \bar{S}[17]$ computation
6: `LDI` $R24, 112$ // $S[14]$	14: `LDI` $R24, 120$ // $S[15]$
7: `LDI` $R25, 32$ // $D[4]$	15: `EOR` $R25, R25$ // $D[0]$
8: `load_state`	16: `load_state`
//R8-R15 : $S'[14] \leftarrow (S[14] \oplus D[4])$	//R8-R15 : $S'[15] \leftarrow (S[15] \oplus D[0])$
9: `rotate_store_t` // $S[14] \leftarrow \bar{S}[4]$	17: `rotate_store_t` // $S[15] \leftarrow \bar{S}[17]$

to $\bar{S}[0]$ using the offset stored in register R24. Here, where π process is executed implicitly. In the standard method, the implementation order is $\theta \rightarrow \pi \rightarrow \rho$. In proposed method, π process is executed through the step of store in memory; thus, the offset used in ρ process, which is executed before π process, is the offset where π process should be applied. Our implementation is actually implemented in $\theta \rightarrow \rho \rightarrow \pi$ order. Since the offset in table applied π process can be used rather than the offset in Table 2, no additional memory cost and generating cost is incurred. When computation of line 5 is completed, $S[4]$ is updated to $\bar{S}[0]$ with θ, π, ρ processes applied. The proposed implementation repeats this sequence. In lines 6–17, the proposed implementation repeats same logic of lines 2–5. If Algorithm 2 is within same logic, executing π process implicitly, and $\theta \sim \rho$ process for all lanes of *state* is completed.

The advantage of the proposed implementation on the 8-bit AVR MCUs is that there is no additional memory cost, which is significant for resource-limited environments. In other words, with the proposed implementation, memory accesses are reduced without having to generate a lookup table. In addition, no additional clock cycles occur in the proposed implementation. As with line 2 and 3, the cost of setting offsets is the same as the offset setup costs required by $\theta \sim \rho$ process in the standard implementation method [15,16].

5   Performance Analysis

We compared the performance of the proposed implementation method to that of existing implementation method on 8-bit AVR MCUs. In this evaluation, we used the following of SHA-3 (256-bit) parameters: $b = 1600$, $r = 1088$, and $c = 512$. Thus, in the sponge structure absorbing process, the message block was divided into 136 bytes. Therefore, if the length of the message was a multiple of 136 bytes, the maximum performance relative to clock cycles per byte can be achieved. The performance was measured relative to execution time (CPB) by hash rate when hashing a byte of various messages in the 8-bit AVR microcontroller. The software was implemented using **Atmel Studio 7**, and the code was compiled using the **-O2** option in **avr-gcc version 5.4.0**. All reported execution times were determined using the cycle-accurate instruction set simulator in **Atmel Studio 7**, where we used the ATmega128 microcontroller as the target device.

Table 6. Performance of proposed SHA-3 Implementations by hash rate when hashing a byte of various message in 8-bit AVR microcontroller, hash rate represent cyc/byte (CPB)[15,16]

Reference	Algorithm	Language	Length of message byte		
			50 byte	100 byte	500 byte
This Paper	SHA-3(256-bit)	Asm	**2667** (+25.1%)	**1333** (+25.7%)	**1073** (+25.0%)
Otte et al. [15]	SHA-3 (256-bit)	C	12854	6427	5142
Balasch et al. [16]	SHA-3 (256-bit)	Asm	3560 (-)	1795 (-)	1432 (-)
Balasch et al. [16]	SHA-256	Asm	672	668	532
Balasch et al. [16]	Blake (256-bit)	Asm	714	708	562
Balasch et al. [16]	Grøstl (256-bit)	Asm	1220	1012	686
Balasch et al. [16]	Photon (256-bit)	Asm	9723	7982	4788

Table 6 compares the proposed implementation to previous SHA-3 implementations, e.g., the implementations proposed by Otte et al. [15] and Balasch et al. [16]. In addition, we compared the proposed implementation to a previous SHA-2 implementation and three SHA-3 candidates. When hashing a 500 bytes message, we obtained an overall execution time of 1073 CPB for the Assembly implementation. The proposed implementation obtained a 79.1% performance improvement over the implementation method proposed by Otte et al. and a 25.0% performance improvement over the method proposed by Balasch et al. The existing implementation [16] incurs nearly three times the CPB cost for SHA-256. However, the proposed implementation requires the CPB by nearly two times to SHA-256 [16], compared to Balash et al.'s implementation [16].

6   Concluding Remarks

In this paper, we have proposed software optimization method for SHA-3 on 8-bit AVR MCUs. The proposed method optimizes SHA-3 process without requiring a lookup table and additional operations from a memory access perspective. With the proposed optimization method, our implementation on an 8-bit AVR microcontroller obtained 25.0% performance improvement over the previous best results on the same condition [16]. The proposed method is applicable to a variety of algorithms using SHA-3, e.g., HASH_DRBG, HMAC_DRBG, PBKDF2, and digital signature algorithms. Furthermore, our method is a generic optimization method that can be applied to various platforms, e.g., both high-end and low-end processor environments.

References

1. Park, S.-E., Hwang, C.-G., Park, D.-C.: Internet of Things (IoT) on system implementation with minimal Arduino based appliances standby power using a smartphone alarm in the environment. JKIECS **10**, 1175–1182 (2015)
2. Stevens, M., Bursztein, E., Karpman, P., Albertini, A., Markov, Y.: The first collision for full SHA-1. In: Katz, J., Shacham, H. (eds.) CRYPTO 2017. LNCS, vol. 10401, pp. 570–596. Springer, Cham (2017). https://doi.org/10.1007/978-3-319-63688-7_19
3. Wang, X., Yin, Y.L., Yu, H.: Finding collisions in the full SHA-1. In: Shoup, V. (ed.) CRYPTO 2005. LNCS, vol. 3621, pp. 17–36. Springer, Heidelberg (2005). https://doi.org/10.1007/11535218_2
4. Rijmen, V., Oswald, E.: Update on SHA-1. IACR Cryptology ePrint Archive 2005:10 (2005)
5. De Cannière, C., Rechberger, C.: Finding SHA-1 characteristics: general results and applications. In: Lai, X., Chen, K. (eds.) ASIACRYPT 2006. LNCS, vol. 4284, pp. 1–20. Springer, Heidelberg (2006). https://doi.org/10.1007/11935230_1
6. Manuel, S.: Classification and generation of disturbance vectors for collision attacks against SHA-1. Des. Codes Cryptogr. **59**(1–3), 247–263 (2011)
7. Khovratovich, D., Rechberger, C., Savelieva, A.: Bicliques for preimages: attacks on Skein-512 and the SHA-2 family. In: Canteaut, A. (ed.) FSE 2012. LNCS, vol. 7549, pp. 244–263. Springer, Heidelberg (2012). https://doi.org/10.1007/978-3-642-34047-5_15
8. Lamberger, M., Mendel, F.: Higher-order differential attack on reduced SHA-256. IACR Cryptology ePrint Archive 2011:37 (2011)
9. Mendel, F., Nad, T., Schläffer, M.: Improving local collisions: new attacks on reduced SHA-256. In: Johansson, T., Nguyen, P.Q. (eds.) EUROCRYPT 2013. LNCS, vol. 7881, pp. 262–278. Springer, Heidelberg (2013). https://doi.org/10.1007/978-3-642-38348-9_16
10. Dobraunig, C., Eichlseder, M., Mendel, F.: Analysis of SHA-512/224 and SHA-512/256. IACR Cryptology ePrint Archive 2016:374 (2016)
11. Sasaki, Y., Wang, L., Aoki, K.: Preimage attacks on 41-step SHA-256 and 46-step SHA-512. IACR Cryptology ePrint Archive 2009:479 (2009)
12. Dworkin Morris, J.: SHA-3 standard: permutation-based hash and extendable-output functions (2015). https://doi.org/10.6028/NIST.FIPS.202

13. Lee, H.-W., Hong, D., Kim, H., Seo, C., Park, K.: An implementation of an SHA-3 hash function validation program and hash algorithm on 16bit-UICC. J. Korea Inst. Inf. Secur. Cryptol. **41**, 885–891 (2014)

14. Kang, M., Lee, H., Hong, D., Seo, C.: Implementation of SHA-3 algorithm based on arm-11 processors. J. Korea Inst. Inf. Secur. Cryptol. **25**, 749–757 (2015)

15. Otte et al.: AVR-crypto-lib (2015). https://wiki.das-labor.org/w/-AVR-Crypto-Lib/en

16. Balasch, J., et al.: Compact implementation and performance evaluation of hash functions in ATtiny devices. In: Mangard, S. (ed.) CARDIS 2012. LNCS, vol. 7771, pp. 158–172. Springer, Heidelberg (2013). https://doi.org/10.1007/978-3-642-37288-9_11

17. Keccack Team. Extended Keccack code package (2018). https://keccak.team/index.html

18. KISA. SHA-3 source code manual (2020). https://seed.kisa.or.kr/kisa/kcmvp/EgovVerification.do

19. Keccack Team. The extended Keccak code package (open-source implementations of the cryptographic schemes defined by the Keccak team). https://github.com/XKCP/XKCP

20. Korea internet & security agency open cryptography algorithms. https://seed.kisa.or.kr/kisa/reference/EgovSource.do

21. Sanadhya, S.K., Sarkar, P.: New collision attacks against up to 24-step SHA-2. IACR Cryptology ePrint Archive 2008:270 (2008)

22. Biryukov, A., Lamberger, M., Mendel, F., Nikolić, I.: Second-order differential collisions for reduced SHA-256. In: Lee, D.H., Wang, X. (eds.) ASIACRYPT 2011. LNCS, vol. 7073, pp. 270–287. Springer, Heidelberg (2011). https://doi.org/10.1007/978-3-642-25385-0_15

23. Atmel. AVR instruction set manual (2012). http://ww1.microch-ip.com/downloads/en/devicedoc/atmel-0856-avr-instruction-set-manual.pdf

24. Kwon, H., Kim, H., Choi, S.J., Jang, K., Park, J., Kim, H., Seo, H.: Compact implementation of CHAM block cipher on low-end microcontrollers. In: You, I. (ed.) WISA 2020. LNCS, vol. 12583, pp. 127–141. Springer, Cham (2020). https://doi.org/10.1007/978-3-030-65299-9_10

25. Kim, Y.B., Seo, S.C.: An efficient implementation of AES on 8-Bit AVR-based sensor nodes. In: You, I. (ed.) WISA 2020. LNCS, vol. 12583, pp. 276–290. Springer, Cham (2020). https://doi.org/10.1007/978-3-030-65299-9_21

26. Liu, Z., Seo, H., Großschädl, J., Kim, H.: Efficient implementation of NIST-compliant elliptic curve cryptography for 8-bit AVR-based sensor nodes. IEEE Trans. Inf. Forensics Secur. **11**(7), 1385–1397 (2016)

27. Guo, X., Huang, S., Nazhandali, L., Schaumont, P.: Fair and comprehensive performance evaluation of 14 second round SHA-3 ASIC implementations, January 2010

28. ARM Coporation. ARM architecture reference manual Armv8 (2010). https://www.scss.tcd.ie/~waldroj/3d1/arm_arm.pdf

29. Cheng, H., Dinu, D., Großschädl, J.: Efficient implementation of the SHA-512 hash function for 8-Bit AVR microcontrollers. In: Lanet, J.-L., Toma, C. (eds.) SECITC 2018. LNCS, vol. 11359, pp. 273–287. Springer, Cham (2019). https://doi.org/10.1007/978-3-030-12942-2_21

30. KISA. KCMVP manual for cryptography (2020). https://seed.k-isa.or.kr/kisa/Board/79/detailView.do

Security Analysis

Can a Differential Attack Work for an Arbitrarily Large Number of Rounds?

Nicolas T. Courtois[1]([⊠])[iD] and Jean-Jacques Quisquater[2]

[1] University College London, Gower Street, London, UK
n.courtois@ucl.ac.uk
[2] Université Catholique de Louvain, Louvain-la-Neuve, Belgium
jjq@uclouvain.be

Abstract. Differential cryptanalysis is one of the oldest attacks on block ciphers. Can anything new be discovered on this topic? A related question is that of backdoors and hidden properties. There is substantial amount of research on how Boolean functions affect the security of ciphers, and comparatively, little research, on how block cipher wiring can be very special or abnormal. In this article we show a strong type of anomaly: where the complexity of a differential attack does not grow exponentially as the number of rounds increases. It will grow initially, and later will be lower bounded by a constant. At the end of the day the vulnerability is an ordinary single differential attack on the full state. It occurs due to the existence of a hidden polynomial invariant. We conjecture that this type of anomaly is not easily detectable if the attacker has limited resources.

Keywords: Feistel ciphers · Boolean functions · Multivariate polynomials · T-310 · Generalized linear cryptanalysis · Polynomial invariants · Hidden polynomial problems · Annihilators · Markov ciphers · k-normality · Algebraic cryptanalysis

1 Introduction

Differential Cryptanalysis (DC) is a well-known basic attack on block ciphers [4,32] and it may seem that what remains to study are just some fine details, cf. [28]. In order to improve DC, researchers have considered various ways to aggregate a larger number of differences [31], for example with truncated differentials of Knudsen [40]. We can hardly just combine two truncated differentials and expect that the propagation probabilities would just be multiplied [9,29,30]. There is a hidden complexity and a lot of non-uniformity: probabilities of individual differentials may differ very substantially. However we do not expect anything special to happen with just old ordinary DC with single differentials. In fact we do: it is the well-known question of Markov ciphers [42]. In this paper we study cases where this property is violated, and DC does not work as expected because relevant events are not independent. What is interesting is showing that this can advantage the attacker in a substantial way.

© Springer Nature Switzerland AG 2021
D. Hong (Ed.): ICISC 2020, LNCS 12593, pp. 157–181, 2021.
https://doi.org/10.1007/978-3-030-68890-5_9

This paper is also about backdooring and hidden properties. Here we will have a hidden polynomial equation in a similar way as in certain public key cryptosystems. It is not apparent for the attacker, even if the attacker knows another, related set of polynomials, which they use for encryption. Our hidden property is going to be a non-linear invariant property, a topic which has attracted considerable attention in recent years [12–19,43,51]. A wider fundamental open problem is the very existence of new attacks on block ciphers, of any sort, such that their complexity would not grow exponentially with the number of rounds. Even though such attacks exist, they seem extremely complex. Surely this would not be possible with good old differential cryptanalysis? In this paper we show that this is actually possible. A similar result for a truncated differential attack was presented at Crypto 2011, cf. Section 3 of [43]. This earlier result worked only for some weak keys. Our attack works with a single differential, which is harder, as probabilities are lower. It is uniform and works for all 2^{240} keys without any exception. It also works in spite of the presence of round constants in T-310.

In this article these considerations come together. We show how to design an anomalous differential attack. The only thing that the attacker observes, will be that a certain differential propagates with a probability which will be bounded by a constant for any number of rounds. This is quite surprising and hides the existence of a hidden polynomial invariant property, the existence of which the attacker could potentially ignore forever; even if they know about the (derived) differential property. Sometimes, differential cryptanalysis does not work as predicted by a "naive" theory and the events in different rounds are not independent. However, this is not just an annoying discrepancy; a bug which was typically ignored by researchers until now. We discover that an anomaly of this sort conceals another strong property extremely useful for the attacker.

This article is organised as follows. In Sect. 2 we explain the philosophy of what we do. In Sect. 3 we study the T-310 cipher. In Sect. 4 we present some older examples of invariant attacks on T-310. In Sect. 5 we describe our attack with one main theorem and 3 technical lemmas. In Sect. 6 we show what happens in practice. In Sect. 7 we discuss several future cryptanalysis research ideas and we wonder if some sort of converse result could be true. Then comes the Conclusion. In Appendix A we look at vulnerability of Boolean functions against our attacks. In Appendix B we consider how invariant properties we study can be used for key recovery.

2   Background: Markov Ciphers and Nonlinear Invariants

The notion of Markov ciphers was introduced at Eurocrypt 1991 by Lai, Massey, and Murphy, see [42]. Probably these questions were already studied earlier, in the Eastern Bloc, cf. [23,24,41]. In short, we have a Markov cipher when the probability that a certain output difference is obtained, does not depend on the input value (but depends on the input difference), when the round key is chosen random. This formulation ignores the question of how the probability depends also on the key, and therefore, our current understanding is yet greatly simplified (we refer to [28,41] to see why this matters). In short, in [42] it is simply assumed that the keys are chosen uniformly at random, similar to averaging probabilities over all possible keys. Many known ciphers are Markov ciphers, for example DES, FEAL, LOKI and IDEA, [42]. Other ciphers such as GOST behave as Markov ciphers with some degree of approximation [9,28].

The importance of Markov ciphers is explained in page 24 of [42]: in a Markov cipher "every differential will be roughly equally likely" after sufficiently many rounds, cf. also [47]. The main goal of the present article is to show that there exists a block cipher **violating** this exact long-term derived property of Markov ciphers in an extremely strong way. Here all differentials will vanish progressively, with probability being zero in practical terms, except with very few special differentials. These differences are able to survive for an arbitrarily large number of rounds. If so, not being a Markov cipher degrades the security of our cipher in a very substantial way. Compared to earlier results in [43], our attack cf. Theorem 5.1.1 works for any key, 100% of keys. Moreover, it works with round constants in T-310. Eliminating the round constants and the key bits alike are hard problems in non-linear cryptanalysis. Many known attacks only work for some keys, not all, see [43,51], or only for some round constants, see Section 7.4 in [14].

What we study in this paper is very much like a backdoor, a hidden unexpected property leading to a strong attack. We emphasise the fact that events of this kind can be easily overlooked. There is an exponential number of differences to study and specific events are detectable only if we have sufficient computing power and a sufficient number of Plaintext/Ciphertext (P/C) pairs. They could also be detected if a specific difference with abnormal propagation is already known, or we are able to characterize some specific input states on n bits where the propagation behaves in an unusual way. Researchers who study this on the experimental side might also discard this result as an outlier. We found it very hard to believe that this is real. Therefore, it is important that in the present article we establish our result through rigourous mathematical proof, see Thoerem 5.1.1 page 10. It is also confirmed by computer simulations in Sect. 6.

2.1 Weak Keys and Weak Components - Long Term Key

There is a substantial amount of research on how non-linear components (Boolean functions and S-boxes) affect the security of ciphers and comparatively little research, on how the block cipher wiring can be special or weak, for example with DES P-box, see [5] or the long-term key LZS in T-310 [21]. In cryptanalysis, we always look for special or even abnormal cases, for example, the block cipher KeeLoq can be broken in an extremely short time of type only 2^{23} for 15 % of keys, cf. [2]. A fair assessment of weakness requires the assumption that weak keys occur at random, with their "natural" probability; see the "multiple random key scenario" in Section 29 in [10]. Here we study the probability for a Boolean function that a certain product of polynomials is zero, see

Appendix A. An essential observation is that in the ring of Boolean polynomials, factorization is not unique and there are typically numerous solutions to such problems, see [15], and one may eventually lead to an attack [17,19].

2.2 Nonlinear Cryptanalysis and Higher Order Nonlinear Cryptanalysis

In recent years many authors show how to construct attacks where a certain non-linear polynomial is invariant [12–19,43,51]. Following ICISC 2019, a good way to study these attacks is a white box method [19]. We formulate our attacks using Boolean polynomial arithmetic. As such, the whole attack could potentially apply to another cipher modulo renaming of variables and we do not use the full specification of the cipher, see [19] and [14]. If a cipher satisfies a certain number of initial conditions on some basic polynomials, then our attack works for an arbitrarily large number of rounds. If a property involves just one encryption, we say it is a property of order 1. The invariants in [16,17,19] and a majority of other recent works on non-linear invariants are of order one (for one single encryption). In this paper a property of order 1 will be used to alter the behaviour of a differential attack. Overall we get an invariant property of order 2.

An important family of invariant attacks are product attacks: the invariant is a product of polynomials. Constructing a non-linear invariant attack is a difficult combinatorial problem. At Eurocrypt'96 Knudsen and Robshaw claimed that this cannot work for Feistel ciphers [44]. Initially, attempts to find a non-linear invariant attack on DES have failed, or produced a tiny improvement compared to Matsui's Linear Cryptanalysis (LC), cf. Crypto 2004 in [20]. Certain block ciphers such as T-310 use only very few key bits in each round, cf. [23], and are particularly vulnerable to this type of attack. Consequently, we have a plethora of attacks of this type [12,14,51] with increasing degrees [13,16,19], which is expected to make the attack increasingly powerful. More general attacks can work with sums of two or more products. For T-310 this is shown in [18], with an example of type $AC + BD$ which we reproduce below in Sect. 4.3. An example with DES is found in Remark 2 in Section 10 of [19].

2.3 On Success Probability and Annihilation Degree in Previous Attacks

In ICISC 2019 the best attack on T-310 was such that if our Boolean function is such that $(Z+e)(a+b)(c+d) = 0$ then a certain product of 8 linear polynomials is an invariant working for any number of rounds, any key, and any choice of round constants. A Boolean function with this type of annihilation with 2 factors is called 4-weakly-normal, where $4 = 6 - 2$, cf. Appendix A and [7]. This notion was earlier studied by Dobbertin [34]. It is easy to see that a Boolean function Z chosen at random will be 4-weakly-normal with very high probability of $2^{-0.68}$, cf. Table 4. A yet stronger or more realistic attack which would only require that Z is 3-normal with $Z(a + d)(b + e)(c + f) = 0$ was described in [16], and a similar attack will be studied inside this paper as a technical Lemma 5.3.1 page

13. The degree of freedom for the attacker increases at last. 100% of all Boolean functions on 6 bits are 3-normal, see Section 5 in [19] and [35]. Moreover, several methods to annihilate with a product of 3 factors exist typically. A recent paper shows that a similar attack with 3 factors exists also with the original Boolean function used during the Cold War to protect government communications [17].

3   Short Description of T-310

We recall the definition of T-310 block cipher from [50]. T-310 operates on 36 bit blocks and a secret key on 240 bits. Each round involves two key bits K, L and one round constant bit F, which is derived from a fixed IV of 61 bits which is transmitted in clear text. The secret key of 240 bits is stored on a paper punch card and is reused after every 120 rounds. The actual encryption is done in a peculiar stream cipher mode which we will ignore here. We refer to [49,50] and [21] for more details. In this paper we only study the underlying block cipher (a keyed permutation on 36 bits).

The wiring or the long term key in T-310, is the equivalent of the P-box in DES, and it is known under the name of LZS or *Langzeitschlüssel*, which means a long-term key. It is changed once per year typically. Formally the LZS wiring is defined by two functions: $D : \{1 \ldots 9\} \to \{0 \ldots 36\}$, $P : \{1 \ldots 27\} \to \{1 \ldots 36\}$ which are typically injective. We need to specify which input state bits are connected to contacts named D1-D9 and v1-v27 in Fig. 2. For example $D(5) = 36$ is about what happens inside the small square box with letter D in Fig. 1. $D(5) = 36$ means that input bit x_{36} is connected to the wire called $D5$

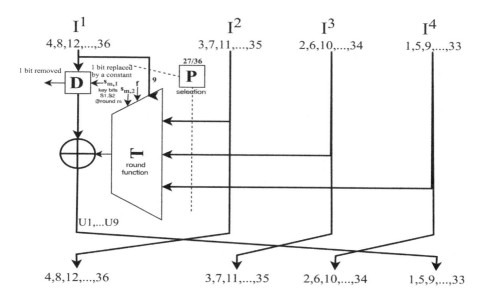

Fig. 1. High-level overview of one round of T-310.

in Fig. 2 which then becomes $U5 = y_{17}$ after XOR with bit $g4$. Then $P(1) = 25$ refers the content of the square box with letter P in Fig. 1. It means that input x_{25} is connected to v1 or the 2nd input of Z_1 in Fig. 2.

In each round only 2 key bits K, L are used. The secret key is defined as $s_{1...120,1...2} \in \{0,1\}^{240}$ which is 240 bits. The same 2 bits are repeated after 120 rounds with

$$K = s_{m,1} \text{ and } L = s_{m,2}$$

In addition each round has a round constant called F, which is derived from the public IV value. In all, for any $F, K, L \in \{0,1\}^3$ one round of this block cipher is a permutation on 36 bits. This requirement is not obvious and it requires some complex technical conditions on the cipher wiring, see [22].

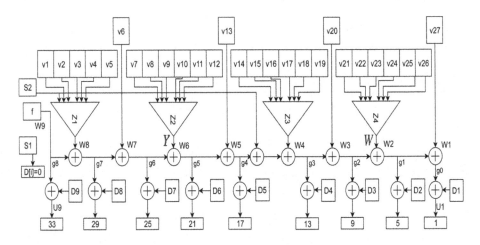

Fig. 2. The internal structure of one round of T-310 block cipher.

In Fig. 3 we give a set of closed formulas to compute the output bits y_{1-36} in each round from the input bits x_{1-36}. These formulas are self contained, i.e. everything can be derived just from these formulas. In one round 9 new bits are created and $36 - 9 = 27$ bits are shifted by one position. The cipher uses 4 identical Boolean functions of 6 bits which are denoted by Z_1, Z_2, Z_3, Z_4 on Fig. 2. A common convention is to rename these 4 Boolean functions and use 1-letter notations $Z(), Y(), X(), W()$ respectively (backwards naming convention).

$$y_{i+1} = x_i \text{ for any } i \neq 4k \qquad (\text{ with } 1 \leq i \leq 36) \qquad (\text{r0})$$

$$y_{33} = F + x_{D(9)} \qquad (\text{r1})$$

$$Z_1 \overset{def}{=} Z(L, x_{P(1)}, \ldots, x_{P(5)}) \qquad (\text{z1})$$

$$y_{29} = F + Z_1 + x_{D(8)} \qquad (\text{r2})$$

$$y_{25} = F + Z_1 + x_{P(6)} + x_{D(7)} \qquad (\text{r3})$$

$$Z_2 \overset{def}{=} Y(x_{P(7)}, \ldots, x_{P(12)}) \qquad (\text{z2})$$

$$y_{21} = F + Z_1 + x_{P(6)} + Z_2 + \qquad x_{D(6)} \qquad (\text{r4})$$

$$y_{17} = F + Z_1 + x_{P(6)} + Z_2 + \qquad x_{P(13)} + x_{D(5)} \qquad (\text{r5})$$

$$Z_3 \overset{def}{=} X(x_{P(14)}, \ldots, x_{P(19)}) \qquad (\text{z3})$$

$$y_{13} = F + Z_1 + x_{P(6)} + Z_2 + \qquad x_{P(13)} + L + Z_3 + x_{D(4)} \qquad (\text{r6})$$

$$y_9 = F + Z_1 + x_{P(6)} + Z_2 + \qquad x_{P(13)} + L + Z_3 + x_{P(20)} + x_{D(3)} \qquad (\text{r7})$$

$$Z_4 \overset{def}{=} W(x_{P(21)}, \ldots, x_{P(26)}) \qquad (\text{z4})$$

$$y_5 = F + Z_1 + x_{P(6)} + Z_2 + \qquad x_{P(13)} + L + Z_3 + x_{P(20)} + Z_4 + x_{D(2)} \qquad (\text{r8})$$

$$y_1 = F + Z_1 + x_{P(6)} + Z_2 + \qquad x_{P(13)} + L + Z_3 + x_{P(20)} + Z_4 + x_{P(27)} + x_{D(1)} \qquad (\text{r9})$$

$$x_0 \overset{def}{=} K \qquad (\text{s1})$$

$$F \in \{0,1\} \text{ is a round constant} \quad \text{depending on a (public) IV} \qquad (f1)$$

$$K = s_{m \bmod 120, 1} \qquad (\text{in encryption round } m = 0,1,2,\ldots) \qquad (k1)$$

$$L = s_{m \bmod 120, 2} \qquad (\text{in encryption round } m = 0,1,2,\ldots) \qquad (k2)$$

Fig. 3. The specification of one round of T-310.

Notation. When we work on invariant attack, we use more compact notations. and the 36 bits $x_1, \ldots, x_{36}$ are replaced by single letters, cf. Fig. 4.

Numbers	1	2	3	4	5	6	7	8	9	10	11	12	13	14	15	16	17	18	19	20	21	22	23	24	25	26	27	28	29	30	31	32	33	34	35	36
Letters	V	U	T	S	R	Q	P	O	N	M	z	y	x	w	v	u	t	s	r	q	p	o	n	m	l	k	j	i	h	g	f	e	d	c	b	a

Fig. 4. Variable naming conventions.

We work on invariants, and variables y_1 and x_1 will be treated likewise and denoted by the same letter (!). Letters were chosen to avoid certain letters like F or W used for a different purpose. Traditionally, if we want to avoid ambiguity, we will distinguish between the variable a at input denoted by a^i and the same variable at output denoted by a^o. Moreover later inside this paper we study two distinct encryptions, in which case we can distinguish the two instances of a by a^1 or a^2 added in the exponent.

4   Some Early Attacks on T-310 and Related Questions

The T-310 block cipher is a good target for cryptanalysis with non-linear invariants. The key reason for this is that extremely few key bits and other round constants are used in each round. This is a crucial property, which distinguishes block ciphers made in the West, typically stronger, and weaker block ciphers made in the Eastern Bloc, a question which was discussed in [23,24]. For this reason, DES is substantially more secure than T-310, even though apart from this property, both ciphers are extremely similar, and can be attacked in the same way. The difference is mainly quantitative: many more key bits are involved in each round of DES. Consequently, attacks on DES typically only work for a small fraction of the key space. This was shown very clearly in ICISC 2019 [19] where two ciphers are studied side-by-side, and earlier in [16].

In Section 7 of [27] the authors propose to look for a non-linear invariant property for T-310, yet at the time no such property was known. For many decades researchers knew about this type of attack [20,38,39], and yet failed to find convincing examples, except for contrived ciphers [25]. More recently, only with T-310 we get powerful invariants working for any number of rounds, any key, and any choice of round constants.

4.1   Linear and Non-Linear Invariants and Phase Transitions

A good way to study such attacks, is the so called "white box" algebraic approach [14,19]. We operate in the cipher specification space and we characterise exactly in which cases the attack works by formal polynomial algebra. The goal of the attacker is to find an invariant and eliminate all the internal state bits, this including the key bits and round constants. As a toy example, we consider the cipher wiring known as 847 in [12].

847: P=32,22,26,14,21,36,30,17,15,29,27,13,4,23,1,8,35,20,
5,16,24,9,10,6,7,28,12 D=24,12,8,16,36,4,20,28,32

We consider two cycles shown in Fig. 5, which show the group action of one encryption round, cf. Fig. 3, instantiated with wiring 847 above, on some very basic polynomials:

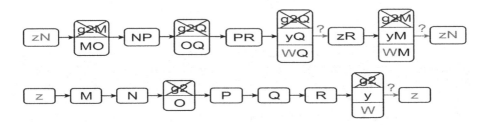

Fig. 5. Transitions between polynomials in an older attack from [12].

Let polynomial $\mathscr{P}$ be the addition of all polynomials of degree 1 and 2 in Fig. 5, excluding those with W, which represents the Boolean function. Here $g2$ depends on the cipher state and the key in a complex way, see Fig. 2, and yet all terms with $g2$ appear an even number of times modulo 2 and are cancelled. Then, it is easy to see that this $\mathscr{P}$ will be a degree 2 polynomial invariant for our cipher, IF the Boolean function W satisfies the following equation:

$$W(1 + M + Q) = 0$$

This is known in general as the Fundamental Equation (FE). Terms W, WM and WQ are eliminated when we add them, not individually. This was not a very good attack. Extremely few Boolean functions satisfy this equation, cf. Table 3 in Appendix A, and our Boolean function cannot be balanced, cf. Theorem 6.4 in [19].

4.2   Phase Transitions or How Impossible Becomes Possible

Here the crucial question is the one of phase transition, cf. Section 2.4 in [16]. This is how ciphers with stronger components can eventually be attacked. The idea is that a spectacular improvement can occur as the degree of the invariant polynomial grows. The paper [14] contains a large body of examples with growing degree and effectively demonstrates this. This leads to the methodology of attack "hopping" or/and attack "lifting". Sometimes the cipher can be modified and fundamental equation does not change. In [18] the attacker modifies a cycle and adds additional polynomials to it. Finally, we also can add one more cycle to our attack, while avoiding our invariant polynomial becoming zero, cf. [19]. We can then hope to obtain a Fundamental Equation which has more roots, or to find an attack which will work for a larger set of Boolean functions, or even find an attack in a real-life setting, cf. Section 3 in [15] and [17].

4.3   Invariant Hopping and Attack Lifting - Example

A short self-contained introduction which shows this process at work can be found in [18]. For example in Section 7.1. and Thm. 7.3. in [18], we find that for a certain cipher wiring known as 551, if we have

$$(Z(a, b, c, d, e, f) + f)(d + e) = 0$$

then the polynomial

$$\mathscr{P} = (e + m) \cdot (g + o) + (f + n) \cdot (h + p)$$

is an invariant for our cipher where $e = x_{32}$ etc, which is different than input e of Z above, following the cipher state variable naming convention of Fig. 4.

In contrast a better product invariant attack of degree 4 can be constructed with

$$\mathscr{P} = (e + m)(f + n)(g + o)(h + p)$$

which invariant works for a substantially larger proportion of Boolean functions. In this case it was shown in [18] that we only need something like:

$$(Z + f)(d + e)(a + b)(c + f) = 0$$

and this happens for any Boolean function with large probability of 2^{-8}, cf. Appendix A or Appendix C in [16]. In general as the degree grows, it becomes easier to find a Boolean function where our polynomial invariant actually works. At this stage, if for a particular function our cipher is still not broken, this is rather accidental than deliberate. Eventually it can also be made to work with a real-life Boolean function, see [17].

Note. All the attacks above were invariant attacks of order 1, dealing with just one encryption. In this paper we will construct an invariant attack of order 2.

5   Constructing An Anomalous Differential Invariant Attack

We define the following 8 basic polynomials:

$$\begin{cases} A \stackrel{def}{=} (m + i) & \text{which is bits } 24, 28 \text{ cf. Fig. 4.} \\ B \stackrel{def}{=} (n + j) & \text{which is bits } 23, 27 \\ C \stackrel{def}{=} (o + k) & \text{which is bits } 22, 26 \\ D \stackrel{def}{=} (p + l) & \text{which is bits } 21, 25 \\ E \stackrel{def}{=} (O + y) & \text{which is bits } 8, 12 \\ F \stackrel{def}{=} (P + z) & \text{which is bits } 7, 11 \\ G \stackrel{def}{=} (Q + M) & \text{which is bits } 6, 10 \\ H \stackrel{def}{=} (R + N) & \text{which is bits } 5, 9. \end{cases}$$

These polynomials allow to greatly simplify our attack. We start by observing that we have the following incomplete cycle, or pseudo-cycle, also shown in later Fig. 7:

$$H \to G \to F \to E \to ? \; D \to C \to B \to A \to ? \; H$$

Here six transitions are completely trivial for example $H \to G$ and due to the internal wiring: these bits are just shifted inside this cipher. Two other transitions, namely $E \to ? \; D$ and $A \to ? \; H$ are in contrast just impossible. They would be true only if certain complex Boolean functions namely $W()$ and $Y()$ were equal to zero for every input, which is not the case and will not be the case. However certain multiples of these polynomials will be annihilated (i.e. 0 for every input, and formally 0 as a polynomial). For example, the attacker discovers that under certain conditions a certain polynomial such as $\mathscr{P} = ABCD$ or $AC + BD$ will be invariant, and its value will not change cf. [14–19].

5.1 Our Main Theorem - An Order Two Invariant Property

Theorem 5.1.1. (An Anomalous Differential Attack). Given the eight basic polynomials $A - H$ defined as above and reproduced also in Fig. 7, AND for each cipher wiring for T-310 s.t.

$$\begin{cases} \{D(2), D(3)\} = \{6 \cdot 4, 7 \cdot 4\} \\ \{D(6), D(7)\} = \{2 \cdot 4, 3 \cdot 4\} \end{cases}$$

AND and if these four multiples of four being 8,12,24,28 are absent from the set of 27 inputs in $\{P(1) \ldots P(27)\}$, where $P : \{1 \ldots 27\} \to \{1 \ldots 36\}$ is an injective wiring, AND for any[1] Boolean function[2] which is such, that we have:

$$Z(a + d)(b + e)(c + f) = 0$$

AND if the 6 inputs of $W()$ defined by integers $P(21), \ldots, P(26)$, are mapped to any 3 out of 6 polynomials B, C, D, F, G, H, in a way which preserves[3] the partitioning in three sets or pairs in $(a+d)(b+e)(c+f)$, for example the inputs of W can be $5, 22, 7, 9, 26, 11$ AND the 6 inputs of $Y()$ defined by integers $P(7), \ldots, P(12)$ are the mapped to remaining 3 out of 6 polynomials B, C, D, F, G, H, while also preserving a partitioning in 3 sets of pairs in $(a+d)(b+e)(c+f)$, for example in order $25, 10, 27, 21, 6, 23$,

THEN for any short term key of 240 bits, and for any initial state on 36 bits, and for any IV, the input difference $[7, 11]$ corresponding to F, i.e. we flip both bits 7 and 11, will be preserved at the output after any number of rounds being a multiple of 8 with probability of at least 2^{-8}.

Remark. This theorem can be transposed by considering arbitrary permutations of 6 inputs a, b, c, d, e, f. These do not need to be applied consistently at both $W()$ and $Y()$, for example inputs 5 and 9 could be exchanged. However, we need to get the same partitioning of 6 inputs into 3 sets of 2 which needs to be consistent in W, in Y and with the partitioning which actually annihilates our Boolean function. We can also consider an arbitrary choice of 3 out of 6 polynomials in $BCDFGH$ to split between W and Y. In our example D, G, B are 3×2 inputs of $W()$ and the remaining 3 go to $Y()$, but it could be any choice of 3 out of 6. For the sake of simplicity and to make our theorem and its proof shorter and easier to follow, we work with a fixed mapping of these 12 variables (Fig. 6).

---

[1] This happens with probability at least 2^{-8} for any Boolean function, see Appendix A.
[2] This function is used twice as W and as Y for 2 disjoints sets of 6 inputs.
[3] For example if one input A is b the other must be e.

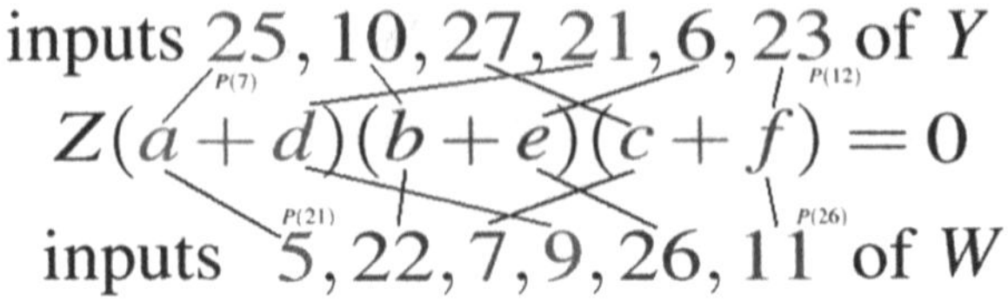

$$\text{inputs } 25, 10, 27, 21, 6, 23 \text{ of } Y$$
$$Z(a + d)(b + e)(c + f) = 0$$
$$\text{inputs } 5, 22, 7, 9, 26, 11 \text{ of } W$$

Fig. 6. For both W and Y we divide inputs in 3 sets of 2 variables in a consistent way.

5.2 A Concrete Example

This is for example achieved for the following full cipher wiring:

268: P=1,20,33,34,15,13,25,10,27,21,6,23,16,14,2,4,3,19,
35,29,5,22,7,9,26,11,17 D=16,28,24,20,32,8,12,4,36

and the following Boolean function $Z(a, b, c, d, e, f) = 1 + a + b + bc + d+$

$abd + cd + acd + bcd + e + ae + abe + ce + ace + de + ade + abde + af + bf + abf + acf + df +$

$bdf + abdf + cdf + bcdf + ef + abef + bcef + adef + abdef + acdef + abcdef$

In Table 1 page 16 we show what happens as the number of round grows. This choice of Boolean function is in no way special: any Boolean function chosen at random will work with high probability of at least 2^{-8}, see Appendix A, or Appendix C in [16].

5.3 Proof of Thoerem 5.1.1

We will show that a certain polynomial expression is invariant for any number of rounds. This for each of two encryptions we consider. The difference we study, $[7, 11]$, is the same as flipping both bits active in our polynomial F. If we have $A^1 = c_A, B^1 = c_B, \ldots H^1 = c_H$ for the first encryption, for some constants $c_A, \ldots c_H \in \{0, 1\}^8$, then we also have $A^2 = c_A, B^2 = c_B, \ldots H^2 = c_H$ for the second encryption. Since our hidden polynomial is built from $A, B, C, D, \ldots$ flipping bits $[7, 11]$ will also preserve this invariant, see Lemma 5.3.1 below. Two invariants will remain linked together for any number of rounds.

The fact that we have two invariants propagating for any number of rounds, which remains yet to be shown, makes that the difference (a bitwise XOR) between both encryptions is mapped to zero, through the linear application $\psi : \{0, 1\}^{36} \rightarrow \{0, 1\}^8$. Here ψ is defined precisely by the set of 8 linear polynomials $A \ldots H$ we defined earlier. This polynomial invariant attack is yet a weak constraint in itself. The fact that $\mathscr{P} = ABCDEFGH$ is an invariant in both encryptions makes that the output difference Δ after any number of rounds can take only 2^{28} possible values with $\psi(\Delta) = 0^8$, on 8 bits. It remains therefore quite surprising that one of these values, namely exactly $F = [7, 11]$ on 28+8

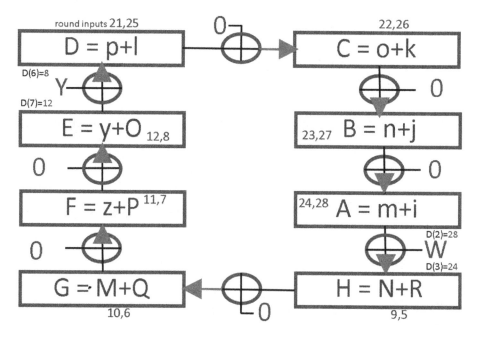

Fig. 7. A cycle on 8 basic polynomials used in our attack with LZS 268 which conceals the existence of a hidden polynomial invariant with $\mathscr{P} = ABCDEFGH$.

bits, is reproduced after any multiple of 8 rounds. This is 28 bits more than expected. Additional things must happen here for our theorem to be true. There is limited diffusion for a few rounds, and these is a finite number of possible output differences Δ which can at all be obtained from the initial difference $F = [7, 11]$. Since the image of the difference $\psi(\Delta)$, is fixed and strongly constrained, we expect that Δ takes fewer values than expected. In fact will show that Δ is fixed, only one value is possible. A rigourous proof with some technical lemmas is given below.

We will first prove that $\mathscr{P} = ABCDEFGH$ an invariant in both/any encryption.

Lemma 5.3.1. The polynomial $\mathscr{P} = ABCDEFGH$ is a non-zero polynomial and under conditions of Thoerem 5.1.1 it is invariant after 1 round of encryption $\forall F, K, L \in \{0, 1\}^3$.

Proof: We distinguish input and output-side polynomials by an index in the exponent such as A^o vs. A^i. We try to eliminate all output-side variables and express everything in input-side polynomials only. Later when there is no ambiguity we will just write A again instead of A^i.

By following the (shortest) path from output 9 to 5 in Fig. 8, or by XORing together the equations (r7) and (r8) in Fig. 3 we get:

$$H^o = y_9 + y_5 = x_{D(3)} + W(.) + x_{D(2)} = W(.) + x_{6\cdot4} + x_{7\cdot4} = W(.) + A^i$$

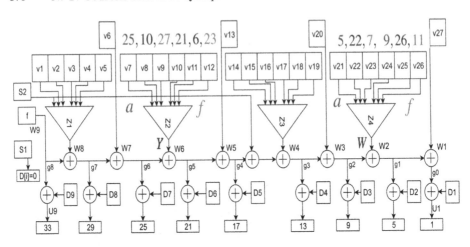

Fig. 8. Internal structure of one round of T-310 block cipher with focus on W and Y in our attack.

then following the path from output 25 to 21 in Fig. 8, or by XORing together the equations (r3) and (r4) in Fig. 3 we get:

$$D^o = y_{25} + y_{21} = x_{D(7)} + Y(.) + x_{D(6)} = Y(.) + x_{2.4} + x_{3.4} = Y(.) + E^i$$

At the input side $\mathscr{P}$ is equal to $\mathscr{P}^i = ABCDEFGH$ and at the output of our cipher

$$\mathscr{P}^o = A^o B^o C^o D^o E^o F^o G^o H^o = B^i C^i D^i (Y(.) + E^i) F^i G^i H^i (W(.) + A^i) =$$

at this moment only input variables are left and we can drop the exponents ^{i} and we have:

$$\mathscr{P}^o = BCD(Y(.) + E)FGH(W(.) + A) =$$

Now we observe that the inputs of $W()$ are $5, 22, 7, 9, 26, 11$, and our assumption $Z(a + d)(b + e)(c + f) = 0$ translated to $W(H)(C)(F) = 0$. Since HCF is a factor of $BCDFGH$ here, we can simply erase $W()$ as it is annihilated, and we get:

$$\mathscr{P}^o = BCD(Y(.) + E)FGH(A) =$$

Likewise, we recall our input mappings on Fig. 9 below, inputs of $Y()$ are $25, 10, 27, 21, 6, 23$, and therefore $Z(a + d)(b + e)(c + f) = 0$ translates to $Y(D)(G)(B) = 0$. Therefore we can also erase $Y()$ and we get:

$$\mathscr{P}^o = ABCDEFGH$$

which is the same as $\mathscr{P}^i$ and hence $\mathscr{P}$ is an invariant after 1 round of encryption. which ends the proof the our invariant work for any input and any F, K, L and any number of rounds. We have a formal equality of two polynomials.   □

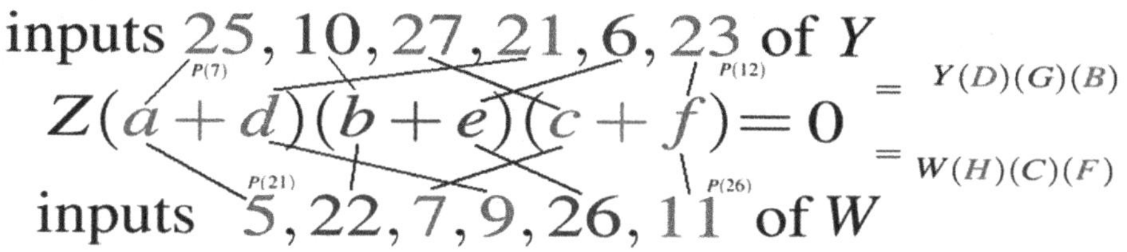

$$\text{inputs } 25, 10, 27, 21, 6, 23 \text{ of } Y$$

$$Z(a+d)(b+e)(c+f) = 0$$

$$\text{inputs } 5, 22, 7, 9, 26, 11 \text{ of } W$$

$$= Y(D)(G)(B)$$
$$= W(H)(C)(F)$$

(with labels $P(7)$, $P(12)$, $P(21)$, $P(26)$)

Fig. 9. We map inputs of W and Y to 3 sets with 2 variables in a way consistent with our annihilation property. In some sense we get two annihilations for the price of one (amplification).

This was just a proof of our lemma. We need yet to show that $F = [7, 11]$ propagates in a certain way which implies our Thoerem 5.1.1. In order to lower bound the propagation probability in general, we need to show that the propagation is special in some cases, so that the invariant $F = [7, 11]$ will be reproduced after 8 rounds, and we can ignore all other cases. It is easy to see that we have for any input I on 36 bits:

$$\mathscr{P}(I) = \prod_{i=1}^{8} (\psi(I))_i$$

which simply means that $\mathscr{P}$ is the same as applying a single 8-ary multiplication $\prod$ to the 8 outputs of ψ. More precisely we are going to show that:

Lemma 5.3.2. If for 2 different encryptions with $I_1 \oplus I_2 = [8, 12]$ of E we have

$$\mathscr{P}(I_1) = 1$$

then we have $O_1 \oplus O_2 = [21, 25]$ a.k.a. D after one round of encryption.

Proof: If $I_1 \oplus I_2 = [8, 12]$ and $\mathscr{P}(I_1) = 1$ then we also have $\mathscr{P}(I_2) = 1$ due to the fact that flipping both bits of $F = [8, 12]$ preserves all the values of $\psi()$ including the E coordinate, which is also unchanged due to double negation. We can then apply Lemma 5.3.1 and we obtain that $\mathscr{P}(O_1) = 1$ and $\mathscr{P}(O_2) = 1$ after one round for each respective encryption. If $\Delta = O_1 \oplus O_2$ we already know that $\psi(\Delta) = 0$. However Δ has 36 bits, not only 8.

We now observe that flipping 8,12 changes nothing else from the equations (r3) and (r4) in Fig. 3 we have

$$y_{25} = F + Z() + x_{P(6)} + x_{D(7)}$$

$$y_{21} = F + Z() + x_{P(6)} + Y() + x_{D(6)}$$

and that outputs of (r5) and all further equations in Fig. 3 are unchanged because, actually all the g_i in Fig. 8 are the same in both encryptions and the inputs $8, 12$ are used only once with $D(6)$ and $D(7)$, due to the fact that in Theorem 5.1.1 we assume that 8,12 are absent from the set of 27 outputs $\{P(1) \ldots P(27)\}$. Thus the only effect of flipping bits $E = 8, 12$ and is to flip bits $D = 21, 25$ in the next round. Similarly we have:

Lemma 5.3.3. If for 2 different encryptions with $I_1 \oplus I_2 = [24, 28]$ from A we have

$$\mathscr{P}(I_1) = 1$$

then we have $O_1 \oplus O_2 = [5, 9]$ a.k.a. H after one round of encryption.

Proof. If $I_1 \oplus I_2 = [24, 28]$ and $\mathscr{P}(I_1) = 1$ then we also have $\mathscr{P}(I_2) = 1$ due to the fact that flipping both bits of $F = [24, 28]$ preserves all the values of $\psi()$ including the A coordinate. We now observe that flipping 24,28 changes nothing else from the equations (r7) and (r8) in Fig. 3 we have

$$y_9 = F + Z() + x_{P(6)} + Y() + x_{P(13)} + L + X() + x_{P(20)} + x_{D(3)}$$

$$y_5 = F + Z() + x_{P(6)} + Y() + x_{P(13)} + L + X() + x_{P(20)} + W() + x_{D(2)}$$

and that outputs of all other equations in Fig. 3 are unchanged because and all internal values in Fig. 8 are the same in both encryptions except y_9 and y_5. This is because inputs 24, 28 are used only once with $D(2)$ and $D(3)$, due to the fact that in Theorem 5.1.1 we assumed that 24,28 are absent from the set of 27 outputs $\{P(1) \ldots P(27)\}$. Thus the only effect of flipping bits of $A = 24, 28$ is to flip just bits of $H = 5, 9$ in the next round.

So far, Lemmas 5.3.2 and 5.3.3 only cover 2 transitions out of 6 for 8 rounds. What if both bits of $F = [7, 11]$ are flipped? Do they flip only $E = [8, 12]$ inside the next round? This is not so obvious as these bits are inputs c, f of W and the output of W could change if we flip both. Now we have twice $Z(a + d)(b + e)(c + f) = 0$ in each encryption, which was already shown to imply $W(H)(C)(F) = 0$ and $Y(D)(G)(B) = 0$. in each encryption. Now if at the input side all the polynomials $ABCDEFGH$ are at 1, due to $ABCDEFGH = 1$, we conclude that outputs of W and Y must be zero. This carries on forever, again assuming $\mathscr{P}(I_1) = 1$ for the beginning round input. This also implies $\mathscr{P}(I_2) = 1$, as already seen in Lemma 5.3.2. If the value of W is zero in both encryptions, flipping two bits of $F = [7, 11]$ has no effect on both Boolean functions W, Y.

Likewise, flipping bits $[21, 25]$ has no effect, and likewise, for all the 6 possibilities corresponding to B, C, D and F, G, H knowing that cases of $E = [8, 12]$ and $A = [24, 28]$ were already covered by Lemmas 5.3.2 and 5.3.3 respectively. Overall we see that we can do a full circle, exactly as in Fig. 7), and the difference $F = [7, 11]$ will after 8 round will become $F = [7, 11]$ again, and all this because the polynomial invariant propagates and remains valid at each round input. More precisely we have in order

$$F \to E \to D \to C \to B \to A \to H \to G \to F$$

This ends the proof that the difference $[7, 11]$ propagates with probability at least 2^{-8}. $\qquad \square$

Linear Spaces. It is easy to see that the same result holds for any linear combination of 8 basic differences $A = [24, 28]$ to $H = [5, 9]$ shown in Fig. 7. The set of anomalous differentials forms a linear space of dimension 8.

6   Computer Simulations and the Choice of the Boolean Function

Is our Theorem 5.1.1 confirmed by computer simulations? The question is really whether our cipher behaves like a typical Markov cipher (in approximation) outside of the proportion of 2^{-8} anomalous input states with $\mathscr{P} = 1$. The answer is yes as it seems. We show two "typical" cases, essentially chosen at random. Our first table is obtained with the exact Boolean function listed as an example after Thoerem 5.1.1 in page 10.

Table 1. Probabilities observed with our Boolean function as the number of rounds grows.

Rounds	8	16	24	32	40	48	56	64
Proba	$2^{-2.40}$	$2^{-4.82}$	$2^{-6.74}$	$2^{-7.71}$	$2^{-7.95}$	$2^{-7.99}$	$2^{-8.00}$	$2^{-8.00}$

We see very clearly that, at the beginning, the probability grows exponentially, $2^{-4.82}$ is almost exactly the square of $2^{-2.40}$. Then, however, for 24 rounds there is already a substantial deviation: we would predict $2^{-3 \cdot 2.40} = 2^{-7.20}$ and we obtain $2^{-6.74}$, a substantially lower result. The results vary very substantially for other Boolean functions which satisfy $Z(a + d)(b + e)(c + f) = 0$. For example, it is easy to see that if we add $(a + d + 1)b$ to our Boolean function which works, we also obtain a function which works. In this case, the cipher is stronger and our differential property less visible, see Table 2.

Table 2. Probabilities observed with a stronger Boolean function.

Rounds	8	16	24	32	40	48	56	64
Proba	$2^{-4.53}$	$2^{-7.51}$	$2^{-7.98}$	$2^{-8.00}$	$2^{-8.00}$	$2^{-8.00}$	$2^{-8.00}$	$2^{-8.00}$

6.1   On Hiding Differentials

We conjecture that this sort of anomaly is not detectable if we have limited computing power or a limited number of samples. There are countless works about backdoors in block ciphers. In 1990s, authors typically concluded that this was infeasible and "hiding differentials" was claimed particularly difficult, Section 3.4. in [48]. The main idea in our work is that we do **not** need to hide high probability events. We hide low probability differentials, the probability of which can be as low as we want, if our invariant polynomial $\mathscr{P}$ had more than 8 factors. Therefore, it appears that we have discovered a valid method of concealing an attack inside a block cipher so that it is not easily detected. In our 2 examples above, we also see that the number of rounds where the propagation will stop decaying exponentially, and the anomaly becomes visible, is not constant and depends on the exact Boolean function used.

7   The Reciprocal Question, Nash Postulate, and Future Research

In this article, we show that with a well-chosen invariant property we can have a strong anomaly in the propagation of ordinary differentials in Differential Cryptanalysis (DC). The key observation is that the complexity of our attack does NOT tend to zero and remains constant for any number of rounds. The probability success first decreases, but eventually it becomes constant. In contrast, with ordinary DC, typically and in the "regular" Markov cipher case, we expect that the complexity will grow exponentially and eventually every differential will be "roughly equally likely" following [42]. An interesting question is whether intermediate cases are possible: where the probability of a single differential in a block cipher is **not constant** but grows polynomially or sub-exponentially with the number of rounds. This would violate the postulate of exponential complexity proposed by John Nash in his letter from 1955 exposed at NSA crypto museum, cf. [46]. More precisely Nash postulated that "For almost all sufficiently complex types of enciphering" where "different portions of the key interact complexly with each other in the determination of their ultimate effects" the computation cost should increase "exponentially with the length of the key". The words of Nash from 1955 are substantially older than modern block ciphers which were invented in 1970s, cf. [24, 27, 32, 36]. However, very clearly these words are what block cipher designers have been aiming at ever since. John Nash also had an intuition that this sort of strong or absolute security claim or result cannot be taken for granted, nor it can be proven in mathematics (today most security results are relative). He wrote: "The nature of this conjecture is such that I cannot prove it, even for a special type of ciphers. Nor do I expect it to be proven." In this article, we suggest that the Nash and many cipher designers were very optimistic and their security will sometimes increase at a slower rate than expected.

7.1   Some Conjectures - Differential Anomalies Vs. Invariants

Moreover, we conjecture that there exists a third possibility, e.g. sub-exponential curve. In present work we show that sometimes, the success probability of a plain ordinary differential attack, does not decrease exponentially, when the number of rounds tends to infinity. The main reason for this is that there is more than just one property. A non-linear invariant property is present, and is acting behind the scenes distorting the input probability distribution forever, each time the differential property propagates. We can then wonder if some sort of reciprocal result exists. Maybe each time when a differential propagates with a probability which does not depend on the number of rounds, some sort of a non-linear invariant would be always present.

This conjecture seems quite strong. However, we do not see any other reason why differential cryptanalysis would behave in such a strongly anomalous way. The space of non-linear invariant attacks is in fact extremely large, and in this way maybe we can **efficiently discover** further new invariant attacks such as $\mathscr{P}$ in present article and possibly attacks more complex than just product attacks.

There is abundant literature about differential cryptanalysis, and it may seem that this topic is well understood. In this article, we show that this topic is not yet well understood and some major questions regarding how the attack could behave asymptotically, when the number of rounds grows, and why this happens, remain actually widely open.

7.2 Related Research - Special Contrived Ciphers

In [26] a toy cipher is presented which is not secure for as many as 2^n rounds, yet it is provably secure if we further increase the number of rounds. We generate the group of all possible permutations on n bits, cf. Appendix A and B in the extended version of [25]. In contrast, in our Theorem 5.1.1 the differential never vanishes, the cipher is not secure no matter how large is the number of rounds.

7.3 Weak Is Beautiful - The World of Periodic Attacks and Weak Keys

It is a major misconception in cryptography research that the interesting attacks to study are those which work for every key. We claim that the special cases are the most interesting ones. Sometimes, they lead to spectacular improvements w.r.t. best attack known in the general case. In addition all differential and polynomial invariant properties we study here are periodic (with a period of 8).

This is particularly interesting in the context of block ciphers when the key scheduling is also periodic. In this (very common) case the key question is to exploit this periodicity and show that in some cases a large number of rounds can be broken for the price of breaking fewer rounds. In this precise sense, a periodic key schedule is a tremendous weakness with T-310, KeeLoq in [1,2], in GOST, but not in DES. The best known single key attack on GOST with truncated differentials has a running time of 2^{179} in [29]. Now, if we study anomalous events with data encrypted with multiple random keys, the (imperfect) periodic structure of GOST is exploited better, and there exists an attack with total running time of 2^{101} in [33]. A wider comprehensive picture is shown in Section 29 in [10]. We see a near-continuous space with various attacks, improving as the proportion of weak keys goes down. Many of these attacks involve differentials. In T-310 the period in the key scheduling is 120, cf. Fig. 3, and our differential property of Theorem 5.1.1 has a period of 8 which divides 120. Unhappily keystream for encryption is extracted in T-310 with a different prime period, cf. Section 3 in [15] and key recovery could be difficult, see Appendix B. Interestingly, previous research has not exhibited differential anomalies as strong as in the present paper for ordinary single differentials. Overall, it appears that the question of weak keys in periodic block ciphers, and in particular the question of anomalous choice of constants (a weak long term key question), has yet not received sufficient attention.

8   Conclusion

In this paper we have demonstrated that the propagation of differentials inside a block cipher can in some cases be truly pathological. This is to the point that the complexity of the attack does not grow exponentially with the number of rounds, and that an arbitrarily large number of rounds can be attacked. After an initial period of quasi-exponential growth, which does not at all look unusual, cf. Table 1, the anomaly begins.

We see that block ciphers can become extremely weak due to a weaker cipher wiring. Interestingly, such modifications are officially allowed, in the sense of being 100 % compatible with the original T-310 encryption hardware. The long term key in T-310 took the form of a printed board, and was changed every few years [21]. This result is particularly significant for T-310, a government encryption system, the hardware implementation cost of which is very large; thousands of times larger than with modern ciphers such as DES or AES, see [23]. However, increasing the number of rounds does not help if the complexity of an attack is constant and it works for an arbitrarily large number of rounds.

This paper is a proof of concept in just one case. We make the unthinkable happen, and show that this works beyond any doubt with a mathematical proof. We conjecture that this sort of anomaly is not detectable, if we have limited computing power or a limited quantity of encrypted data. We conjecture that this kind of Non-Markovian vulnerability exists also in other ciphers. If the hidden polynomial has a higher degree, it will become very hard to know if such a property is present or not, in any given cipher.

In comparison to an earlier result of this type presented at Crypto 2011, see [43], our Theorem 5.1.1 works with ordinary single differentials, for any key, and in spite of the presence of round constants in T-310. The vulnerability is principally a question of cipher wiring, which is without doubt very special. In contrast, no Boolean function should be considered to be resistant to our attack. Our vulnerability works with any Boolean function chosen at random with a probability of 2^{-8}, which is not at all small. Several works such as [16] and [17] show, that 100 % of Boolean functions are vulnerable against polynomial invariant attacks. Now we also have a similar result for ordinary differential cryptanalysis. The security of the whole block cipher cannot be taken for granted, cf. [47], just on the basis of avoiding high probability iterative differentials.

Appendix A   On Boolean Function Vulnerability

It is possible to see that a Boolean function chosen at random will satisfy our exact property $Z(a + d)(b + e)(c + f) = 0$ with probability 2^{-8}, cf. Section 5 in [13] and/or Appendix C in [16]. The result is the same as long as we have three linear factors which are linearly independent. In general, Boolean functions which are constant over large affine spaces are not an exception, it is systematic. 100% of Boolean functions in 6 variables are 3-normal and can be annihilated by a product of 3 affine polynomials. cf. Section 5 in [19] and [35]. We use another method to obtain the same result. It is sufficient to check all the 150357 classes of Boolean functions based on a database of Boolean functions of [6] based on earlier work by Maiorana [45].

Moreover, our experience shows that typically (when the Boolean function is balanced) both Z or $Z + 1$ will admit numerous solutions of this type, some of which could work with an attack such as described in this paper.

Table 3. Classes of Boolean Functions with 6 Variables w.r.t. k-normality

total ↓ (any k)	k-normal Boolean functions			
k value →	6	≥ 5	≥ 4	≥ 3
150357	1	205	47446	150357
100%	$2^{-17.2}$	$2^{-9.52}$	$2^{-1.66}$	$2^{-0.0}$

Table 4. Classes of Boolean Functions with 6 Variables w.r.t. k-weak-normality

total ↓ (any k)	k-weakly-normal B. functions			
k value →	6	≥ 5	≥ 4	≥ 3
150357	1	205	93760	150357
100%	$2^{-17.2}$	$2^{-9.52}$	$2^{-0.68}$	$2^{-0.0}$

No Boolean function whatsoever should be assumed to be secure against the attacks such as described in this paper. For example with the original Boolean function used in T-310 we have $Zc(b + d)f = 0$ and $Z(a + b)c(1 + e) = 0$ and many other relations of this type. From here it is possible to construct a product invariant attack on demand, using exactly one single relation like this, see [17]. In other words, just one such annihilation equation, which was not chosen by the attacker, can lead to an attack on T-310 working for any number of rounds. This is already for an invariant attack at order 1. Properties which involve two encryptions like in our Theorem 5.1.1 and the existence of multiple ways to annihilate polynomials further increase the freedom for the attacker.

Appendix B   The Key Recovery Question

There exists multiple ways in which non-linear invariant attacks can be exploited in cryptanalysis in order to decrypt actual encrypted communications. This question was already studied in Section 9 in [16] and Section 6 in [12] and Section 6 in [13] and there are several distinct ways to approach this problem. Some invariants (not all) introduce pervasive biases made of higher order correlation properties which do not degrade as the number of rounds increases. Other invariants do directly involve some key bits. In some sense we expect that most invariants are NOT suitable for actual attacks, in the sense that other invariants are more suitable for various technical reasons.

Appendix B.1   New Ways to Exploit Polynomial Invariants

In this paper we discover a possibility to convert a non-linear invariant attack into a differential attack. This opens new possibilities for key recovery in 3 steps as follows. First, we guess some key bits, then, determine some internal values, finally, confirm through a statistical distinguisher. It is important to note that the question of which key bits should be guessed and which ones are determined, is a major practical combinatorial optimization problem in cryptanalysis. It leads to interesting security "metric" notions such as SAT immunity and UNSAT immunity, cf. [11].

Appendix B.2   Multiple Simultaneous Differentials and Cube Attacks

A more advanced method to enable key recovery would be to explore the rich world of cube attacks which is a form of a higher order differential attack. This type of discrete differential properties is much older than it is usually assumed, it was studied since at least 1976, cf. [24], and there are many flavours of cube attacks [52,53]. It is quite rare that several differential properties can work simultaneously and that the overall combined probability remains very high. One example of this is with MiFare classic in [8,37], and it happens again here. Our attack has 8 differences which form a linear space and could be used simultaneously in a variety of combined differential, invariant or/and cube attacks. An interesting question is then how quickly the complexity of such attacks increases as the number of rounds grows. Here we need to look at a new type of conditional cube attack: when a certain product of polynomials is at 1. We need to focus on cube properties which involve key bits, which cannot be taken for granted in general, cf. Section 4.1. in [3]. The space of possible attacks is enormous and we leave this for future research.

References

1. Courtois, N.T., Bard, G.V., Wagner, D.: Algebraic and slide attacks on KeeLoq. In: Nyberg, K. (ed.) FSE 2008. LNCS, vol. 5086, pp. 97–115. Springer, Heidelberg (2008). https://doi.org/10.1007/978-3-540-71039-4_6
2. Courtois, N.T., Bard, G.V.: Random permutation statistics and an improved slide-determine attack on KeeLoq. In: Naccache, D. (ed.) Cryptography and Security: From Theory to Applications. LNCS, vol. 6805, pp. 35–54. Springer, Heidelberg (2012). https://doi.org/10.1007/978-3-642-28368-0_6
3. Bard, G.V., Courtois, N.T., Nakahara, J., Sepehrdad, P., Zhang, B.: Algebraic, AIDA/Cube and side channel analysis of KATAN family of block ciphers. In: Gong, G., Gupta, K.C. (eds.) INDOCRYPT 2010. LNCS, vol. 6498, pp. 176–196. Springer, Heidelberg (2010). https://doi.org/10.1007/978-3-642-17401-8_14
4. Biham, E., Shamir, A.: Differential cryptanalysis of DES-like cryptosystems. J. Cryptol. **4**, 3–72 (1991). https://doi.org/10.1007/BF00630563

5. Brown, L., Seberry, J.: On the design of permutation P in des type cryptosystems. In: Quisquater, J.-J., Vandewalle, J. (eds.) EUROCRYPT 1989. LNCS, vol. 434, pp. 696–705. Springer, Heidelberg (1990). https://doi.org/10.1007/3-540-46885-4_71

6. Çalık, Ç., Sönmez Turan, M., Peralta, R.: The multiplicative complexity of 6-variable Boolean functions. Cryptogr. Commun. **11**(1), 93–107 (2018). https://doi.org/10.1007/s12095-018-0297-2. https://ia.cr/2018/002.pdf

7. Charpin, P.: Normal Boolean functions. J. Complex. **20**(2–3), 245–265 (2004)

8. Courtois, N.T.: The dark side of security by obscurity and cloning MiFare classic rail and building passes anywhere, anytime. In: SECRYPT 2009, pp. 331–338. INSTICC Press (2009). ISBN 978-989-674-005-4

9. Courtois, N.T., Mourouzis, T.: Propagation of truncated differentials in GOST. In: SECURWARE (2013). http://www.thinkmind.org/download.php?articleid=securware_2013_7_20_30119

10. Courtois, N.T.: Algebraic complexity reduction and cryptanalysis of GOST. Monograph study on GOST cipher, 224 p. https://ia.cr/2011/626

11. Courtois, N., Gawinecki, J.A., Song, G.: Contradiction immunity and guess-then-determine attacks on GOST. In: CECC 2912, Tatra Mt. Math. Publ. vol. 53, no. 3, pp. 65–79 (2012). http://www.sav.sk/journals/uploads/0114113604CuGaSo.pdf

12. Courtois, N.T., Georgiou, M.: Variable elimination strategies and construction of nonlinear polynomial invariant attacks on T-310. Cryptologia **44**(1), 20–38 (2020). https://doi.org/10.1080/01611194.2019.1650845

13. Courtois, N.T., Patrick, A., Abbondati, M.: Construction of a polynomial invariant annihilation attack of degree 7 for T-310. Cryptologia **44**(4), 289–314 (2020)

14. Courtois, N.T.: On the existence of non-linear invariants and algebraic polynomial constructive approach to backdoors in block ciphers. https://ia.cr/2018/807. Accessed 27 Mar 2019

15. Courtois, N.T., Patrick, A.: Lack of unique factorization as a tool in block cipher cryptanalysis. https://arxiv.org/abs/1905.04684. Accessed 12 May 2019

16. Courtois, N.T.: Structural nonlinear invariant attacks on T-310: attacking arbitrary Boolean functions. https://ia.cr/2018/1242. Accessed 12 Sept 2019

17. Courtois, N.T.: A nonlinear invariant attack on T-310 with the original Boolean function. Cryptologia, 23 Apr 2020. https://www.tandfonline.com/doi/full/10.1080/01611194.2020.1736207. to appear also in paper version in 2020

18. Courtois, N.T.: Invariant hopping attacks on block ciphers. In: Presented at WCC 2019, Abbaye de Saint-Jacut de la Mer, France, 31 March–5 April 2019. https://arxiv.org/pdf/2002.03212.pdf. Accessed 8 Feb 2020

19. Courtois, N.T., Abbondati, M., Ratoanina, H., Grajek, M.: Systematic construction of nonlinear product attacks on block ciphers. In: Seo, J.H. (ed.) ICISC 2019. LNCS, vol. 11975, pp. 20–51. Springer, Cham (2020). https://doi.org/10.1007/978-3-030-40921-0_2

20. Courtois, N.T.: Feistel schemes and bi-linear cryptanalysis. In: Franklin, M. (ed.) CRYPTO 2004. LNCS, vol. 3152, pp. 23–40. Springer, Heidelberg (2004). https://doi.org/10.1007/978-3-540-28628-8_2

21. Courtois, N.T., et al.: Cryptographic security analysis of T-310. Monography study on the T-310 block cipher, 132 p. 20 May 2017. https://ia.cr/2017/440.pdf. Accessed 29 June 2018

22. Courtois, N.T., Oprisanu, M.-B.: Ciphertext-only attacks and weak long-term keys in T-310. Cryptologia, **42**(4), 316–336 (2018). http://www.tandfonline.com/doi/full/10.1080/01611194.2017.1362065

23. Courtois, N., Drobick, J., Schmeh, K.: Feistel ciphers in East Germany in the communist era. Cryptologia **42**(6), 427–444 (2018)
24. Courtois, N.T.: Block ciphers: lessons from the cold war. In: Slides presented at 2019 biennial Symposium on Cryptologic History, Laurel, Maryland, US, October 2019. http://www.nicolascourtois.com/papers/Feistel_East_Cold_War_US_Oct2019.pdf
25. Courtois, N.T.: The inverse S-Box, non-linear polynomial relations and cryptanalysis of block ciphers. In: Dobbertin, H., Rijmen, V., Sowa, A. (eds.) AES 2004. LNCS, vol. 3373, pp. 170–188. Springer, Heidelberg (2005). https://doi.org/10.1007/11506447_15. https://www.researchgate.net/publication/221005723_The_Inverse_S-Box_Non-linear_Polynomial_Relations_and_Cryptanalysis_of_Block_Ciphers
26. Courtois, N.: The inverse S-box and two paradoxes of whitening. Long extended version of the Crypto 2004 rump session presentation, Whitening the AES S-box. http://www.nicolascourtois.com/papers/invglc_rump_c04.pdf
27. Courtois, N., Oprisanu, M.-B., Schmeh, K.: Linear cryptanalysis and block cipher design in East Germany in the 1970s. Cryptologia (2018). https://www.tandfonline.com/doi/abs/10.1080/01611194.2018.1483981
28. Courtois, N.: The best differential characteristics and subtleties of the Biham-Shamir attacks on DES. https://ia.cr/2005/202
29. Courtois, N.T.: An improved differential attack on full GOST. In: Ryan, P.Y.A., Naccache, D., Quisquater, J.-J. (eds.) The New Codebreakers. LNCS, vol. 9100, pp. 282–303. Springer, Heidelberg (2016). https://doi.org/10.1007/978-3-662-49301-4_18
30. Courtois, N.: An improved differential attack on full GOST. Cryptology ePrint Archive, Report 2012/138, 15 March 2012. https://ia.cr/2012/138. Accessed Dec 2015
31. Courtois, N., Misztal, M.: Aggregated differentials and cryptanalysis of PP-1 and GOST. Periodica Mathematica Hungarica **65**(2), 11–26 (2012). https://doi.org/10.1007/s10998-012-2983-8. In CECC 2011, 11th Central European Conference on Cryptology
32. Courtois, N.T., Mourouzis, T., Misztal, M., Quisquater, J.J., Song, G.: Can GOST be made secure against differential cryptanalysis? Cryptologia **39**(2), 145–156 (2015)
33. Courtois, N.: On multiple symmetric fixed points in GOST. Cryptologia **39**(4), 322–334 (2015)
34. Dobbertin, H.: Construction of bent functions and balanced Boolean functions with high nonlinearity. In: Preneel, B. (ed.) FSE 1994. LNCS, vol. 1008, pp. 61–74. Springer, Heidelberg (1995). https://doi.org/10.1007/3-540-60590-8_5
35. Dubuc, S.: Etude des propriétés de dégénérescence et de normalité des fonctions booléennes et construction de fonctions q-aires parfaitement non-linéaires, Ph.D. thesis, Université de Caen (2001)
36. Feistel, H., Notz, W.A., Smith, J.L.: Cryptographic techniques for machine to machine data communications, 27 Dec 1971, Report RC-3663, IBM T. J. Watson Research (1971)
37. Golić, J.D.: Cryptanalytic attacks on MIFARE classic protocol. In: Dawson, E. (ed.) CT-RSA 2013. LNCS, vol. 7779, pp. 239–258. Springer, Heidelberg (2013). https://doi.org/10.1007/978-3-642-36095-4_16
38. Harpes, C., Kramer, G.G., Massey, J.L.: A generalization of linear cryptanalysis and the applicability of Matsui's Piling-up lemma. In: Guillou, L.C., Quisquater,

J.-J. (eds.) EUROCRYPT 1995. LNCS, vol. 921, pp. 24–38. Springer, Heidelberg (1995). https://doi.org/10.1007/3-540-49264-X_3

39. Harpes, C., Massey, J.L.: Partitioning cryptanalysis. In: Biham, E. (ed.) FSE 1997. LNCS, vol. 1267, pp. 13–27. Springer, Heidelberg (1997). https://doi.org/10.1007/BFb0052331

40. Knudsen, L.R.: Truncated and higher order differentials. In: Preneel, B. (ed.) FSE 1994. LNCS, vol. 1008, pp. 196–211. Springer, Heidelberg (1995). https://doi.org/10.1007/3-540-60590-8_16

41. Kovalchuk, L.V.: Generalized Markov ciphers: evaluation of practical security against differential cryptanalysis. In: Proceedings of 5th All-Russian Scientific Conference MaBIT-06, 25–27 Oct 2006, MGU, Moscow, pp. 595–599 (2006). (in Russian)

42. Lai, X., Massey, J.L., Murphy, S.: Markov ciphers and differential cryptanalysis. In: Davies, D.W. (ed.) EUROCRYPT 1991. LNCS, vol. 547, pp. 17–38. Springer, Heidelberg (1991). https://doi.org/10.1007/3-540-46416-6_2

43. Leander, G., Abdelraheem, M.A., AlKhzaimi, H., Zenner, E.: A cryptanalysis of PRINTCIPHER: the invariant subspace attack. In: Rogaway, P. (ed.) CRYPTO 2011. LNCS, vol. 6841, pp. 206–221. Springer, Heidelberg (2011). https://doi.org/10.1007/978-3-642-22792-9_12

44. Knudsen, L.R., Robshaw, M.J.B.: Non-Linear Characteristics in Linear Cryptoanalysis. In: Maurer, U.M. (ed.) EUROCRYPT 1996. LNCS, vol. 1070, pp. 224–236. Springer, Heidelberg (1996)

45. Maiorana, J.A.: A classification of the cosets of the Reed-Muller code R(1,6). Math. Comput. **57**(195), 403–414 (1991)

46. John Nash, handwritten letters and documents relating to their evaluation, available at NSA crypto museum, January-March 1955. cryptologicfoundation.org. https://www.nsa.gov/news-features/declassified-documents/nash-letters/assets/files/nash_letters1.pdf. declassified in 2012

47. Nyberg, K., Knudsen, L.R.: Provable security against differential cryptanalysis. In: Brickell, E.F. (ed.) CRYPTO 1992. LNCS, vol. 740, pp. 566–574. Springer, Heidelberg (1993). https://doi.org/10.1007/3-540-48071-4_41

48. Peyrin, T., Wang, H.: The MALICIOUS framework: embedding backdoors into tweakable block ciphers. In: Micciancio, D., Ristenpart, T. (eds.) CRYPTO 2020. LNCS, vol. 12172, pp. 249–278. Springer, Cham (2020). https://doi.org/10.1007/978-3-030-56877-1_9

49. Referat 11: Kryptologische Analyse des Chiffriergerätes T-310/50. Central Cipher Organ, Ministry of State Security of the GDR, document referenced as 'ZCO 402/80', a.k.a. MfS-Abt-XI-594, Berlin, 123 p. (1980)

50. Schmeh, K.: The East German encryption machine T-310 and the algorithm it used. Cryptologia **30**(3), 251–257 (2006)

51. Todo, Y., Leander, G., Sasaki, Y.: Nonlinear invariant attack: practical attack on full SCREAM, iSCREAM and Midori 64. J. Cryptol. **32**, 1–40 (2018)

52. Vielhaber, M.: AIDA Breaks BIVIUM (A&B) in 1 Minute Dual Core CPU Time. https://ia.cr/2009/402

53. Winter, R., Salagean, A., Phan, R.C.-W.: Comparison of cube attacks over different vector spaces. In: Groth, J. (ed.) IMACC 2015. LNCS, vol. 9496, pp. 225–238. Springer, Cham (2015). https://doi.org/10.1007/978-3-319-27239-9_14

Key Mismatch Attack on ThreeBears, Frodo and Round5

Jan Vacek and Jan Václavek[(✉)]

Thales DIS, Prague, Czech Republic
{jan.vacek,jan.vaclavek}@thalesgroup.com

Abstract. In the last years, several key reuse attacks were proposed against Round 2 candidates of the NIST Post-Quantum Cryptography Standardization Process. In these attacks, the adversary has access to the key mismatch oracle which tells her if a given ciphertext decrypts to a given message under the targeted secret key. One of the so far non-targeted candidates is ThreeBears, which is a key encapsulation mechanism based on the integer module learning with errors (I-MLWE) problem. In this paper, we present a first key mismatch attack against the ThreeBears cryptosystem. Our attack recovers the whole secret key with probability of 100% and requires about 2^{11} queries on average. Besides that, we use our technique to target other Round 2 candidates Frodo and Round5, and we improve the state-of-the-art results for them.

Keywords: ThreeBears · Frodo · Round5 · Key mismatch attack · Post quantum cryptography · Cryptanalysis · Attack

1 Introduction

The increasingly relevant threat of a large-scale quantum computer has motivated the community to base cryptosystems on problems believed to be resistant even against quantum computers. One of the second-round candidates of the NIST Standardization Process is ThreeBears which is a key exchange algorithm based on the integer module learning with errors (I-MLWE) problem. Among other candidates belong Frodo and Round5, which are schemes based on learning with errors (LWE) and learning with rounding (LWR) problems, respectively.

In the last years, several key mismatch attacks were proposed against Round 2 candidates [3, 5, 6, 8, 11–14]. In these attacks, the adversary has access to the key mismatch oracle which tells her if a given ciphertext decrypts to a given message under the targeted secret key. Such an attack model was originally proposed in [8] and is relevant in scenarios when the same secret key is reused for several key exchanges since a lot of matches/mismatches are usually needed to recover the secret key. One of the candidates without a proposed key mismatch attack against it is ThreeBears.

If a secret key is reused in a passively secure (IND-CPA) encryption based KEM, then it is quite straightforward for the attacker to access the key mismatch

© Springer Nature Switzerland AG 2021
D. Hong (Ed.): ICISC 2020, LNCS 12593, pp. 182–198, 2021.
https://doi.org/10.1007/978-3-030-68890-5_10

oracle. The attacker just sends the chosen ciphertext to the party doing the decryption and tries to communicate with this party using the chosen message (which plays the role of the shared key). If the attacker is able to communicate, they must have the same key, which corresponds to a match from the oracle. On the other hand, if the attacker cannot communicate with the second party, they do not have the same key, which corresponds to a mismatch from the oracle.

For the actively secure (IND-CCA) variants, it is necessary to use some side-channel or fault injection attacks to get access to the oracle. For example, in [15], they observed that a hash function operating on the decrypted message within the Fujisaki-Okamoto transformation exhibits a differential behavior based on the value of the decrypted message. They used the fact that this behavior can be observed over the electromagnetic side-channel to instantiate the key mismatch oracle.

Despite the fact that a key reuse is considered as a misuse by the specification of the targeted schemes, we think that the condition of the attack is still relevant since being considered as a misuse does not prevent it from happening. We suppose that it can still easily happen either as a result of misinterpreting the specification of the scheme or by deliberately reusing the secret key for efficiency reasons due to the lack of understanding of possible attacks and their complexities in this case.

1.1 Our Contribution

In this paper, we mainly focus on ThreeBears [10] and we present the first key mismatch attack against it. First, we recall the attack model and formally define the key mismatch oracle. After that, we show the main idea of the attack and then we provide the details about the choice of the concrete queries to the oracle.

Besides that, we target other Round 2 candidates, Frodo [1] and Round5 [2]. The only key mismatch attacks on Frodo and Round5 we are aware of were published in [3,6,15]. We improve and extend the attacks on Frodo and provide the first complete results for all variants of Round5 since the attack from [6] is actually against a predecessor of Round5, HILA5, and the attack from [15] targets only the variant with the error-correcting code. Our method shares some similarities with the attack from [15]. We both try to reduce the number of possibilities for the targeted secret coefficient, but in a different manner. The concrete differences are described in more details later at the end of Subsect. 4.1. We developed our method without being aware of the method presented in [15] and we analyze the similarities between the two methods a posteriori.

All three targeted schemes use an encryption based approach and we target the underlying public key encryption schemes. Our attacks against all three above mentioned candidates recover the whole secret key with success probability of 100% and the required number of queries to recover the whole secret key is shown in Table 1 for the NIST security level 5 variants. The complete and detailed results are discussed in Subsect. 3.4 and at the end of Subsects. 4.1 and 4.2. We implemented the attacks in Python[1] and they confirm the given results.

Table 1. Results for NIST security level 5 variants

	Name of the variant	Average number of queries to recover the whole secret key
ThreeBears	PAPABEAR	2847
Frodo	FRODOPKE-1344	28 008
Round5	R5N1_5CPA_0D	15 005
Round5	R5ND_5CPA_0D	1525
Round5	R5ND_5CPA_5D	1551

1.2 Outline of the Paper

In Sect. 2, we introduce the notation, define the key mismatch oracle and briefly describe the ThreeBears cryptosystem. In Sect. 3, we describe the attack against ThreeBears. In Sect. 4, we sketch the attacks against Frodo and Round5. In Sect. 5, we conclude the paper. Moreover, we provide some missing proofs in the Appendices.

2 Preliminaries

2.1 Notations

For a positive integer q, we denote by $\mathbb{Z}_q$ the quotient ring $\mathbb{Z}/q\mathbb{Z}$, where we take the elements of $\mathbb{Z}_q$ to be the integers between 0 and $q - 1$. For an integer x and a positive integer q, we define the $x \bmod q$ operation in a standard way to always produce an integer between 0 and $q - 1$.

For a set A, we denote by $\xleftarrow{\$} A$ sampling an element uniformly random from the set A. Similarly, for a probability distribution ψ, we denote by $\xleftarrow{\$} \psi$ picking an element according to ψ. For $x \in \mathbb{R}$, we define $\lceil x \rfloor = \lfloor x + \frac{1}{2} \rfloor \in \mathbb{Z}$, where $\lfloor y \rfloor$ is the greatest integer not exceeding y.

2.2 Key Mismatch Oracle

In the targeted underlying public key encryption schemes, Alice first generates her secret key $\mathbf{S}$ and computes the public key pk, which she sends to Bob. Bob encrypts the message m and sends the ciphertext ct back to Alice. Alice then decrypts the ciphertext and gets the message m'.

---

[1] https://github.com/Mismatch-attack-threebears-frodo-round5/Attacks.

In the key mismatch attack, the adversary Eve, who is acting as Bob, wants to recover Alice's secret key S. She does not compute the ciphertext according to the specification, but she chooses a message m_E and an arbitrary ciphertext ct, which she sends back to Alice. The key mismatch oracle tells her if the ciphertext ct decrypts to m_E.

We define the key mismatch oracle formally in the next definition.

Definition 1 (Key mismatch oracle). *Let S be the secret key of Alice. On input of ct and m_E, the output of the key mismatch oracle $\mathcal{O}$ is defined as follows:*

$$\mathcal{O}(ct, m_E) = \begin{cases} + & \text{if } Decryption(ct, S) = m_E \\ - & otherwise. \end{cases} \tag{1}$$

We already discussed ways to access key mismatch oracle in the introduction.

2.3 Description of ThreeBears

ThreeBears is based on a non-standard variant of LWE called integer module learning with errors (I-MLWE) which was first introduced in 2017 by Chunsheng in [9]. By the words of its author, ThreeBears is based on KYBER at a high level. The difference is that instead of working in some polynomial ring $\mathbb{Z}_q[x]/f(x)$, the ring $\mathbb{Z}_{f(q)}$ is used by substituting x with q. In ThreeBears, all parameter sets use $f(x) = x^{312} - x^{156} - 1$ and $q = 2^{10}$. The degree of the polynomial f is denoted by D, i.e., $D = 312$. Furthermore, $f(q)$ is denoted by N, i.e., $N = 2^{3120} - 2^{1560} - 1$. A special multiplication is used, for $a, b \in \mathbb{Z}_N$ it is

$$clar \cdot a \cdot b \bmod N$$

instead of a standard $a \cdot b \bmod N$, where $clar = q^{156} - 1$. This extra factor, $clar$, is called a clarifier and the purpose of this clarifier is to distort the noise and so to decrease the failure probability. The dimension of the underlying module is $2, 3$ or 4 and is denoted by d.

To sample from the noise distribution, an auxiliary distribution ψ with very small support, which is either $[-1, 1]$ or $[-2, 2]$, is used. Sampling noise elements from $\mathbb{Z}_N$ is then described in Algorithm 1. We denote this noise distribution on $\mathbb{Z}_N$ by χ.

As in other LWE based schemes, ENCODE and DECODE functions are used, see Algorithms 3 and 4. These functions make use of another function called EXTRACT, which is described in Algorithm 2. The whole simplified version of ThreeBears is depicted in Fig. 1. We omitted some parts which are unnecessary for the attack, such as using seeds.

Error correcting code. Moreover, a forward error-correcting code is applied to the message before encrypting it in order to reduce the failure probability.[2]

---

[2] Which means that during decryption after I-MLWE decoding, the ECC decoding is performed.

The authors chose a Melas-type BCH code that corrects 2 errors. We described the scheme without this error-correcting code and in the next section, we first describe the attack on the version of the scheme without the code as it is simpler and more straightforward. Then we comment on how to target the variant with the error-correcting code in Subsect. 3.3. We point out that targeting the variant with the error-correcting code does not require any additional queries to the oracle compared to targeting the same variant without the error-correcting code.

Algorithm 1. Sampling noise

1: **function** χ
2:     $s \leftarrow 0$
3:     **for** i from 0 to $D - 1$ **do**
4:         $digit \xleftarrow{\$} \psi$
5:         $s \leftarrow s + digit \cdot q^i$
6:     **end for**
7:     **return** $s \bmod N \in \mathbb{Z}_N$
8: **end function**

Algorithm 2. Extracting bits

1: **function** $\text{EXTRACT}_b(S, i)$
2:     **if** $i \bmod 2 == 0$ **then**
3:         $j \leftarrow i/2$
4:     **else**
5:         $j \leftarrow D - (i + 1)/2$
6:     **end if**
7:     $S \leftarrow S \bmod N$
8:     **return** $\lfloor S \cdot 2^b / q^{j+1} \rfloor$
9: **end function**

Algorithm 3. Encoding message

1: **function** $\text{ENCODE}(C \in \mathbb{Z}_N, \text{mes-sage } m \in \{0,1\}^L)$
2:     $encr \leftarrow 0$
3:     **for** i from 0 to $L - 1$ **do**
4:         $encr[i] \leftarrow (\text{EXTRACT}_4(C, i) + 8 \cdot m[i]) \bmod 16$
5:     **end for**
6:     **return** $encr$
7: **end function**

Algorithm 4. Decoding message

1: **function** $\text{DECODE}(C \in \mathbb{Z}_N, encr \in [0,15]^L)$
2:     $m' \leftarrow 0$
3:     **for** i from 0 to $L - 1$ **do**
4:         $m'[i] \leftarrow \lfloor \frac{2 \cdot encr[i] - \text{EXTRACT}_5(C,i)}{16} \rceil \bmod 2$
5:     **end for**
6:     **return** m'
7: **end function**

---

3 Key Mismatch Attack on ThreeBears

A secret key is an element of $\mathbb{Z}_N^d$, which is formed by $d \cdot 312$ numbers from the distribution ψ, where each 312 numbers of these $d \cdot 312$ determine one noise element in $\mathbb{Z}_N$ (see Algorithm 1). We target each of these $d \cdot 312$ small numbers separately and we call them as secret coefficients $s_{i,j}$, where $i \in [0, d-1]$ and $j \in [0, 311]$ based on to which part and by which power of q it contributes to the secret key.

First, we need to take care of the special form of multiplication using the so-called clarifier. We observe that

$$clar \cdot q^{156} = (q^{156} - 1) \cdot q^{156} = 1 \pmod{N}.$$

If we want a standard multiplication by $\mathbf{C}_1'$, we can simply set $\mathbf{C}_1 = q^{156} \cdot \mathbf{C}_1'$, because

$$clar \cdot \mathbf{C}_1 \mathbf{S} = clar \cdot q^{156} \cdot \mathbf{C}_1' \mathbf{S} = \mathbf{C}_1' \mathbf{S} \pmod{N}.$$

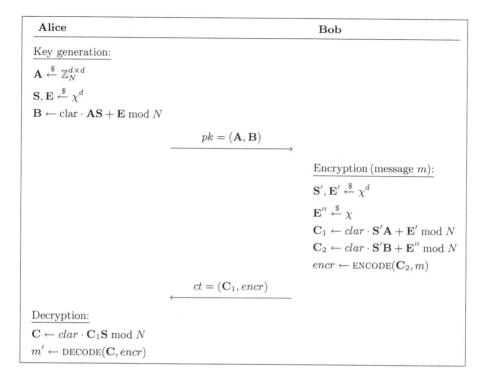

Fig. 1. Simplified version of ThreeBears

Another difficulty is that the length of the message, denoted by L, is less than D, which means that not every part of $\mathbf{C}$ is used within the DECODE function. For that reason, we cannot use exactly the same method to target all secret coefficients. We use three sets of queries with small differences to the oracle based on the second index j in a secret coefficient $s_{i,j}$. First group are coefficients with $j \in [0, 127] \cup [184, 311]$, second group are coefficients with $j \in [128, 155]$ and the third group are coefficients with $j \in [156, 183]$.

3.1 High Level Description of the Attack

As mentioned above, the secret coefficients $s_{i,j}$ are targeted one by one. The main idea of the attack is to gradually reduce the possibilities for the targeted coefficient.

We are interested in queries for which the output from the oracle depends only on one secret coefficient from the secret key $\mathbf{S}$. Then the output from the oracle can give us useful information on this concrete coefficient, because the output is independent from the other coefficients.

Moreover, we prefer queries such that for some values of the targeted coefficient, the output from the oracle would be $+$, and for some of them would be $-$. Then, using this query, we will know in which group the targeted coefficient lies.

By repeating this process, we can further reduce the number of possibilities for the targeted coefficient. Then, if there is only one possibility left, it must be the targeted coefficient.

We construct suitable queries which can recover each possible coefficient. The actual attack then just consists of using these already defined queries and of reducing the possibilities for the targeted coefficient. The gradual splitting of possible values for the targeted coefficient into two smaller disjoint groups naturally corresponds to a binary tree. Hence, the queries form a binary tree and the actual attack consists of following a path within this tree to some leaf corresponding to the value of the targeted coefficient.

We keep in mind two objectives. First, we have to choose the queries such that it is possible to recover each possible value of the secret coefficient. Otherwise, the attack does not work with a success probability of 100%. Second, we minimize the expected number of queries to recover the secret coefficient, which equals the weighted average of depths[3] of the leaves and the probabilities of values corresponding to these leaves.

3.2 Choice of Queries

We use the same guessed message $m_E = 0^L = (0, \ldots, 0) \in \{0, 1\}^L$ for all queries to the oracle. We choose the ciphertext $ct = (\mathbf{C}_1, encr)$ to the oracle depending on which group we target. Recall that $\mathbf{C}_1 \in \mathbb{Z}_N^d$. First, based on the index j, we set

$$
g = \begin{cases} 55 \cdot q^{156} & \text{if } j \in [0, 127] \cup [184, 311] \\ 55 & \text{if } j \in [128, 155] \\ 55 \cdot q^{212} & \text{if } j \in [156, 183] \end{cases} \tag{2}
$$

and then, we set

$$
\mathbf{C}_1[k] = \begin{cases} g & \text{if } k = i \\ 0 & \text{if } k \neq i. \end{cases} \tag{3}
$$

This $\mathbf{C}_1$ is then multiplied with the secret vector $\mathbf{S}$, so by setting all entries in $\mathbf{C}_1$ except one to 0 ensures that only one part of $\mathbf{S}$ appears in the result, i.e., only coefficients with the same index i. The reason for different powers of q is that not all parts of $\mathbf{C}$ are used during the decryption as mentioned earlier. For that reason, by using these different powers we shift the secret key $\mathbf{S}$ correspondingly so that we can recover all its parts at the end.

The constant 55 is chosen such that we can have control over decryption (m' will have a specific form, see Lemma 1) but still we get some useful information from the key mismatch oracle. With values larger than 55, we start losing control over decryption. With values smaller than 55, we start losing useful information from the oracle due to the compression.

---

[3] The number of queries to recover some coefficient equals the depth of the leaf corresponding to the value of this coefficient.

In the following lemma, we justify the mentioned control over decryption using constant 55. In the paper of Bauer et al. [5], the assumption of having such control was called Hypothesis 1. We sketch the proof in the Appendix A.

Lemma 1. *Let C_1 have the form as in Eq. 3 and set $encr[l] = 0$ for some $l \in [0, \ldots, L-1]$. Then, in decryption, $m'[l] = 0$.*

Choice of encr. Above, we have defined the choice of the first part of the cipher-text ct (i.e. C_1) queried to the oracle. This part is fixed during targeting some coefficient $s_{i,j}$. Now we define the second part, i.e. $encr$, which will vary for each oracle query depending on the position in the binary tree. Motivated by Lemma 1, all parts of $encr$ except one are set to zero. First, based on the index j, we set

$$k = \begin{cases} 2j & \text{if } j \in [0, 127] \\ 2(311 - j) + 1 & \text{if } j \in [184, 311] \\ 2(155 - j) + 1 & \text{if } j \in [128, 155] \\ 2(255 - j) + 1 & \text{if } j \in [156, 183] \end{cases} \tag{4}$$

and then we set

$$encr[l] = \begin{cases} node.value & \text{if } l = k \\ 0 & \text{otherwise.} \end{cases} \tag{5}$$

The values of k are chosen to be in accordance with the choice of C_1, i.e., to correspond to particular shifts of the secret key S.

Concrete Binary Tree. Recall that $encr \in [0, 15]^L$, so we are choosing $node.value$ from $[0, 15]$. Because the range of the targeted coefficients is very small, i.e. $[-1, 1]$ or $[-2, 2]$, we were even able to construct the binary trees manually instead of using some recursive search. We used values $2, 4, 5$ and 11 in the trees. We show the trees in Figs. 2 and 3.[4] In the leaves there are the possible values of the secret coefficient. For the inner nodes, the first part is the remaining set of possibilities, the second part is the value substituted to $encr[k]$ for this particular query. We denoted this value as $node.value$ above.

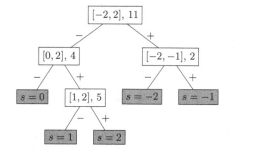
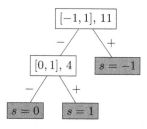

Fig. 2. T_1, search tree for ThreeBears, range $[-2, 2]$

Fig. 3. T_2, search tree for ThreeBears, range $[-1, 1]$

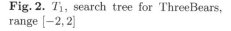

[4] We point out that it is possible to deduce the smaller tree from the larger one.

We show the pseudocode for the attack against ThreeBears in Algorithms 5 and 6. The function *setup* sets the $\mathbf{C}_1[k]$ and k depending on the index j as in Subsect. 3.2. The input is the tree which we are using, i.e. either T_1 or T_2, which depends on the support of the distribution ψ.

Algorithm 6. Recover one secret coefficient in ThreeBears

```
1: function COEFF(i, j, Tree)
2:      C₁ ← 0ᵈ ∈ Z_N^d
3:      encr ← 0^L ∈ [0, 15]^L
4:      m_E = 0^L ∈ {0, 1}^L
5:      C₁[i], k ← setup(j)
6:      Node ← Tree.root
7:      while Node ≠ leaf do
8:          encr[k] ← Node.value
9:          ct = (C₁, encr)
10:         b ← O(ct, m_E)
11:         if b = '+' then
12:             Node ← Node.right
13:         else
14:             Node ← Node.left
15:         end if
16:     end while
17:     s_{i,j} ← Node.coefficient
18:     return s_{i,j}
19: end function
```

Algorithm 5. Key mismatch attack against ThreeBears

```
1: function RECOVER(Tree)
2:      S ← 0ᵈ ∈ Z_N^d
3:      for i from 0 to d − 1 do
4:          for j from 0 to 311 do
5:              s_{i,j} ← COEFF(i, j, Tree)
6:              S[i] ← S[i] + s_{i,j} · qʲ
7:          end for
8:      end for
9:      return S
10: end function
```

3.3 Error-Correcting Code

In this subsection, we describe how to adjust the attack to work also for the variants with the error-correcting code. So far, we chose the ciphertext such that after the decryption, it results in zero bits except for one position which could be possibly 1.

The problem with the error-correcting code is that both decryption variants, i.e. all zeroes or all zeroes except one 1, result into all zeroes after performing ECC decoding.[5] This means that we get only matches from the oracle which gives us no useful information. Because the error-correcting code corrects two errors, we change the ciphertext such that after decryption, it results into zero bits except for three positions:

- First of these three positions is the same as before and the resulting bit is zero or one depending on the targeted secret coefficient $s_{i,j}$.
- The resulting bit for the other two positions is always one. These positions are fixed at indices 256 and 257, because these correspond to the correcting bits and hence are not used in the original attack.

---

[5] All zeroes are encoded to all zeroes using this error-correcting code and the code correct up to two errors.

The guessed message remains the same, i.e. all zeroes. Now we realize that the output from the oracle is the same as in the original attack. If the resulting bit on the position we are interested in (corresponding to the targeted coefficient $s_{i,j}$) is 0, then there are 2 ones which are corrected to all zeroes and we get a match from the oracle. On the other hand, if this bit is 1, then there are 3 ones which are no longer corrected to all zeroes (code corrects up to 2 errors, not 3) and we get a mismatch as before.

We know that if we choose $encr[l] = 0$, then $m'[l] = 0$ no matter what the secret key $\mathbf{S}$ is. Hence, we can set $encr[l] = 8$ and we get that $m'[l] = 1$. The full justification of this fact is proved in Appendix B. This concludes the required modifications to target also the full variants with the error-correcting code. We can see that there are no extra queries needed to handle the error-correcting code.

3.4   Results

Using this method, the number of queries required to recover some value is equal to the depth of the leaf corresponding to this value in the binary tree. The expected number of queries to recover one coefficient is then computed as a weighted average of these depths by probabilities of the corresponding values.

We use the probabilities of values from ψ to compute the expected number of queries to recover one secret coefficient and to recover the whole secret key $\mathbf{S}$. Results are shown in Table 2. We denote by # coeffs the total number of secret coefficients, by $\mathbb{E}[\text{one coeff}]$ the expected number of queries to recover one secret coefficient and by $\mathbb{E}[\text{whole key}]$ the expected number of queries to recover the whole secret key.

4   Key Mismatch Attack on Frodo and Round5

In this section, we apply the same method, i.e. gradually reduce the possibilities for the targeted coefficients, against other candidates Frodo and Round5.

Except for targeting coefficients one by one as before, we target them now also by pairs, triplets and possibly by quadruplets[6] using the exactly same method via the binary trees. Again, we choose the queries such that the output from the oracle depends only on the targeted pair/triplet/quadruplet of coefficients. The guessed message again consists only of zero bits.

4.1   Frodo

Frodo was first introduced in [7] in 2016 and is based on the standard LWE problem. The overall structure of the scheme is similar to all other LWE-based schemes including ThreeBears (see Fig. 1). In Frodo, all the underlying objects

---

[6] We call this the dimension of the attack. For example, dimension 2 means targeting pairs of coefficients.

are matrices and operations are matrix multiplications modulo some integer. Range of the secret coefficients is larger than in ThreeBears, it is $[-12, 12]$, $[-10, 10]$ or $[-6, 6]$ depending on the concrete parameter set. For all details, we refer the reader to the specification of Frodo [1].

The Key Mismatch Attack. The key mismatch attack is simpler against Frodo, because we do not have to distinguish any cases depending on the indices of the targeting coefficients. Secret key $\mathbf{S}$ is an element of $\mathbb{Z}_q^{n \times 8}$, where the particular entries of this matrix, called secret coefficients $s_{i,j}$, come from the error distribution with the above mentioned range. Now, we define the structure of the queried ciphertext[7] such that the output from the oracle depends only on the targeted coefficients and hence it is possible to use our method using the binary trees.

In order to target the secret coefficient $s_{i,j}$, we set

$$\mathbf{C}_1[0, i] = 1 \quad \text{and} \quad \mathbf{C}_2[0, j] = node.value$$

and the remaining entries are set to zeroes.

In order to target the pair $s_{i,j}$, $s_{i+1,j}$, we set

$$\mathbf{C}_1[0, i] = node.\alpha \quad \text{and} \quad \mathbf{C}_1[0, i+1] = node.\beta,$$

the remaining entries of $\mathbf{C}_1$ are set to zeroes and the choice of $\mathbf{C}_2$ remains unchanged.

Having the structure of the queries, we have to choose the values of $node.value$, $node.\alpha$ and $node.\beta$ which then define the binary trees. As mentioned earlier, we want to be able to recover each possible coefficient (or pair of coefficients) and simultaneously minimize the expected number of queries.

When targeting coefficients one by one, it was again possible to construct the trees manually. For the pairs, there is a huge number of possibilities and we had to use a recursive search to find a tree as optimal as possible. Nevertheless, the number of possible trees is too huge and it is not computationally feasible to try each of them. Hence, we used the following heuristic: we were splitting the sets of remaining possibilities into the two disjoint subsets in a way that these two subsets have similar probability.

Results and Comparison. The only key mismatch attacks on Frodo we are aware of are the attacks from [3] and [15]. In [3], they use a different approach. Using the key mismatch oracle, they recover linear equations with the secret coefficients being the unknown. From these equations, they compute the secret key. The authors target only the NIST security level 1 variant and need 2^{16} queries.

In [15], they find a sequence of queries G such that they can associate each value of the secret coefficient with a unique output sequence from G. Then, during the attack, they perform this sequence of queries G and determine the

---

[7] We use the notation from the specification of Frodo.

value of the secret coefficient from the output sequence. Unfortunately, they have not provided the required number of queries for their attack against Frodo.

Nevertheless, we think that the attack from [15] requires less queries than the attack from [3], but requires more queries than our attack. The reason why we think that the attack from [15] requires more queries than our attack is twofold. First, our attack uses the binary tree, which allows for a flexible number of queries to recover one secret coefficient compared to [15], where a fixed number of queries is used to recover one secret coefficient. Second, using our method, we know precisely what the expected number of queries to recover a secret coefficient is. Hence, we choose the queries not randomly but in order to minimize the expected number of queries.

We target all three variants of Frodo and for the NIST security level 1 variant, we need about 2^{14} queries using our method. Complete results are presented in Table 2.

4.2   Round5

Round5 is a merger of Round 1 candidates HILA5 and Round2. Round5 is based on the learning with rounding (LWR) and on the ring learning with rounding (RLWR) problems. These two problems were first introduced by Banerjee et al. in 2011 in [4]. The authors propose ring parameter sets with error-correction, ring parameter sets without error-correction and non-ring parameter sets without error-correction.

The non-ring variants are similar to Frodo, but instead of adding errors from some noise distribution, rounding is used. In the ring variants, the underlying structures are polynomials and operations are computed in some polynomial ring.

Key Mismatch Attack Against LWR Variant. We recall that the secret key $\mathbf{S}$ is an element of $\{-1, 0, 1\}^{n \times \overline{n}}$. Due to the above mentioned similarity with Frodo, we use analogous structure of queries to the oracle as in the attack against Frodo. Because the range of the secret coefficients is only $[-1, 1]$, it was possible to target also triplets and quadruplets of secret coefficients. When targeting coefficients individually, we again constructed binary trees manually. For pairs, triplets and quadruplets, we had to use a recursive search to find suitable binary trees. We used the same heuristic as for Frodo described above.

Key Mismatch Attack Against RLWR Variant. We recall that the secret key $\mathbf{S}$ is a polynomial of degree at most $n - 1$ with coefficients from $\{-1, 0, 1\}$. We use similar structure of queries as in the non-ring case, but the situation is more complicated now. The first problem is that, similarly to ThreeBears, only a part of the polynomial is used within the decode function. Hence, we need to shift the polynomials accordingly, which adds extra noise to the polynomials and makes the attack more difficult. Moreover, when targeting pairs (or triplets) it is more difficult to combine two (three) coefficients into one entry of a polynomial. To resolve these problems, we need to have several binary trees and we choose a suitable tree during the attack based on the previous recovered coefficients.

Error-Correcting Code. Actually, it is easier to target the variants using the error-correcting code. The reason is that these variants are working in a different polynomial ring where it is easier to shift polynomials and to combine several coefficients into one entry. Otherwise, the structure of queries is similar, but it is enough to have only one binary tree. We use the same trick as in the attack against ThreeBears to handle the issue with the error-correcting code.

Results and Comparison. In [6], a key mismatch attack against a predecessor of Round5, HILA5, was published. Despite the fact that it was referenced as the attack against Round5 in [11], we think that the two schemes are quite different and hence we do not compare our attack with the attack from [6].

In [15], they use the same attack methodology used against Frodo also against Round5. They target the ring variants with the error-correcting code. For these variants, our attack requires about 16–33% less queries to recover the secret key.

Moreover, in [15] they argue that their method is also applicable to other variants of Round5, but they do not define the structure of queries nor provide any concrete results.

We want to point out that in our opinion, the attack against the variants of Round5 which do not use the error-correcting code is more difficult because of the different polynomials used in these variants. Hence, the queries must be chosen more carefully which is also the reason why we use more binary trees for these variants.

5   Conclusion

In this paper, we presented the first (to our best knowledge) key mismatch attack against ThreeBears and we improved the key mismatch attacks on Frodo and Round5.

We found out that it is often more efficient to target coefficients in tuples instead of just one by one. The biggest difference was for the variant R5ND_1CPA_0d of Round5, where targeting triplets of secret coefficients saved more than 40% of queries to the oracle compared to targeting the coefficients individually. On the other hand, it is more difficult to construct the binary trees for bigger tuples since the trees are also bigger and at some point, it starts to be infeasible to improve the results by targeting bigger tuples.

Our method using the binary trees provides a general technique to target LWE-based schemes. We applied the same method against NewHope [16], where we improved the state-of-the-art by a wide margin, and also against Kyber and Saber, where we achieved minor improvements, usually between 3–10%.

Table 2. Comparison of our results, d is the dimension of the attack

	d	Security level	# coeffs	$\mathbb{E}$[one coeff]	$\mathbb{E}$[whole key]
BABYBEAR	1	1	624	2.31	1443
FRODOPKE-640	2	1	5120	3.56	18 239
R5N1_1CPA_0D	4	1	4158	1.39	5790
R5ND_1CPA_0D	3	1	618	1.11	687
R5ND_1CPA_5D	4	1	490	1.34	656
MAMABEAR	1	3	936	2.30	2150
FRODOPKE-976	2	3	7808	3.29	25 672
R5N1_3CPA_0D	4	3	7048	1.20	8436
R5ND_3CPA_0D	2	3	786	1.55	1221
R5ND_3CPA_5D	4	3	756	1.69	1277
PAPABEAR	1	5	1248	2.28	2847
FRODOPKE-1344	2	5	10752	2.60	28 008
R5N1_5CPA_0D	3	5	9488	1.58	15 005
R5ND_5CPA_0D	2	5	1018	1.50	1525
R5ND_5CPA_5D	4	5	940	1.65	1551

A   Proof of Lemma 1

Proof. Let start with the first case from Eq. 2, i.e.

$$\mathbf{C}_1[k] = \begin{cases} 55 \cdot q^{156} & \text{if } k = i \\ 0 & \text{if } k \neq i. \end{cases} \tag{6}$$

Then

$$\mathbf{C} = clar \cdot \mathbf{C}_1 \mathbf{S} \bmod N = 55 \cdot \mathbf{S}[i] \bmod N.$$

We call $\mathbf{S}[i]$ just $\mathbf{S}$ from now on. Let first assume that $\mathbf{S} \geq 0$. Then

$$\mathbf{C} = 55 \cdot \mathbf{S} \bmod N = 55 \cdot \mathbf{S}.$$

So according to the definition of the DECODE function, we have that

$$m'[l] = \left\lfloor \frac{2 \cdot encr[l] - \text{EXTRACT}_5(\mathbf{C}, l)}{16} \right\rceil \bmod 2 = \left\lfloor \frac{2 \cdot 0 - \text{EXTRACT}_5(\mathbf{C}, l)}{16} \right\rceil \bmod 2 \tag{7}$$

According to the definition of EXTRACT function, we have

$$\text{EXTRACT}_5(\mathbf{C}, l) = \left\lfloor \frac{\mathbf{C} \cdot 32}{q^{j'+1}} \right\rfloor \tag{8}$$

for some j' depending on l. We know that

$$\mathbf{C} = 55 \cdot \mathbf{S} = 55 \cdot \sum_{i=0}^{311} s_i \cdot q^i,$$

which we substitute back to the Eq. 8, which leads to

$$
\begin{aligned}
\text{EXTRACT}_5(\mathbf{C}, l) &= \left\lfloor \frac{32 \cdot 55 \cdot \sum_{i=0}^{311} s_i \cdot q^i}{q^{j'+1}} \right\rfloor \\
&= \left\lfloor 32 \cdot 55 \cdot \sum_{m=0}^{j'} \frac{s_m}{q^{j'+1-m}} + 32 \cdot 55 \cdot \sum_{m=j'+1}^{311} s_m \cdot q^{m-j'-1} \right\rfloor \quad (9) \\
&= \epsilon + 32 \cdot 55 \cdot \sum_{m=j'+1}^{311} s_m \cdot q^{m-j'-1}
\end{aligned}
$$

with $\epsilon \in [-4, 3]$, because s_m is small (see definition of χ and ψ). Substituting this to the Eq. 7, we have

$$
\begin{aligned}
m'[l] &= \left\lfloor \frac{\epsilon - 32 \cdot 55 \cdot \sum_{m=j'+1}^{311} s_m \cdot q^{m-j'-1}}{16} \right\rceil \bmod 2 \\
&= \left(\left\lfloor \frac{\epsilon}{16} \right\rceil - 2 \cdot 55 \cdot \sum_{m=j'+1}^{311} s_m \cdot q^{m-j'-1} \right) \bmod 2 \quad (10) \\
&= \left\lfloor \frac{\epsilon}{16} \right\rceil \bmod 2 = 0.
\end{aligned}
$$

For the case $\mathbf{S} < 0$, we have $\mathbf{S} \bmod N = N + \mathbf{S} = q^{312} - q^{156} - 1 + \mathbf{S}$ and the rest of the proof is similar. For the $\mathbf{C}_1$ having the other forms, we have to take into account the multiplication by other powers of q and the proof is more technical and shows actually the motivation for the value 55.[8]

B   Proof of $m'[l] = 1$

Proof.

$$
\begin{aligned}
m'[l] &= \left\lfloor \frac{2 \cdot encr[l] - \text{EXTRACT}_5(\mathbf{C}, l)}{16} \right\rceil \bmod 2 \\
&= \left\lfloor \frac{16 - \text{EXTRACT}_5(\mathbf{C}, l)}{16} \right\rceil \bmod 2 \\
&= \left(1 + \left\lfloor \frac{-\text{EXTRACT}_5(\mathbf{C}, l)}{16} \right\rceil \right) \bmod 2 \\
&= 1,
\end{aligned}
$$

---

[8] This part of the proof would work with any constant $\in [-127, 127]$ instead of 55.

where we used that

$$\left\lceil \frac{-\text{EXTRACT}_5(\mathbf{C}, l)}{16} \right\rceil \bmod 2 = 0,$$

which we know from Lemma 1.

References

1. Alkim, E., et al.: FrodoKEM - specifications and supporting documentation (2019)
2. Baan, H., et al.: Round5: KEM and PKE based on (Ring) Learning with Rounding (2020)
3. Băetu, C., Durak, F.B., Huguenin-Dumittan, L., Talayhan, A., Vaudenay, S.: Misuse attacks on post-quantum cryptosystems. In: Ishai, Y., Rijmen, V. (eds.) EUROCRYPT 2019. LNCS, vol. 11477, pp. 747–776. Springer, Cham (2019). https://doi.org/10.1007/978-3-030-17656-3_26
4. Banerjee, A., Peikert, C., Rosen, A.: Pseudorandom functions and lattices. In: Pointcheval, D., Johansson, T. (eds.) EUROCRYPT 2012. LNCS, vol. 7237, pp. 719–737. Springer, Heidelberg (2012). https://doi.org/10.1007/978-3-642-29011-4_42
5. Bauer, A., Gilbert, H., Renault, G., Rossi, M.: Assessment of the key-reuse resilience of NewHope. In: Matsui, M. (ed.) CT-RSA 2019. LNCS, vol. 11405, pp. 272–292. Springer, Cham (2019). https://doi.org/10.1007/978-3-030-12612-4_14
6. Bernstein, D.J., Groot Bruinderink, L., Lange, T., Panny, L.: HILA5 Pindakaas: on the CCA security of lattice-based encryption with error correction. In: Joux, A., Nitaj, A., Rachidi, T. (eds.) AFRICACRYPT 2018. LNCS, vol. 10831, pp. 203–216. Springer, Cham (2018). https://doi.org/10.1007/978-3-319-89339-6_12
7. Bos, J., et al.: Frodo: take off the ring! practical, quantum-secure key exchange from LWE. In: Proceedings of the 2016 ACM SIGSAC Conference on Computer and Communications Security, CCS 2016, pp. 1006–1018. Association for Computing Machinery, New York (2016). https://doi.org/10.1145/2976749.2978425
8. Fluhrer, S.: Cryptanalysis of Ring-LWE based key exchange with key share reuse. Cryptology ePrint Archive, Report 2016/085 (2016). https://eprint.iacr.org/2016/085
9. Gu, C.: Integer version of Ring-LWE and its applications. In: Meng, W., Furnell, S. (eds.) SocialSec 2019. CCIS, vol. 1095, pp. 110–122. Springer, Singapore (2019). https://doi.org/10.1007/978-981-15-0758-8_9
10. Hamburg, M.: Post-quantum cryptography proposal: threebears (2019)
11. Huguenin-Dumittan, L., Vaudenay, S.: Classical misuse attacks on NIST round 2 PQC. In: Conti, M., Zhou, J., Casalicchio, E., Spognardi, A. (eds.) ACNS 2020. LNCS, vol. 12146, pp. 208–227. Springer, Cham (2020). https://doi.org/10.1007/978-3-030-57808-4_11
12. Okada, S., Wang, Y., Takagi, T.: Improving key mismatch attack on NewHope with fewer queries. In: Liu, J.K., Cui, H. (eds.) ACISP 2020. LNCS, vol. 12248, pp. 505–524. Springer, Cham (2020). https://doi.org/10.1007/978-3-030-55304-3_26
13. Qin, Y., Cheng, C., Ding, J.: A complete and optimized key mismatch attack on NIST candidate NewHope. In: Sako, K., Schneider, S., Ryan, P.Y.A. (eds.) ESORICS 2019. LNCS, vol. 11736, pp. 504–520. Springer, Cham (2019). https://doi.org/10.1007/978-3-030-29962-0_24

14. Qin, Y., Cheng, C., Ding, J.: An efficient key mismatch attack on the NIST second round candidate Kyber. Cryptology ePrint Archive, Report 2019/1343 (2019). https://eprint.iacr.org/2019/1343
15. Ravi, P., Sinha Roy, S., Chattopadhyay, A., Bhasin, S.: Generic side-channel attacks on CCA-secure lattice-based PKE and KEMs. IACR Trans. Cryptogr. Hardw. Embed. Syst. **2020**(3), 307–335 (2020). https://doi.org/10.13154/tches.v2020.i3.307-335. https://tches.iacr.org/index.php/TCHES/article/view/8592
16. Vacek, J., Václavek, J.: Key mismatch attack on newhope revisited. Cryptology ePrint Archive, Report 2020/1389 (2020). https://eprint.iacr.org/2020/1389

A New Non-random Property
of 4.5-Round PRINCE

Bolin Wang[1,2(✉)], Chan Song[1,2], Wenling Wu[1,2(✉)], and Lei Zhang[1,2]

[1] TCA Laboratory, SKLCS, Institute of Software, Chinese Academy of Sciences,
Beijing 100190, People's Republic of China
{wangbolin,songchan,wwl,zhanglei}@tca.iscas.ac.cn
[2] University of Chinese Academy of Sciences,
Beijing 100049, People's Republic of China

Abstract. PRINCE is a widely analyzed block cipher proposed in 2012. Subspace trail cryptanalysis is a new cryptanalytic technique to generalize the invariant subspace attack. So far, two subspace trails that exist with probability 1 are known for 2.5 rounds of PRINCE. In this paper, we first describe a new non-random property for 4.5 rounds of PRINCE based on subspace trail with certain probability, which is independent of the secret key, the details of the Linear layer and of the S-Box layer. Then, we obtain that by appropriate choices of difference for a number of input pairs, it is possible to make sure that the number of times that the difference of the resulting output pairs lie in a particular subspace is always a multiple of 8. Later, a detailed proof is given as why it has to exist. Relying on this property, a new distinguisher can be set up to distinguish the 4.5-round PRINCE from a random permutation.

Keywords: PRINCE · Subspace trail · Block cipher · Structural property · Permutation

1 Introduction

The area of lightweight cryptography, i.e., ciphers with particularly low implementation costs, has drawn considerable attention over the last years. Most of the existing lightweight block cipher algorithms are influenced by the design principles of DES and AES. PRINCE is a low-latency block cipher proposed at ASIACRYPT 2012 [1]. It is an iterated block cipher structured as a substitution-permutation network (SPN). And a round of PRINCE is similar to a round of AES, with one exception: the MixColumns operation of the forward round function is before the ShiftRows operation. PRINCE has a new, original feature called the α-reflection property that involves a specific fixed parameter α. Because of this property, decryption with round key K is identical to encryption with round key $K \oplus \alpha$, which significantly reduces the cost of implementation of decryption.

Over the past few years, with the development of lightweight block cipher algorithms, some new cryptanalytic methods have been proposed. Invariant

© Springer Nature Switzerland AG 2021
D. Hong (Ed.): ICISC 2020, LNCS 12593, pp. 199–212, 2021.
https://doi.org/10.1007/978-3-030-68890-5_11

attack method is one of them. Invariant subspace attack was first proposed by Leander et al. [2] in CRYPTO 2011, which was used to analyze the block cipher PRINT [3]. Later, Bulygin et al. [4] conducted a complete study on the invariant subspace attack of PRINT. In EUROCRYPT 2015, Leander et al. [5] proposed a generic algorithm for detecting invariant subspaces and applied it to iSCREAM [6], Robin [7] and Zorro [8]. In 2016, Guo et al. [9] applied this method to conduct a full-round attack on Midori64 [10]. In CRYPTO 2017, Beierle et al. [11] analyzed the influence of the selection of round constants and linear layer on resisting invariant subspace attack. In ASIACRYPT 2016, Todo et al. [12] proposed nonlinear invariant attack. This new method is able to distinguish the full versions of the (tweakable) block ciphers SCREAM, iSCREAM and Midori64 in a weak-key setting. In order to eliminate the influence of round constants, Wei et al. [13] proposed a generalized nonlinear invariant attack using a pair of constants in the input of nonlinear invariant functions. The subspace trail cryptanalysis is a generalization of invariant subspace attack, which was introduced by Grassi et al in [14]. The authors implemented competitive key-recovery attack with very low data complexity on 2, 3 and 4 rounds of AES. In EUROCRYPT 2017, Grassi et al. [15] described a new structural property for up to 5 rounds of AES.

PRINCE has received considerable attention with many different attacks on round-reduced version since it has been proposed. In [16], Derbez and Perrin described attacks relying on a Meet-in-the-Middle approach, applicable (theoretically) up to 10 rounds. In [17], Morawiecki introduced up to 7 rounds of integral and higher-order differential cryptanalysis based on a 3.5-round distinguisher with one active nibble. In [18], Posteuca and Negara found a 4.5-round integral distinguisher which needs three (not arbitrary) active nibbles instead of one. Due to the involution structure of PRINCE, an improved differential attack was proposed in [19]. A related work on truncated differentials was presented in [20], which showed the existence of 5- and 6-round truncated difference distinguishers. In [21], the first application of reflection cryptanalysis on PRINCE-like ciphers is presented. The authors showed that there exist values of α which would allow a key-recovery attack on the full 12-round cipher. Moreover, because of new cryptanalysis method, new design criteria concerning the selection of the value of α for PRINCE-like ciphers are obtained.

In [22], Grassi et al. set up truncated differential attacks on round-reduced PRINCE that exploit subspace trails. However, the subspace trails used in their paper exist with probability 1. As we will argue below, it is possible to find a structural property of PRINCE which exploits subspace trails with certain probability.

This paper is organized as follows. In Sect. 2, we start by recalling the block cipher PRINCE and the subspace trail cryptanalysis, and then continue to introduce the existing subspaces of PRINCE. In Sect. 3, we illustrate our new subspace trails for 4.5 rounds of PRINCE. Later, we present a new structural property of PRINCE using the two 4.5-round subspace trails. In Sect. 4 we present our proof details. Finally, we conclude in Sect. 5.

2   Preliminaries

2.1   The Block Cipher PRINCE

The lightweight PRINCE cipher [1] uses a block size of 64 bits and a key size of 128 bits. The cipher state is conceptually arranged in a 4×4 grid where every cell represents a nibble. PRINCE is based on the so-called FX construction, where one part of the key is used for a core cipher F, which contains the major encryption process, and the remaining parts are used for whitenings before and after the core: $FX_{k,k_1,k_2} = k_2 \oplus F_k(x \oplus k_1)$. The key is first split into two parts of 64 bits each (i.e. $k = k_0||k_1$), and then it is expanded into 192 with a simple linear transformation:

$$(k_0||k_1) \rightarrow (k_0||k_0'||k_1) := (k_0||(k_0 \ggg 1) \oplus (k_0 \lll 63)||k_1)$$

The 64-bit subkeys k_0 and k_0' are used as whitening keys to the underlying block cipher called PRINCEcore, while the 64-bit key k_1 is used for the 12-round PRINCEcore.

Each round of PRINCEcore consists of an S-box layer, a linear layer, a ShiftRows operation, a key addition and the addition of a round constant.

S-Box Layer. The cipher uses a 4-bit S-Box. The action of the S-Box in hexadecimal notation is given by the following table (Table 1).

Table 1. S-Box of PRINCE

x	0	1	2	3	4	5	6	7	8	9	A	B	C	D	E	F
$S[x]$	B	F	3	2	A	C	9	1	6	7	8	0	E	5	D	4

Linear Layer M': In the linear layer, the 64-bit state is multiplied with a 64×64 matrix. More precisely, two 16×16 submatrices $\hat{M}^{(0)}$ and $\hat{M}^{(1)}$ are arranged on the diagonal of a bigger matrix, where every submatrix affects a 16-bit chunk x_i of the 64-bit state $x = (x_1||x_2||x_3||x_4)$:

$$M' \cdot x = (\hat{M}^{(0)} \cdot x_1 || \hat{M}^{(1)} \cdot x_2 || \hat{M}^{(1)} \cdot x_3 || \hat{M}^{(0)} \cdot x_4).$$

ShiftRows Operation SR: Equal to the one in the AES.

RC_i**-add:** A 64-bit round constant is xored with the state.

k_i**-add:** The 64-bit state is xored with the 64-bit subkey.

In the last 5 rounds (the backward rounds), the order of operations is inverse with respect to the first 5 rounds (the forward rounds), where only the round constants differ. The middle rounds consist of three key-less operations: an S-Box layer, a matrix multiplication with M' and an inverse S-Box layer. The difference between RC_i and RC_{11-i} is always equal to a constant α and since M' is self-inverting (i.e. $M' = M'^{-1}$), the core cipher has the so called α-reflection property, i.e. the core cipher is such that the inverse of PRINCEcore parametrized with k is equal to PRINCEcore parametrized with $k \oplus \alpha : D_{(k_0||k_0'||k_1)}(\cdot) = E_{(k_0'||k_0||k_1 \oplus \alpha)}(\cdot)$.

2.2   Subspace Trails

Let F denote a round function in an iterated block cipher $E_K(\cdot)$. Assume there exists a coset $V \oplus a$ such that $F(V \oplus a) = V \oplus b$, then if the round key $K \in V \oplus a \oplus b$, it follows that $F(V \oplus a) \oplus K = V \oplus a$. We say that $V \oplus a$ is an invariant coset of the subspace V for the function F. In [14], the authors generalized the above concept to subspace trails. The specific definition is as follows.

Definition 1. *Let* $(V_1, V_2, \cdots, V_{r+1})$ *denote a set of* $r + 1$ *subspaces with* $dim(V_i) \leq dim(V_{i+1})$. *If for each* $i = 1, 2, \cdots, r$ *and for each* $a_i \in V_i^{\perp}$, *there exists (unique)* $a_{i+1} \in V_{i+1}^{\perp}$ *such that*

$$F(V_i \oplus a_i) \subseteq V_{i+1} \oplus a_{i+1},$$

then $(V_1, V_2, \cdots, V_{r+1})$ *is a subspace trail of length* r *for the function* F. *If the previous relation holds with equality, then the trail is called a constant-dimensional subspace trail.*

2.3   Subspaces of PRINCE

In this section, we recall the subspace trails of PRINCE presented in [22]. We denote five families of subspaces essential to PRINCE: the column subspaces C_I, the diagonal subspaces D_I, the inverse-diagonal subspaces ID_I, the mixed subspaces M_I and the inverse-mixed subspaces IM_I. Moreover, let $E = \{e[0], \cdots, e[15]\}$ denote the unit vectors of $F_{2^4}^{16}$ (e_i has a single 1 in position i).

Definition 2. *(Column subspaces) The column subspaces* C_i *are defined as:*

$$C_i = \langle e[4 \cdot i], e[4 \cdot i + 1], e[4 \cdot i + 2], e[4 \cdot i + 3] \rangle.$$

For instance, C_0 corresponds to matrix representation:

$$C_0 = \left\{ \begin{bmatrix} x & 0 & 0 & 0 \\ y & 0 & 0 & 0 \\ z & 0 & 0 & 0 \\ w & 0 & 0 & 0 \end{bmatrix} \middle| \forall x, y, z, w \in F_{2^4} \right\} \equiv \begin{bmatrix} x & 0 & 0 & 0 \\ y & 0 & 0 & 0 \\ z & 0 & 0 & 0 \\ w & 0 & 0 & 0 \end{bmatrix}.$$

Definition 3. *(Diagonal subspaces, Inverse-diagonal subspaces) The diagonal subspaces* D_i *and inverse-diagonal subspaces* ID_i *are defined as*

$$D_i = SR(C_i), ID_i = SR^{-1}(C_i).$$

For instance, D_0 and ID_0 correspond to matrix representations:

$$D_0 = \begin{bmatrix} x & 0 & 0 & 0 \\ 0 & 0 & 0 & y \\ 0 & 0 & z & 0 \\ 0 & w & 0 & 0 \end{bmatrix}, ID_0 = \begin{bmatrix} x & 0 & 0 & 0 \\ 0 & y & 0 & 0 \\ 0 & 0 & z & 0 \\ 0 & 0 & 0 & w \end{bmatrix}.$$

Definition 4. *(Mixed subspaces, Inverse-mixed subspaces) The mixed subspaces* M_i *and inverse-mixed subspaces* IM_i *are defined as:*

$$M_i = M'(D_i), \quad IM_i = M'(ID_i).$$

For instance, M_0 and IM_0 correspond to matrix representations:

$$M_0 = \begin{bmatrix} \alpha_3(x) & \alpha_3(w) & \alpha_0(z) & \alpha_2(y) \\ \alpha_2(x) & \alpha_2(w) & \alpha_3(z) & \alpha_1(y) \\ \alpha_1(x) & \alpha_1(w) & \alpha_2(z) & \alpha_0(y) \\ \alpha_0(x) & \alpha_0(w) & \alpha_1(z) & \alpha_3(y) \end{bmatrix}, IM_0 = \begin{bmatrix} \alpha_3(x) & \alpha_1(y) & \alpha_0(z) & \alpha_0(w) \\ \alpha_2(x) & \alpha_0(y) & \alpha_3(z) & \alpha_3(w) \\ \alpha_1(x) & \alpha_3(y) & \alpha_2(z) & \alpha_2(w) \\ \alpha_0(x) & \alpha_2(y) & \alpha_1(z) & \alpha_1(w) \end{bmatrix}$$

where $\alpha_i(\cdot)$ are defined as

$$\alpha_i(x) = x \wedge (0x2^i \oplus 0xf)$$

and where $\wedge$ is the *and (logic) operator*.

Given $I \subseteq \{0, 1, 2, 3\}$, subspaces $C_I, D_I, ID_I, M_I, IM_I$ are defined as:

$$C_I = \bigoplus_{i \in I} C_i, \quad D_I = \bigoplus_{i \in I} D_i, \quad ID_I = \bigoplus_{i \in I} ID_i, \quad M_I = \bigoplus_{i \in I} M_i, \quad IM_I = \bigoplus_{i \in I} IM_i$$

In this paper, we are working over the field $GF(2)$. The dimension of any of the subspaces $C_I, D_I, ID_I, M_I, IM_I$ is $16 \cdot |I|$.

In the following, we present two subspace trails for 2.5 rounds of PRINCE in [22]. Let R denote one round of PRINCE, $ARK(\cdot)$ means a bit-wise XOR with the secret key k_1. To simplify the notation, we denote by *super-SBox* the middle two rounds.

Theorem 1. *Let* $I \subseteq \{0, 1, 2, 3\}$. *For each* $a \in C_I^\perp$, *there exists unique* $b \in M_I^\perp$ *such that* $R^{1+1.5}(C_I \oplus a) = M_I \oplus b$, *where* b *depends on* a *and on the secret key. Equivalently:*

$$Prob(R^{(1+1.5)}(x) \oplus R^{(1+1.5)}(y) \in M_I | x \oplus y \in C_I) = 1.$$

This means that a coset of C_I is certainly mapped into a coset of M_I after 2.5 rounds:

$$C_I \oplus a \xrightarrow{R \circ ARK(\cdot)} D_I \oplus b \xrightarrow{M' \circ S - Box(\cdot)} M_I \oplus c.$$

The middle rounds without the final S-Box are denoted by 1.5 rounds.

Theorem 2. *Let* $I \subseteq \{0, 1, 2, 3\}$. *For each* $a \in C_I^\perp$, *there exists unique* $b \in IM_I^\perp$ *such that* $R^{2+0.5}(C_I \oplus a) = IM_I \oplus b$, *where* b *depends on* a *and on the secret key. Equivalently:*

$$Prob(R^{(2+0.5)}(x) \oplus R^{(2+0.5)}(y) \in IM_I | x \oplus y \in C_I) = 1.$$

This means that a coset of C_I is mapped into a coset of IM_I after 2.5 rounds:

$$C_I \oplus a \xrightarrow{super - SBox \circ ARK(\cdot)} C_I \oplus b \xrightarrow{M' \circ SR^{-1} \circ ARK(\cdot)} IM_I \oplus c.$$

The linear part of the next round is defined as 0.5 rounds.

These two theorems state that the two 2.5-round subspace trails exist with probability 1 and both start with a coset of C_I. Therefore, we consider two elements that belong to the same coset of C_I, which belong to the same coset of D_I with probability 1 after one round, and belong to the same coset of C_Q with certain probability after one more round. In the following section, we will continue with these properties of PRINCE subspace trails.

3   New Structural Property of 4.5-Round PRINCE

In this section, we present two subspace trails for 4.5 rounds of PRINCE with certain probability. Moreover, we propose a new property for PRINCE based on the two 4.5-round subspace trails.

3.1   4.5-Round Subspace Trails for PRINCE

Using the two subspace trails of PRINCE in Sect. 2, it is possible to extend backward by two rounds to obtain two 4.5-round subspace trails. Before we go on, we have the following proposition that is analogous in [14].

Proposition 1. *For any D_I and C_J, we have that*

$$Prob(x \in C_J | x \in D_I) = 2^{-16|I|+4|I| \cdot |J|}.$$

Proof. Let $Y \in D_I \cap C_J$. We have that dimension$(Y) =$ dimension$(D_I \cap C_J) = 4|I| \cdot |J|$. Let Z be the subspace of dimension $16|I| - 4|I| \cdot |J|$ such that $D_I = Y \oplus Z$, and let π_Y and π_Z be the projection of D_I on Y and Z respectively. That is

$$\pi_Y : D_I \to Y, \qquad \pi_Y(x) = x_y,$$
$$\pi_Z : D_I \to Z, \qquad \pi_Z(x) = x_z.$$

It follows $\forall x \in D_I$, there exists unique $x_y \in Y$ and $x_z \in Z$ such that $x = x_y \oplus x_z$. Thus, we can obtain that $Pr(x \in C_J | x \in D_I) = Pr(\pi_z(x) = 0 | x \in D_I)$. Since Z has dimension $16|I| - 4|I| \cdot |J|$, we get that:

$$Pr(x \in C_J | x \in D_I) = Pr(\pi_z(x) = 0 | x \in D_I) = 2^{-16|I|+4|I| \cdot |J|}.$$

According to this proposition, given two texts in the same coset of C_I, then they belong to the same coset of D_I after one forward round, while they are in the same coset of $D_I \cap C_J$ with probability $2^{-16|I|+4|I| \cdot |J|}$. Because of $D_I \cap C_J \subseteq C_J$, a coset of C_J is mapped into a coset of D_J after one more round. Considering the intersection of D_J and C_Q again, we can obtain that if two elements belong to the same coset of D_J, then they belong to the same coset of C_Q with probability $2^{-16|J|+4|J| \cdot |Q|}$.

As we have just seen, considering the first subspace trail for $1 + 1.5$ rounds of PRINCE, a coset of C_Q is mapped into a coset of M_Q after 2.5 rounds. Using

Proposition 1 twice, we can set up a 4.5-round subspace trail for PRINCE. It follows that if two texts belong to the same coset of C_I, then the probability that they belong to the same coset of M_Q after 4.5 rounds is equal to $2^{-16|I|+4|I|\cdot|J|} \cdot 2^{-16|J|+4|J|\cdot|Q|}$.

Same analysis can be applied to the subspace trail for $2 + 0.5$ rounds of PRINCE. If two texts belong to the same coset of C_I, then the probability that they belong to the same coset of IM_Q after 4.5 rounds is equal to $2^{-16|I|+4|I|\cdot|J|} \cdot 2^{-16|J|+4|J|\cdot|Q|}$. As a result, the following theorems hold.

Theorem 3. *Let $I, J, Q \subseteq \{0, 1, 2, 3\}$ where $0 < |I| \leq 3, 0 < |J| \leq 3, 0 < |Q| \leq 3$. For any I, J and Q, we have that $R^{m_1}(C_I \oplus a) = M_Q \oplus e$ with probability $2^{-16|I|+4|I|\cdot|J|} \cdot 2^{-16|J|+4|J|\cdot|Q|}$, where the input and output of second round need to consider the intersection of D_I and C_J, D_J and C_Q respectively. Equivalently:*

$$Prob(R^{m_1}(x) \oplus R^{m_1}(y) \in M_Q | x \oplus y \in C_I) = 2^{-16|I|+4|I|\cdot|J|} \cdot 2^{-16|J|+4|J|\cdot|Q|},$$

$$C_I \oplus a \xrightarrow{R(\cdot)} D_I \oplus b \xrightarrow{R(\cdot)} C_Q \oplus c \xrightarrow{R(\cdot)} D_Q \oplus d \xrightarrow{\Lambda(\cdot)} M_Q \oplus e$$

where $m_1 = 2 + 1 + 1.5$ and $\Lambda(\cdot) = M' \circ S - Box(\cdot)$.

Theorem 4. *Let $I, J, Q \subseteq \{0, 1, 2, 3\}$ where $0 < |I| \leq 3, 0 < |J| \leq 3, 0 < |Q| \leq 3$. For any I, J and Q, we have that $R^{m_2}(C_I \oplus a) = IM_Q \oplus e$ with probability $2^{-16|I|+4|I|\cdot|J|} \cdot 2^{-16|J|+4|J|\cdot|Q|}$, where the input and output of second round need to consider the intersection of D_I and C_J, D_J and C_Q respectively. Equivalently:*

$$Prob(R^{m_2}(x) \oplus R^{m_2}(y) \in IM_Q | x \oplus y \in C_I) = 2^{-16|I|+4|I|\cdot|J|} \cdot 2^{-16|J|+4|J|\cdot|Q|},$$

$$C_I \oplus a \xrightarrow{R(\cdot)} D_I \oplus b \xrightarrow{R(\cdot)} C_Q \oplus c \xrightarrow{\Gamma_1(\cdot)} C_Q \oplus d \xrightarrow{\Gamma_2(\cdot)} IM_Q \oplus e$$

where $m_2 = 2 + 2 + 0.5$, $\Gamma_1(\cdot) = super - SBox \circ ARK(\cdot)$ and $\Gamma_2(\cdot) = M' \circ SR^{-1} \circ ARK(\cdot)$.

As a consequence, we obtain two subspace trails for 4.5 rounds of PRINCE with certain probability.

3.2   New Property of 4.5-Round PRINCE

Using the first 4.5-round subspace trail presented in Theorem 3, it is possible to show a new structural property of 4.5-round PRINCE. Let's consider a set of plaintexts in the same coset of column subspace C_I, that is $C_I \oplus a$ for a certain $a \in C_I^\perp$, and the corresponding ciphertexts after 4.5 rounds. In order to set up the distinguisher on 4.5 even more rounds of PRINCE, the core idea is to count the number of different pairs of ciphertexts that belong to the same coset of M_Q for a fixed Q. Through analysis, we can prove that for 4.5-round PRINCE this number is a multiple of 8 with probability 1. Instead, for a random permutation the same number does not have any special property. Therefore, this allows to distinguish 4.5-round PRINCE from a random permutation. The same property holds for the second 4.5-round subspace trail of PRINCE in Theorem 4. The above property of PRINCE can be summarized as the following theorems.

Theorem 5. *Let C_I and M_Q be the subspaces defined as before for certain fixed I and Q, and assume $|I| = 1$. Given an arbitrary coset of C_I, consider all the 2^{16} plaintexts and the corresponding ciphertexts after 4.5 rounds (three rounds before + middle 1.5 rounds), that is (p^i, c^i) for $i = 0, \cdots, 2^{16} - 1$ where $p^i \in C_I \oplus a$ and $c^i = R^{2+1+1.5}(p^i)$. The number n of different pairs of ciphertexts (c^i, c^j) for $i \neq j$ such that $c^i \oplus c^j \in M_Q$ (i.e. c^i and c^j belong to the same coset of M_Q)*

$$n := |\{(p^i, c^i), (p^j, c^j) | \forall\ p^i, p^j \in C_I \oplus a, p^i < p^j\ and\ c^i \oplus c^j \in M_Q\}|$$

is a multiple of 8, that is $\exists n' \in N$ such that $n = 8 \cdot n'$.

Theorem 6. *Let C_I and IM_Q be the subspaces defined as before for certain fixed I and Q, and assume $|I| = 1$. Given an arbitrary coset of C_I, consider all the 2^{16} plaintexts and the corresponding ciphertexts after 4.5 rounds (two rounds before + middle 2 rounds + 0.5 round), that is (p^i, c^i) for $i = 0, \cdots, 2^{16} - 1$ where $p^i \in C_I \oplus a$ and $c^i = R^{2+2+0.5}(p^i)$. The number n of different pairs of ciphertexts (c^i, c^j) for $i \neq j$ such that $c^i \oplus c^j \in IM_Q$ (i.e. c^i and c^j belong to the same coset of IM_Q)*

$$n := |\{(p^i, c^i), (p^j, c^j) | \forall\ p^i, p^j \in C_I \oplus a, p^i < p^j\ and\ c^i \oplus c^j \in IM_Q\}|$$

is a multiple of 8, that is $\exists n' \in N$ such that $n = 8 \cdot n'$. "$<$" in Theorem 5 and 6 means the partial order [15].

In the following, we will prove Theorem 5 in detail, and Theorem 6 has similar analysis. Since

$$C_I \oplus a \xrightarrow{R(\cdot)} D_I \oplus b \xrightarrow{R(\cdot)} C_Q \oplus c \xrightarrow{R^{1+1.5}(\cdot)} M_Q \oplus d,$$

the main idea is to focus on the second round $D_I \oplus b \xrightarrow{R(\cdot)} C_Q \oplus b$ in order to prove the statement of Theorem 5. Therefore, the proof of Theorem 5 is related to the following lemma on 1-round PRINCE.

Lemma 1. *Let D_I and C_Q be the subspaces defined as before for certain fixed I and Q, and assume $|I| = 1$. Given an arbitrary coset of D_I, consider all the 2^{16} plaintexts and the corresponding ciphertexts after 1 round, that is $(\hat{p}^i, \hat{c}^i)$ for $i = 0, \cdots, 2^{16} - 1$ where $\hat{c}^i = R(\hat{p}^i)$. The number n of different pairs of ciphertexts $(\hat{c}^i, \hat{c}^j)$ for $i \neq j$ such that $\hat{c}^i \oplus \hat{c}^j \in C_Q$ (i.e. $\hat{c}^i$ and $\hat{c}^j$ belong to the same coset of C_Q) is a multiple of 8, that is $\exists n' \in N$ such that $n = 8 \cdot n'$.*

The complete proof is provided in the next section - Sect. 4. The proof of Theorem 5 follows immediately by the proof of Lemma 1.

4   A Detailed Proof of Lemma 1 and Theorem 5

In this section we will give a formal and detailed proof of Theorem 5. As already known, it is sufficient to prove Lemma 1 in order to prove Theorem 5. Firstly, let us concentrate on Lemma 1.

Consider two elements p^1 and p^2 in the same coset of $D_i \oplus a$ for $a \in D_i^\perp$. Without loss of generality (W.l.o.g.), assume $i = 0$ (it is analogous for the other cases). By definition of D_i, there exist $x, y, z, w \in F_{2^4}$ and $x', y', z', w' \in F_{2^4}$ such that:

$$p^1 = a \oplus \begin{bmatrix} x & 0 & 0 & 0 \\ 0 & 0 & 0 & y \\ 0 & 0 & z & 0 \\ 0 & w & 0 & 0 \end{bmatrix}, \quad p^2 = a \oplus \begin{bmatrix} x' & 0 & 0 & 0 \\ 0 & 0 & 0 & y' \\ 0 & 0 & z' & 0 \\ 0 & w' & 0 & 0 \end{bmatrix}.$$

For the following, we say that p^1 is "generated" by the variables $\langle x, y, z, w \rangle$ and that p^2 is "generated" by the variables $\langle x', y', z', w' \rangle$.

First Case. Firstly, we consider the case in which three variables are equal. W.l.o.g. we assume that $y = y', z = z', w = w'$ and $x \neq x'$ (the other cases are analogous). In other words, we suppose that p^1 and p^2 belong to the same coset of $(D_0 \cap C_0) \oplus a$, where $a \in (D_0 \cap C_0)^\perp$. Since $D_0 \cap C_0 \subseteq C_0$, it follows that if $p^1 \oplus p^2 \in C_0$, then $R(p^1) \oplus R(p^2) \in D_0$. In more details, $R(p^1) \oplus R(p^2)$ is given by:

$$(R(p^1) \oplus R(p^2))_{0,0} = \alpha_3(S - Box(x \oplus a_{0,0}) \oplus S - Box(x' \oplus a_{0,0})),$$
$$(R(p^1) \oplus R(p^2))_{1,3} = \alpha_2(S - Box(x \oplus a_{0,0}) \oplus S - Box(x' \oplus a_{0,0})),$$
$$(R(p^1) \oplus R(p^2))_{2,2} = \alpha_1(S - Box(x \oplus a_{0,0}) \oplus S - Box(x' \oplus a_{0,0})),$$
$$(R(p^1) \oplus R(p^2))_{3,1} = \alpha_0(S - Box(x \oplus a_{0,0}) \oplus S - Box(x' \oplus a_{0,0})).$$

Since $S - Box(x \oplus a_{0,0}) \oplus S - Box(x' \oplus a_{0,0})$ is different from zero, it follows that at least three output nibbles (one per column) must be different from zero. In other words, it is possible that p^1 and p^2 exist such that $R(p^1) \oplus R(p^2) \in C_Q$ for $|Q| = 3$. Moreover, observe that $R(p^1) \oplus R(p^2) \in C_Q$ for $|Q| = 3$ if and only if one column of $R(p^1) \oplus R(p^2)$ is equal to zero. Since there are two "free" variables (i.e. x, x') and one equation, such a system can have a non-negligible solution.

Finally, since the previous result is independent of the values of $y = y', z = z', w = w'$, it follows that the number of collisions for this case must be a multiple of 2^{12}.

Second Case. Secondly, we consider the case in which two variables are equal. W.l.o.g. we assume for example that $y = y', z = z'$ while $x \neq x', w \neq w'$ (the other cases are analogous). That is, we suppose that p^1 and p^2 belong to the same coset of $(D_0 \cap C_{0,1}) \oplus a$, where $a \in (D_0 \cap C_{0,1})^\perp$.

Assume that - for certain $y = y'$ and $z = z'$ - there exist two elements p^1 (generated by $\langle x, w \rangle$) and p^2 (generated by $\langle x', w' \rangle$) defined as before in the same coset D_0 that belong to the same coset of C_Q for a certain Q with $|Q| = 3$ after one round. This implies that the two elements $\hat{p}^1$ (generated by $\langle x', w \rangle$) and $\hat{p}^2$ (generated by $\langle x, w' \rangle$)

$$\hat{p}^1 = a \oplus \begin{bmatrix} x^{'} & 0 & 0 & 0 \\ 0 & 0 & 0 & 0 \\ 0 & 0 & 0 & 0 \\ 0 & w & 0 & 0 \end{bmatrix}, \quad \hat{p}^2 = a \oplus \begin{bmatrix} x & 0 & 0 & 0 \\ 0 & 0 & 0 & 0 \\ 0 & 0 & 0 & 0 \\ 0 & w^{'} & 0 & 0 \end{bmatrix}$$

belong to the same coset of C_Q after one round since $R(p^1) \oplus R(p^2) = R(\hat{p}^1) \oplus R(\hat{p}^2)$. Note that the existence of the two elements $\hat{p}^1$ and $\hat{p}^2$ is guaranteed by the fact that we are working with the entire coset of D_0. This implies that the number of collisions must be even, that is a multiple of 2.

Then we have only to prove that it is possible that $x, x^{'}, w, w^{'}$ can exist such that $R(p^1) \oplus R(p^2) \in C_Q$ for $|Q| = 3$. We compute and analyze the first column (the others are analogous):

$$(R(p^1) \oplus R(p^2))_{.,0} = \begin{bmatrix} \alpha_3(S-Box(x \oplus a_{0,0}) \oplus S-Box(x^{'} \oplus a_{0,0})) \\ \alpha_2(S-Box(w \oplus a_{3,1}) \oplus S-Box(w^{'} \oplus a_{3,1})) \\ 0 \\ 0 \end{bmatrix}.$$

Since $S-Box(x \oplus a_{0,0}) \oplus S-Box(x^{'} \oplus a_{0,0})$ and $S-Box(w \oplus a_{3,1}) \oplus S-Box(w^{'} \oplus a_{3,1})$ are different from zero, it follows that at least 3×2 output nibbles must be different from zero. That is to say, it is possible that p^1 and p^2 exist such that $R(p^1) \oplus R(p^2) \in C_Q$ for $|Q| = 3$. Similarly, $R(p^1) \oplus R(p^2) \in C_Q$ for $|Q| = 3$ holds if and only if one column of $R(p^1) \oplus R(p^2)$ is equal to zero. Since there are four "free" variables (i.e. $x, x^{'}, w, w^{'}$) and a system of two equations, such a system can have a non-negligible solution.

Finally, since the previous result is independent of the values of $y = y^{'}, z = z^{'}$, it follows that the number of collisions for this case must be a multiple of 2^9.

Third Case. Thirdly, we consider the case in which only one variable is equal. W.l.o.g. we assume for example that $y = y^{'}$, while $x \neq x^{'}, z \neq z^{'}$ and $w \neq w^{'}$ (the other cases are analogous). That is, we suppose that two texts p^1 and p^2 belong to the same coset of $(D_0 \cap C_{0,1,2}) \oplus a$, where $a \in (D_0 \cap C_{0,1,2})^{\perp}$.

Assume there exist two elements p^1 (generated by $\langle x, z, w \rangle$) and p^2 (generated by $\langle x^{'}, z^{'}, w^{'} \rangle$) defined as before in the same coset D_0 that belong to the same coset of C_Q with $|Q| = 3$ after one round. Similar to before, it follows that the following three pairs of elements in the same coset of D_0 generated by:

- $\langle x^{'}, z, w \rangle$ and $\langle x, z^{'}, w' \rangle$
- $\langle x, z^{'}, w \rangle$ and $\langle x^{'}, z, w' \rangle$
- $\langle x, z, w^{'} \rangle$ and $\langle x', z^{'}, w \rangle$

belong to the same coset of C_Q after one round since $R(p^1) \oplus R(p^2) = R(\hat{p}^1) \oplus R(\hat{p}^2)$, where $\hat{p}^1$ and $\hat{p}^2$ are generated by the previous combinations of variables. Note that the existence of these elements is guaranteed by the fact that we are working with the entire coset of D_0. This implies that the number of collisions must be a multiple of 4.

Then we just need to prove that such x, z, w and x', z', w' can exist. As before we compute and analyze the first column (the others are analogous):

$$(R(p^1) \oplus R(p^2)).,_0 = \begin{bmatrix} a_3(S - Box(x \oplus a_{0,0}) \oplus S - Box(x' \oplus a_{0,0})) \\ a_2(S - Box(w \oplus a_{3,1}) \oplus S - Box(w' \oplus a_{3,1})) \\ a_2(S - Box(z \oplus a_{2,2}) \oplus S - Box(z' \oplus a_{2,2})) \\ 0 \end{bmatrix}.$$

Since $S-Box(x \oplus a_{0,0}) \oplus S-Box(x' \oplus a_{0,0})$, $S-Box(w \oplus a_{3,1}) \oplus S-Box(w' \oplus a_{3,1})$ and $S-Box(z \oplus a_{2,2}) \oplus S-Box(z' \oplus a_{2,2})$ are different from zero, it follows that at least 3×3 output nibbles must be different from zero. This implies that the event $R(p^1) \oplus R(p^2) \in C_Q$ for $|Q| = 3$ is possible. As before, $R(p^1) \oplus R(p^2) \in C_Q$ for $|Q| = 3$ holds if and only if one column of $R(p^1) \oplus R(p^2)$ is equal to zero. Also in this case, variables x, x', z, z', w, w' can exist since the number of equations is less than the number of variables.

Finally, since the previous result is independent of the values of $y = y'$, it follows that the number of collisions for this case must be a multiple of 2^6.

Fourth Case. Fourthly, we consider the case in which all variables are different. W.l.o.g. we assume that $x \neq x', y \neq y', z \neq z'$ and $w \neq w'$. That is, we suppose that two texts p^1 and p^2 belong to the same coset of $D_0 \oplus a$, where $a \in D_0^\perp$.

Assume there exist two elements p^1 (generated by $\langle x, y, z, w \rangle$) and p^2 (generated by $\langle x', y', z', w' \rangle$) defined as before in the same coset of D_0 that belong to the same coset of C_Q with $|Q| = 3$ after one round. Similarly, it follows that the following seven pairs of elements in the same coset of D_0 generated by:

- $\langle x', y, z, w \rangle$ and $\langle x, y', z', w' \rangle$ - $\langle x, y', z, w \rangle$ and $\langle x', y, z', w' \rangle$
- $\langle x, y, z', w \rangle$ and $\langle x', y', z, w' \rangle$ - $\langle x, y, z, w' \rangle$ and $\langle x', y', z', w \rangle$
- $\langle x', y', z, w \rangle$ and $\langle x, y, z', w' \rangle$ - $\langle x', y, z', w \rangle$ and $\langle x, y', z, w' \rangle$
- $\langle x', y, z, w' \rangle$ and $\langle x, y', z', w \rangle$

belong to the same coset of C_J after one round since $R(p^1) \oplus R(p^2) = R(\hat{p}^1) \oplus R(\hat{p}^2)$, where $\hat{p}^1$ and $\hat{p}^2$ are generated by the previous combinations of variables. Note that the existence of these elements is guaranteed by the fact that we are working with the entire coset of D_0 as before. This implies that the number of collisions must be a multiple of 8.

Then we have only to prove that such x, y, z, w and x', y', z', w' can exist. As before we compute and analyze the first column (the others are analogous):

$$(R(p^1) \oplus R(p^2)).,_0 = \begin{bmatrix} a_3(S - Box(x \oplus a_{0,0}) \oplus S - Box(x' \oplus a_{0,0})) \\ a_2(S - Box(w \oplus a_{3,1}) \oplus S - Box(w' \oplus a_{3,1})) \\ a_2(S - Box(z \oplus a_{2,2}) \oplus S - Box(z' \oplus a_{2,2})) \\ a_3(S - Box(y \oplus a_{1,3}) \oplus S - Box(y' \oplus a_{1,3})) \end{bmatrix}.$$

Since $S-Box(x \oplus a_{0,0}) \oplus S-Box(x' \oplus a_{0,0})$, $S-Box(w \oplus a_{3,1}) \oplus S-Box(w' \oplus a_{3,1})$, $S-Box(z \oplus a_{2,2}) \oplus S-Box(z' \oplus a_{2,2})$ and $S-Box(y \oplus a_{1,3}) \oplus S-Box(y' \oplus a_{1,3})$ are different from zero, it follows that at least 3×4 output nibbles must be different from zero. This means that the event $R(p^1) \oplus R(p^2) \in C_Q$ for $|Q| = 3$

is possible. Also in this case, variables $x, x^{'}, y, y^{'}, z, z^{'}, w, w^{'}$ can exist since the number of equations is less than the number of variables.

To sum up, according to the previous analysis, there exist $n_1, n_2, n_3, n_4 \in N$ such that the total number of collisions n is equal to $n = 2^{12} \cdot n_1 + 2^9 \cdot n_2 + 2^6 \cdot n_3 + 8 \cdot n_4 = 8 \cdot (2^9 \cdot n_1 + 2^6 \cdot n_2 + 2^3 \cdot n_3 + n_4)$, i.e. it is a multiple of 8. This proves the lemma.

For completeness, we briefly recall why the proof of Lemma 1 implies Theorem 5. Consider the following description of 4.5-round of PRINCE:

$$C_I \oplus a \xrightarrow{R(\cdot)} D_I \oplus b \xrightarrow{R(\cdot)} C_Q \oplus c \xrightarrow{R(\cdot)} D_Q \oplus d \xrightarrow{M^{'} \circ S - Box(\cdot)} M_Q \oplus e.$$

Combining Lemma 1, we start with the second round and extend it forward by 2.5 rounds, then we obtain that a coset of C_Q is mapped into a coset of M_Q. Let's consider the second round again and extend it backward by one round, then we obtain that a coset of D_I is mapped into a coset of C_I. Since these two events hold with probability 1, this finally proves Theorem 5, and Theorem 6 is the same. For completeness, for the cases $|I| = 2$ and $|I| = 3$, there are similar theorems. Moreover, we have practically verified the property of PRINCE using C/C++ implementation. To verify Theorem 5, we have chosen $|I| = 1$ and $|Q| = 3$ fixed, given plaintexts in the same coset of C_i, the programs counts the number of collisions n among the ciphertexts in the same coset of M_Q and prints the corresponding $n\%8$. The experimental results show that for 4.5-round PRINCE n is a multiple of 8, while it can take any possible value in the case of a random permutation. Finally, note that Theorem 6 holds exactly in the similar way for PRINCE.

5   Conclusion

Over the past several years, we've seen the rapid deployment of secure microcontrollers in the Internet of Things, automotive, and cloud infrastructure. Various areas of technology, including industrial automation, robotics as well as the 5th generation mobile networks, require real-time operation, low latency execution, while preserving the highest level of security. PRINCE is the first publicly known low-latency family of block ciphers that got scrutinized by the cryptographic community. In this paper, we have presented two new subspace trails for 4.5 rounds of PRINCE. Additionally, we showed a new structural property that can be exploited to set up an efficient 4.5-round secret-key distinguisher for PRINCE, which is independent of the secret key, improving the previous results in [21]. Starting from our results, this new 4.5-round property might be applied to set up round-reduced attacks on PRINCE. We leave these questions for further study.

Acknowledgements. The authors would like to thank all anonymous referees for their valuable comments. This work is supported by National Natural Science Foundation of China (No. 61672509, No. 62072445), and the National Cryptography Development Foundation of China (no. MMJJ20170101).

References

1. Borghoff, J., et al.: PRINCE – a low-latency block cipher for pervasive computing applications. In: Wang, X., Sako, K. (eds.) ASIACRYPT 2012. LNCS, vol. 7658, pp. 208–225. Springer, Heidelberg (2012). https://doi.org/10.1007/978-3-642-34961-4_14

2. Leander, G., Abdelraheem, M.A., AlKhzaimi, H., Zenner, E.: A cryptanalysis of PRINTCIPHER: the invariant subspace attack. In: Rogaway, P. (ed.) CRYPTO 2011. LNCS, vol. 6841, pp. 206–221. Springer, Heidelberg (2011). https://doi.org/10.1007/978-3-642-22792-9_12

3. Knudsen, L., Leander, G., Poschmann, A., Robshaw, M.J.B.: PRINTCIPHER: a block cipher for IC-printing. In: Mangard, S., Standaert, F.-X. (eds.) CHES 2010. LNCS, vol. 6225, pp. 16–32. Springer, Heidelberg (2010). https://doi.org/10.1007/978-3-642-15031-9_2

4. Bulygin, S., Walter, M., Buchmann, J.: Full analysis of PRINTcipher with respect to invariant subspace attack: efficient key recovery and countermeasures. Des. Codes Crypt. **73**(3), 997–1022 (2014)

5. Leander, G., Minaud, B., Rønjom, S.: A generic approach to invariant subspace attacks: cryptanalysis of Robin, iSCREAM and Zorro. In: Oswald, E., Fischlin, M. (eds.) EUROCRYPT 2015. LNCS, vol. 9056, pp. 254–283. Springer, Heidelberg (2015). https://doi.org/10.1007/978-3-662-46800-5_11

6. Grosso, V., Leurent, G., Standaert, F.X., et al.: SCREAM & iSCREAM. Entry in the CAESAR competition (2014). http://competitions.cr.yp.to/round1/screamv1.pdf

7. Grosso, V., Leurent, G., Standaert, F.-X., Varıcı, K.: LS-designs: bitslice encryption for efficient masked software implementations. In: Cid, C., Rechberger, C. (eds.) FSE 2014. LNCS, vol. 8540, pp. 18–37. Springer, Heidelberg (2015). https://doi.org/10.1007/978-3-662-46706-0_2

8. Gérard, B., Grosso, V., Naya-Plasencia, M., Standaert, F.-X.: Block ciphers that are easier to mask: how far can we go? In: Bertoni, G., Coron, J.-S. (eds.) CHES 2013. LNCS, vol. 8086, pp. 383–399. Springer, Heidelberg (2013). https://doi.org/10.1007/978-3-642-40349-1_22

9. Guo, J., Jean, J., Nikolic, I., et al.: Invariant subspace attack against Midori64 and the resistance criteria for S-box designs. IACR Trans. Symmetric Cryptol. **2016**(1), 33–56 (2016)

10. Banik, S., et al.: Midori: a block cipher for low energy. In: Iwata, T., Cheon, J.H. (eds.) ASIACRYPT 2015. LNCS, vol. 9453, pp. 411–436. Springer, Heidelberg (2015). https://doi.org/10.1007/978-3-662-48800-3_17

11. Beierle, C., Canteaut, A., Leander, G., Rotella, Y.: Proving resistance against invariant attacks: how to choose the round constants. In: Katz, J., Shacham, H. (eds.) CRYPTO 2017. LNCS, vol. 10402, pp. 647–678. Springer, Cham (2017). https://doi.org/10.1007/978-3-319-63715-0_22

12. Todo, Y., Leander, G., Sasaki, Y.: Nonlinear invariant attack. In: Cheon, J.H., Takagi, T. (eds.) ASIACRYPT 2016. LNCS, vol. 10032, pp. 3–33. Springer, Heidelberg (2016). https://doi.org/10.1007/978-3-662-53890-6_1

13. Wei, Y., Ye, T., Wu, W., Pasalic, E.: Generalized nonlinear invariant attack and the new design criterion for round constants. IACR Trans. Symmetric Cryptol. **2018**(4), 62–79 (2018)

14. Grassi, L., Rechberger, C., Rønjom, S.: Subspace trail cryptanalysis and its applications to AES. IACR Trans. Symmetric Cryptol. **2016**(2), 192–225 (2016)

15. Grassi, L., Rechberger, C., Rønjom, S.: A new structural-differential property of 5-round AES. In: Coron, J.-S., Nielsen, J.B. (eds.) EUROCRYPT 2017. LNCS, vol. 10211, pp. 289–317. Springer, Cham (2017). https://doi.org/10.1007/978-3-319-56614-6_10

16. Derbez, P., Perrin, L.: Meet-in-the-middle attacks and structural analysis of round-reduced PRINCE. In: Leander, G. (ed.) FSE 2015. LNCS, vol. 9054, pp. 190–216. Springer, Heidelberg (2015). https://doi.org/10.1007/978-3-662-48116-5_10

17. Morawiecki, P.: Practical attacks on the round-reduced PRINCE. IET Inf. Secur. **11**(3), 146–151 (2017)

18. Posteuca, R., Negara, G.: Integral cryptanalysis of round-reduced PRINCE cipher. Proc. Rom. Acad. **2015**(16), 265–270 (2015)

19. Abed, F., List, E., Lucks, S.: On the security of the core of PRINCE against biclique and differential cryptanalysis. Cryptology ePrint Archive, Report 2016/712 (2016)

20. Zhao, G., Sun, B., Li, C., Su, J.: Truncated differential cryptanalysis of PRINCE. Secur. Commun. Netw. **8**(16), 2875–2887 (2015)

21. Soleimany, H., et al.: Reflection cryptanalysis of PRINCE-like ciphers. J. Cryptol. **28**(3), 718–744 (2013). https://doi.org/10.1007/s00145-013-9175-4

22. Grassi, L., Rechberger, C.: Practical low data-complexity subspace-trail cryptanalysis of round-reduced PRINCE. In: Dunkelman, O., Sanadhya, S.K. (eds.) INDOCRYPT 2016. LNCS, vol. 10095, pp. 322–342. Springer, Cham (2016). https://doi.org/10.1007/978-3-319-49890-4_18

Artificial Intelligence and Cryptocurrency

Generative Adversarial Networks-Based Pseudo-Random Number Generator for Embedded Processors

Hyunji Kim, Yongbeen Kwon, Minjoo Sim, Sejin Lim, and Hwajeong Seo$^{(\boxtimes)}$ ⓘ

IT Department, Hansung University, Seoul, South Korea
khj1594012@gmail.com, vexyoung@gmail.com, minjoos9797@gmail.com,
dlatpwls834@gmail.com, hwajeong84@gmail.com

Abstract. A pseudo-random number generator (PRNG) is a fundamental building block for modern cryptographic solutions. In this paper, we present a novel PRNG based on generative adversarial networks (GAN). A recurrent neural network (RNN) layer is used to overcome the problems of predictability and reproducibility for long random sequences, which is found in the result of the NIST test suite for the previous method. The proposed design generates a random number of 1,099,200-bits with a 64-bit seed. The proposed method is also efficiently implemented on embedded processors by using the Edge TPU. To support the Edge TPU, the proposed GAN based PRNG is converted to a TensorFlow Lite model. During model training, the number of epochs is significantly reduced with the proposed approach. The PRNG generates random numbers in 13.27 ms using the Edge TPU. Also, our PRNG achieved a speed of 1.0 GB/s, which is about 6.25x compared to the speed of other lightweight PRNG. To the best of our knowledge, this is the first GAN based PRNG for embedded processors. Finally, generated random numbers were tested through the NIST random number test suite. Compared with the previous method, the proposed method reduced the percentage of test failures by 2.85x. The result shows that the proposed GAN-based PRNG achieved high randomness even on embedded processors.

Keywords: Pseudo-random number generator · Generative adversarial networks · Edge TPU · Recurrent neural networks

1 Introduction

Pseudo-random number generators (PRNGs) are widely used in cryptographic applications. For this reason, the implementation of PRNGs on modern computers is important for real-world applications. In the past, a number of PRNG implementations have been investigated [1–4].

Previous PRNG implementations utilized unique hardware features and mathematical functions. Recently, a novel approach to a generative adversarial

ⓒ Springer Nature Switzerland AG 2021
D. Hong (Ed.): ICISC 2020, LNCS 12593, pp. 215–234, 2021.
https://doi.org/10.1007/978-3-030-68890-5_12

network (GAN)-based PRNG was presented. It was proved that the machine-learning algorithm can efficiently generate random sequences [5]. However, previous methods require a number of epochs and achieve low generation performance and randomness.

In this paper, we present the first GAN-based PRNG for embedded processors. The proposed approach improved the previous GAN-based PRNG by modifying the neural network layers and output type, which also reduces the number of epochs. Moreover, the model is tailored to support embedded environments (i.e. Edge TPU). Finally, the generated random number sequence passes the NIST test suite with high randomness and performance.

1.1   Contribution

Novel Generative Adversarial Networks-Based Pseudo-Random Number Generator. We present a novel GAN-based PRNG. Unlike the previous approach, the proposed design generates 1,099,520-bit random numbers with only a 64-bit seed. During the model training session, the number of epochs is significantly reduced with new approach. Finally, the proposed PRNG shows better randomness and performance than previous works.

Lightweight GAN-Based PRNG for Embedded Processors. We tailored the proposed GAN-based PRNG for Edge TPU to ensure high-performance and high entropy. The model was successfully converted to a Tensorflow Light model and uploaded to the Edge TPU. Random number sequences were successfully generated on the embedded processor. Also, the GAN-based PRNG achieved faster speed than other lightweight PRNG.

Randomness Test Based on NIST Suite. The proposed PRNG was evaluated through the NIST suite. The generated random sequence successfully passed the NIST test. The entropy was also higher than that achieved by previous methods.

The remainder of this paper is organized as follows. In Sect. 2, related technologies, such as random number generator, deep learning framework, generative adversarial networks, and previous GAN-based PRNG implementations are presented. In Sect. 3, the proposed GAN-based PRNG implementation is introduced. In Sect. 4, the evaluation of proposed GAN-based PRNG implementation is discussed. Finally, Sect. 5 concludes the paper.

2   Related Works

2.1   Random Number Generator

A random number generator (RNG) produces a sequence of numbers that cannot be predicted better than by random chance. Random number generators are largely divided into true random number generators (TRNGs) and pseudo-random number generators (PRNGs). In the following subsection, we describe both RNG approaches in detail.

True Random Number Generator. A TRNG generates genuinely random numbers. These numbers are non-deterministic. According to Kerchoff's principle, the random number generator must produce unpredictable bits even if every detail of the generator is available [6]. Physical sources, including Johnson's noise, Zener noise, radioactive decay, photon path splitting at a two-way beam splitter, and photon arrival times, have been utilized to achieve randomness [7–11].

Pseudo Random Number Generator. A PRNG, namely, a deterministic random bit generator (DRBG), generates numbers that look random by producing the random sequence with perfect balance between 0's and 1's. However, these numbers are deterministic, periodic, and predictable. For this reason, these numbers can be reproduced when the inner state of the PRNG is available. A PRNG suitable for cryptographic applications is a cryptographically secure PRNG (CSPRNG). Some examples of CSPRNGs include stream ciphers and block ciphers in the counter mode of operation [12].

2.2 Random Number Generator Attack

The inner state of a RNG is updated through the update function with a seed, and it outputs a random number. Because the previous state is not known by using a one-way function, such as the update function, it can prevent predictive attacks. If the length of the inner state is short, an attacker is able to predict the output through a brute force attack on the inner state. Therefore, the length of the inner state should be sufficiently long enough. If the attacker can predict or control even some of the operating conditions used to generate or control the output, it would be relatively easy to carry out a brute force attack on the inner conditions. Therefore, as much noise as possible should be used to generate the inner state. The entropy of the noise must be large enough.

2.3 Deep Learning Framework

The deep learning method is a type of machine learning method based on artificial neural networks with representation learning. The deep learning method uses multiple layers to extract higher-level features from raw input. There are various deep learning structures. There are software (TensorFlow) and hardware (Tensor Processing Unit) frameworks to support the deep learning method.

TensorFlow. TensorFlow is an open-source software library for machine learning applications, such as neural networks [13]. The library is used for both research and production, such as DeepDream, which generates automated image-captioning[1]. The programming language is Python.

---

[1] https://www.vice.com/en_uk/topic/motherboard.

Keras. Keras is a deep learning library for machine learning and artificial intelligence. It provides a high-level API for users to easily build neural networks and runs on TensorFlow. Using the Keras library, we can easily design a model using pre-implemented modules and additionally use TensorFlow for detailed design (low-level). Recently, with the release of TensorFlow 2.0, TensorFlow allows the Keras function to be used through the tf.keras module. This makes TensorFlow's low-level design and Keras's high-level design more flexible than before. We designed the model using Keras in this work.

Edge TPU. In 2018, Google announced the Edge tensor processing unit (TPU), which runs machine learning models for edge computing. It is available as a USB companion or as a self-contained development board. [14] The Edge TPU performs 4 trillion operations per second while using only $2W^2$. In comparison to a floating-point architecture with a similar form factor, the Intel Compute Stick, the Edge TPU has shown superior performance in terms of latency and computational efficiency. The machine learning models on the Edge TPU are based on TensorFlow Lite[3]. Because the Edge TPU is capable of accelerating forward-pass operations, the Edge TPU is efficient for making inferences.

The Edge TPU is mainly used for classification, becuase it provides pre-trained and pre-compiled model detection tasks for image classification and objects[4].

2.4   Generative Adversarial Networks

A generative adversarial network (GAN) is a type of machine learning framework [15]. A GAN is an in-depth neural network structure consisting of two networks, generator and discriminator. Given a training set, a GAN learns to generate new data with the same statistics as the training set. The generator is intended to produce data that is as real as possible, and the discriminator is intended to distinguish between real and fake data. The training course repeats the process of training the discriminator first and then the generator and discriminator exchange data. The discriminator follows two main courses. The first is to enter real data and learn that the network really classifies that data. The second is the process of entering fake data generated by the generator and learning to classify that data as fake. This allows the discriminator to classify real data as real or fake. If it determines that data is real, it outputs 1, and if it determines that data is fake, it outputs 0. After the discriminator has completed this learning process, the generator is trained to deceive the learned discriminator. In other words, the generator gradually develops its output based on the discriminator judgment so that it can judge its output as true data. By repeating the above training process, both the discriminator and the generator will be

---

[2] https://coral.ai/docs/edgetpu/benchmarks/.

[3] https://www.blog.google/products/google-cloud/bringing-intelligence-to-the-edge-with-cloud-iot/.

[4] https://coral.ai/models/.

gradually developed. As a result, the generator will be able to create fake data that is completely identical to real data, and the discriminator will be unable to distinguish between real data and fake data.

GANs have been used for various fields, such as fashion, art, science, and video games [16,17]. In this paper, we used GAN for pseudo-random number generation.

2.5   Previous GAN-Based PRNG Implementations

In [5], the first GAN based PRNG implementation was presented. The GAN's generator is partially hidden, and the adversary is trained to discover a mapping from the overt part to the hidden part. In that study, the generated random numbers achieved randomness.

In general, a GAN is a model consisting of the learning of the generator and the discriminator. However, it was novel that a GAN was designed with the learning of a generator and a predictor rather than a discriminator. The discriminative approach requires an external source of randomness that it attempts to imitate, whereas this predictive approach does not require external inputs. In the predictor approach, the output of the generator cannot be predicted by the improved predictor.

3   Proposed Method

We propose an PRNG based on GAN for embedded processors. In Fig. 1, the proposed system configuration is presented. The basic GAN model consists of a generator and a discriminator. The proposed method uses the predictor introduced in the previous method rather than a discriminator. For each training session, a fresh random seed source is entered into the generator. Then, the generator produces a random bit stream based on the random seed. The generated random bit stream is split into two parts. The predictor is trained to predict the back part by the divided front part. Then, the generator produces a random bit stream so that the predictor cannot predict by reflecting the predictor's training result. Because the inner state of the generator is updated according to the training result of the predictor, different results are generated even if the same random seed input is given. The generator produces a bit stream with high randomness because both models are alternately trained. As shown in Fig. 2, a model trained to generate a random bit stream is deployed to edge devices using Edge TPU and TensorFlow Lite. A random seed is generated from a secure entropy source for the embedded device. The random bit stream generated by the trained generator is converted to random numbers. Finally, a PRNG for an embedded processor is designed.

3.1   Design of Generator Model

As shown in Fig. 3, the generator model consists of four fully connected layers. The generator uses a random seed as input, and a bit stream with a length of

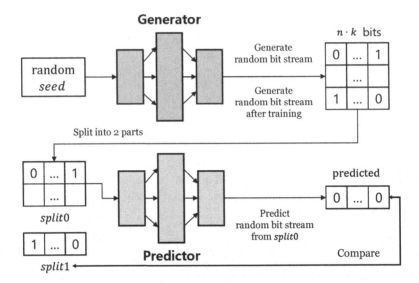

Fig. 1. System configuration for proposed method.

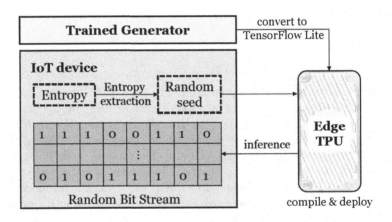

Fig. 2. Configuration of random bit stream generation in embedded processors using Edge TPU.

$n \cdot k$ is generated. In this case, n and k are adjustable hyper-parameters. The generator is trained by using the combined model. Algorithm 1 shows only the process of predicting a random bit stream, not the training process. Because the proposed method learns the bit stream, the sigmoid is used as the activation function. The value of the sigmoid activation function is a floating-point number between 0 and 1. For this reason, we round the value to 0 or 1 to generate the result in a bit stream format. The generated output is used as the input of the predictor model.

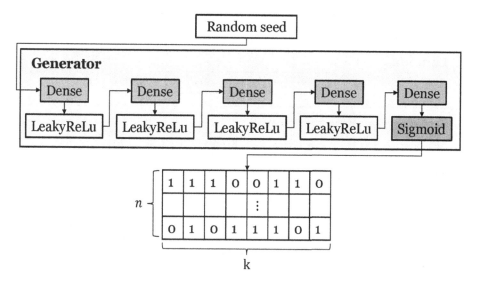

Fig. 3. Architecture of generator.

Algorithm 1. Generator mechanism

Input: Random seed (s), Generator (G)
Output: Random bit stream (RBS)
1: $x \leftarrow Dense(s)$
2: **for** $i = 1$ **to** 4 **do**
3:     $x \leftarrow Dense(x)$
4: **end for**
5: $x \leftarrow Sigmoid(x)$
6: $RBS \leftarrow$ round x into nearest integer (0 or 1)
7: **return** RBS

3.2 Design of Predictor Model

In Algorithm 2, the predictor mechanism is expressed. First, the bit stream received by the generator is split into one for training($split0$) and one for loss($split1$). Figure 4 shows the process of splitting the input data before training. $split0$ is $(n-1) \cdot k\text{-}bits$ and $split1$ is $k\text{-}bits$. The predictor uses $split0$ as the RNN layer's input for training. The predictor uses spilt0 as training data to predict what split1 was. In other words, the predictor does not require training data, unlike the basic GAN model.

As shown in Fig. 5, the predictor model adds a recurrent neural network (RNN) layer to the convolution layer used in the previous method. An RNN learns the correlation of data points in a sequence [18]. Because the past information is stored through hidden states, it is possible to learn even long sequence data. Each row of the generator's output used as the predictor's input becomes time-series data. Each of the n time series data has k features. This feature is suitable for learning and predicting the sequence of bit streams. In particular,

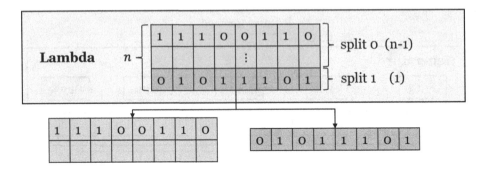

Fig. 4. Split the input bit stream into two parts for training.

by adjusting the k, each time series data can be learned in decimal or hexadecimal units within the desired range. In addition, features of time-series data in a specific range are reflected, and long-term dependencies can be maintained. As in the previous work, using only a convolution layer means that there is less weight to learn. However, there is a tendency to learn regional features, which reduces randomness. In Sect. 4, the NIST test suite result shows how effective the use of RNN is in ensuring the randomness of long bit streams. In the case of the proposed model, using LSTM, a type of RNN, it takes too long to train. In addition, the weight of learning is four times greater than that for a simple RNN, so it is inefficient compared to the randomness being learned. For this reason, we selected a simple RNN layer for training the random bit stream.

Because the predictor also learns and predicts the bit stream, we use the sigmoid activation function. The sigmoid activation function returns a value between 0 and 1, the return value is rounded to the nearest integer and is used as a bit.

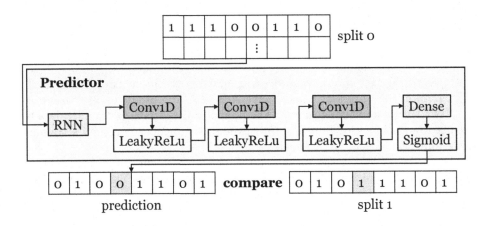

Fig. 5. Architecture of a predictor.

Algorithm 2. Predictor mechanism

Input: Random bit stream (RBS), Length of
Output: Predicted random bit stream (RBS_P)
1: $n \leftarrow$ row size of RBS
2: $k \leftarrow$ column size of RBS
3: **for** $i = 0$, **to** n **do**
4:   **for** $j = 0$, **to** k **do**
5:     **if** $i < n - 1$ **then**
6:       $Split0[i][j] = RBS[i][j]$
7:     **else**
8:       $Split1[i][j] = RBS[i][j]$
9:     **end if**
10:   **end for**
11: **end for**
12: $x \leftarrow RNN(Split0)$
13: **for** $i = 1$, **to** 3 **do**
14:   $x \leftarrow Conv1D(x)$
15: **end for**
16: $x \leftarrow Dense(x)$
17: $x \leftarrow Sigmoid(x)$
18: $RBS_P \leftarrow$ round x into nearest integer (0 or 1)
19: $Loss_p \leftarrow mean(|\ Split1 - RBS_P\ |)$
20: Train to minimize $Loss_p$
21: **return** $RBS_P, Split1$

The loss is calculated as the mean absolute error over $split1$ and RBS_P which is the predicted random bit stream by learning $split0$. If $split1$ and the predicted random bit stream are the same, the predictor has made a correct prediction, and the loss is minimized. Therefore, the predictor performance is improved by training, which reduces the loss.

3.3   Design of GAN-Based PRNG

The GAN-PRNG is the final model that combines two models, including a generator and a predictor. The generator is trained through a combined model to reflect the results of the predictor. Algorithm 3 shows the detailed operation of the proposed GAN-PRNG.

Built-in random functions are used as a random seed in the training process. However, we obtain the entropy from a secure entropy source or generator implemented in hardware to generate the random seed in the inference process. The GAN requires a random seed that is the input of the generator and uses a value randomly extracted from a uniform distribution or a normal distribution. This simple distribution is mapped to a complex distribution through training.

The random seed is 64 bits, and it is used as input to the generator. The random bit stream is generated through the generator and used as the input of the predictor. After the predictor is trained, the predicted bit stream(RBS_P)

Algorithm 3. Proposed RNG based on GAN

Input: Random seed (s), Generator (G), Predictor (P), epochs $(EPOCHS)$, Secure
    parameter (t), Range of random number (r), The number of bits needed to represent
    random number (m)
Output: Random Number (num)
1: **for** $epoch = 1$ **to** $EPOCHS$ **do**
2:    $s \leftarrow$ sample $entropy$ from IoT device
3:    $RBS \leftarrow G(s)$
4:    $RBS_P, Split1 \leftarrow P(RBS)$
5:    $Loss_G \leftarrow mean(abs(1 - Split1 - RBS_P)) \cdot 0.5$
6:    Train G to minimize $Loss_G$
7:    $RBS \leftarrow G(s)$
8: **end for**
9: $c \leftarrow \sum_{i=0}^{m+t-1} 2^i \cdot RBS_i$
10: $num \leftarrow c \ mod \ r$
11: **return** num

---

and the actual bit stream($split1$) are returned. The combined model computes
the loss with two loss values and trains the generator to minimize the loss. If
$split1$ and the predicted random bit stream are different, the loss of the combined
model is minimized to zero. This means that the predictor cannot predict the
random bit stream generated by the generator. In other words, the generator
also generates a better bit stream by reflecting the predictor's output.

This process is repeated for each epoch. A bit stream with high randomness
is generated. Then the random bit stream is converted to a random number.
There are three methods (the simple discard method, complex discard method,
and simple modular method) of converting a random bit stream into a random
number. Among them, we chose the simple modular method. Compared to the
other methods, it does not require a conditional loop; therefore, it is possible to
operate in constant time. Through this entire process, a random number stream
is generated.

Both loss values are calculated as seen in Algorithm 3. Through the training,
the predicted bit stream and the actual bit stream become similar, reducing
the loss of the predictor. The loss of the generator is also calculated using the
predicted bit stream and the actual bit stream. If the predictor fails to predict
the random bit stream, the generator loss is reduced by the calculation formula.
Both models are trained to minimize the loss function. This means that the
generator is trained to generate an unpredictable random bit stream.

3.4   GAN-Based PRNG in Embedded Processors

The predictor is trained from its own model with the output of the generator.
The model that directly generates a random bit stream is a generator. Among
the trained models, only the generator model is converted into a TensorFlow
Lite model. Algorithm 4 shows the process of converting a trained model into a

TensorFlow Lite model. We compiled the TensorFlow Lite model and performed the inference using the Edge TPU. The inference model expressed in Algorithm 5 is to generate random bit stream without the training process through the pre-trained model. The fixed weight means that the inner state of the model is fixed. When the input is exposed, there is a risk of prediction. In the previous work, the random seed extraction and their PRNG model are separated. For that reason, instead of a secure entropy source, the built-in random function is used. In the proposed method, the entropy is obtained on embedded processors as random seeds (e.g. sensor data) to prevent such prediction [19].

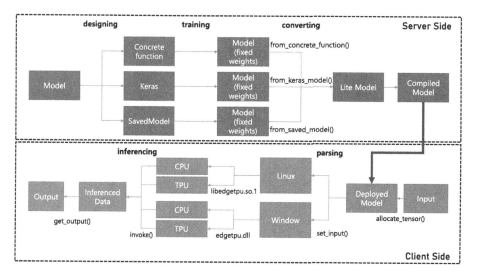

Fig. 6. Flowchart from model design to inference on Edge TPU.

The entire flow chart from the design of the model to inferring a random bit sequence is shown in Fig. 6. The deployment of the proposed GAN based PRNG on embedded processors is carried out as follows. First, the GAN model is set, which includes the generator (G) model and predictor (P) model. Second, the models (G, P) are trained under the GAN framework. Third, the trained models are saved with fixed weights. Fourth, only a G model is converted into a TensorFlow flatbuffer file. This is because we need a random bit generator, not a random bit predictor. It is the generator that generates the random bits directly, and because there is no training process on the embedded processor, only the generator model is converted. The G model has the ability to generate a random bit sequence with random seeds. Three types of models can be converted into a TensorFlow flatbuffer file.

– Using SavedModel Directories
  The first way to save the trained model is to use the 'SavedModel' format. TensorFlow offers an API module to save a trained model named

---

Algorithm 4. Converting algorithm

Input: Trained Combined Model (M), TFLiteConverter
Output: TensorFlow Lite FlatBuffer file
1: $trained\_G \leftarrow M.get\_generator$
2: $converter \leftarrow TFLiteConverter(trained\_G)$
3: $tflite\_model \leftarrow converter.convert()$
4: with $tf.io.gfile.GFile('trained\_G.tflite', 'wb')$ as f:
5: $f.write(tflite\_model)$
6: **return** $trained\_G.tflite$

Algorithm 5. Inference for TPU

Input: TensorFlow Lite Model LM, Entropy, Interpreter I
Output: Inferred Random Bit Stream (RBS)
1: $s \leftarrow sample\ entropy\ from\ IoT\ device$
2: $I.setModel(LM)$
3: $input\_size \leftarrow I.get\_input\_details()$
4: $output\_size \leftarrow I.get\_output\_details()$
5: $s.reshape(input\_size)$
6: $I.setTensor(s)$
7: $I.Invoke()$
8: $results \leftarrow I.getTensor()$
9: $RBS \leftarrow I.get\_Output$
10: **return** RBS

'tf.saved_model', which has the functions of save and load. The save function makes a directory consisting of model weights and functions. It can be loaded and used as a trained model with fixed weights. This model cannot have a specified input shape.

– Using tf.keras models
The second way to save a trained model is to use 'tf.keras' models. 'tf.keras' can support a sequential model, which is constructed by a list of layers. Each layer in a sequential model must have one input and one output. A sequential model can be trained by the inner function 'fit' with epochs. After training, the model has fixed weights. For saving a tf.keras model, all functions should be from a single module. Otherwise, module synchronization problems occur during the conversion process.

– Using concrete functions
The third way to save a trained model is to use concrete functions. Currently, only one concrete function is supported for one model.

Python supports API 'tf.lite.TFLiteConverter' for converting these types of models. Through the API functions 'from_saved_model()', 'from_keras_module()' and 'from_concrete_functions()', each type of models is converted into a Tensor-Flow FlatBuffer file, for which the filename extension is .tflite.

Finally, this TensorFlow FlatBuffer file can be deployed as a PRNG. For using the full potential of Edge TPU, there are some limitations for models and layers.

In our case, we saved only the G model as a tf.keras model. To generate a random bit stream, only the G model is needed. Fourth, a tf.keras model is converted to a TensorFlow Lite FlatBuffer file (.tflite). We converted the G model to a FlatBuffer file. Fifth, we compiled the model and deployed it to the embedded processors.

4   Evaluation

For the experiment, Google Co-laboratory PRO, a cloud-based service, was utilized. It ran on Ubuntu 18.04.3 LTS and consists of an Nvidia GPU (Tesla T4, Tesla P100, or Tesla K80) with 25 GB RAM. In terms of the programming environment, Python 3.6.9, TensorFlow 2.2.0-rc and Keras 2.3.1 version were used.

The GAN-PRNG for the embedded processor was implemented using the TensorFlow Lite model and Google Edge TPU. We saved the trained model in the Colab environment and converted it to a TensorFlow Lite model. Using a TensorFlow Lite file (.tflite), random numbers can be generated on embedded devices without training. In addition, we performed statistical tests on random bit sequences generated by the GAN-PRNG using NIST test suite.

4.1   NIST Test Suite

The randomness was verified for the output generated by the RNG through the NIST test suite. The test suite consists of 188 individual tests. Each test is repeated 10 times, which is called a test instance. For each repetition, 1,000,000 bits are used as input. The p-value is measured for each instance. The ideal random number sequence has a p-value of 1, and the test is passed when the threshold value is greater than α ($\alpha = 0.01$). The final analysis report shows the number of instances passed and the p-value for the distribution of instance p-value. In the proposed method, the execution time on the Edge TPU is measured to measure the rate of random number generation on the embedded processor.

Parameters. For the fair comparison, the hyper parameters were set as follows. Table 1 compares the hyper parameters optimized for each model with the previous method. First, the data types for random number generation in the previous work and this work are decimal and bit, respectively. The activation function must change according to the data type. Therefore, the previous work uses the customized activation function to generate a decimal in the range $[0, 2^{16}-1]$, and this work uses the sigmoid function with the range $[0,1]$. So, the range of data does not exceed 1 in this work. In this case, using the mean squared error loss function is not effective because the distance between the actual value and the predicted value is reduced by the square. So we use mean absolute error to calculate the difference between the two values. The previous method consists of

400 mini-batches of 2,048 input vectors. It predicts and learns 1 integer from 7 integers. Therefore, 112-bits are learned to predict 16-bits, and 262,144-bits are learned in one mini-batch. It trained 104,857,600-bits in 1 epoch. The proposed method consists of 100 mini-batches of 137,440 input vectors, so 1,099,192-bits are learned to predict 8 bits, and 1,099,200-bits are learned in one mini-batch. The total number of bits trained in 1 epoch is 109,920,000-bits. In both methods, a 64-bit seed is the input in one mini-batch. This means that the proposed method generates a longer random number stream in comparison to the previous method using the same length seed. It also shows that our method learns about longer sequences, and this work achieves a similar level of randomness, up to 2.5 million bits per random seed. The unit of the generator's dense layer is 30. The unit of the RNN layer is 8, the filter of the convolution layer is set to 8, and the kernel size is set to 1.

The learning rate of this neural network is 0.02, and an Adam optimizer with a learning rate of 0.0002 is used as an optimization function. In addition, only 30 epochs are used for training, which improves the previous method by 200,000 epochs.

Table 1. Comparison of parameters with the previous works.

	Bernardi et al. [5]	This work
Data type	Decimal	Bit
Activation	Custom (range[0, $2^{16} - 1$])	Sigmoid (0 or 1)
Loss	Mean Square Error	Mean Absolute Error
Seed: Output (bits)	64:262,144	64:1,099,200
Output Length	104,857,600-bits	109,920,000-bits
Optimizer	Adam (lr = 0.02)	Adam (lr = 0.0002)
Learning rate of network	0.02	0.02
Epoch	200,000	30

Results. Table 2 shows the NIST test suite results and inference time. Figure 7 presents the final analysis report of the NIST test suite. For one of the individual tests, the random excursion (variant), each experiment has different numbers. Therefore, T_I is measured differently from the previous work. The results before training show that it passed only 2 out of 188 individual tests; thus, it cannot be used as an PRNG.

The proposed method achieve better randomness for longer sequences than the original method. In 10 experiments on 1794 test instances, 196 test instances failed. There was no case where the p-value did not exceed the minimum pass rate. There was only 1 individual test for which it did not pass in the entire experiment. Because the minimum pass rate is 8 for each individual test, more than 8 test instances passed for all individual tests except 1 individual test. Therefore, this does not indicate that there is a vulnerability for a particular individual test. There were several failed test instances, but for individual tests, the pass criterion was achieved.

The results for p-value and individual tests were reduced by about 2.85 and 45 times, respectively, compared to the previous work. In the previous method, there was no content of time measurement. Also, because it is not an PRNG on an embedded processor, it was measured on a desktop for testing purposes. The proposed method is an PRNG on an embedded processor. Therefore, we use Edge TPU, an ASIC designed to accelerate inference. The result was 14.1 times faster than that achieved by the previous method on the desktop.

The individual tests that the previous method failed to pass were mainly frequency, cumulative sum, run, fast Fourier transform (FFT), and non-overlapping-template. The frequency test concerns the proportion of zeroes and ones for the entire sequence. This means that randomness was not secured due to statistical bias. The cumulative sum test converts 0 to -1 and then calculates the cumulative sum. This is a random walk test and the result of an ideal sequence of random numbers is zero. In the cumulative sum test, if the value is 0, there is randomness, and the farther from 0, the more the test cannot be passed. Consecutive bits of either 0 or 1 are called a run, and the run test checks the probability that a run of 0 will change to a run of 1. For an ideal sequence of random numbers, the probability value is 0.5. The purpose of an FFT test is to detect periodic features using the peak heights in the FFT. This means that the sequence of random numbers has a periodic pattern that is not truly random. Thus, there is a problem that the random number can be reproduced. The non-overlapping-template test checks the number of occurrences of pre-defined target strings. This test rejects sequences that exhibit too many occurrences of a given non-periodic (aperiodic) pattern. The GAN-PRNG learns how to learn the bit stream in the front part to predict the bit stream in the back, and then generate an unpredictable bit stream through the predicted bit stream. In other words, it learns not to repeat a specific bit stream after a specific bit stream, which has the effect of not having a pattern even periodically. Considering such a training process, it indicates that the previous method was not trained enough to achieve randomness. In summary, the fact that the previous method failed these tests indicates that an ideal random number stream cannot be achieved due to frequency problems or the presence of patterns.

However, the proposed method passed most of the tests in the NIST test suite. Using the RNN layer makes it suitable for sequence data that has long-term dependencies, so even longer sequences can learn a previous bit stream and overall features. Compared to the results of the previous method learned with regional features using only the convolution layer, the result of learning the entire sequence has better randomness. Therefore, the proposed method achieves overall improvement in several tests (frequency, cumulative sum, run, FFT and non-overlapping-template) that, the previous method mostly failed. Finally, our method achieved high randomness and overcame the problems of predictability and reproducibility by using time series neural networks (RNN).

Table 2. Comparison of GAN based PRNG, where T, T_I, F_I, $F_I/\ \%$, F_P, F_T, $F\%$ are the number of individual tests, test instances, failed instances, their percentage, individual tests with p-value below the threshold, individual tests that failed, their percentage, respectively. The inference time is the time to generate a random number through trained generator.

	T	T_I	F_I	$F_I/\%$	F_P	F_T	$F\%$	Inference time
Before training	188	1789	1769	98.8	160.8	186	98.9	177.32 ms
Bernardi et al. [5]	188	1830	56	3.0	2.7	4.5	2.5	187.09 ms
Proposed method	188	1794	19.6	1.09	0.00	0.1	0.00	13.27 ms

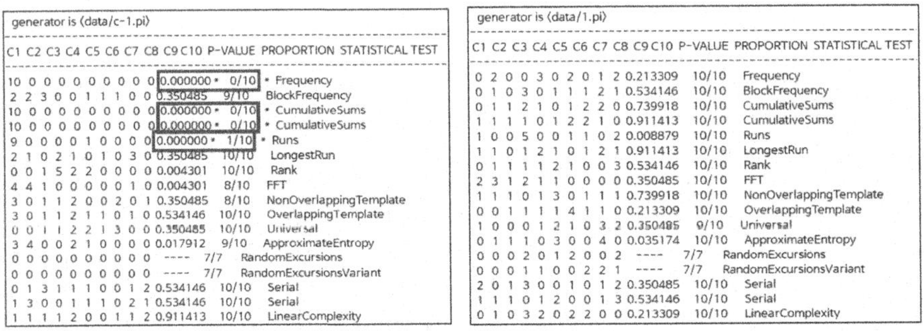

generator is ⟨data/c-1.pi⟩

C1	C2	C3	C4	C5	C6	C7	C8	C9	C10	P-VALUE	PROPORTION	STATISTICAL TEST
10	0	0	0	0	0	0	0	0	0	0.000000 *	0/10	* Frequency
2	2	3	0	0	1	1	1	0	0	0.350485	9/10	BlockFrequency
10	0	0	0	0	0	0	0	0	0	0.000000 *	0/10	* CumulativeSums
10	0	0	0	0	0	0	0	0	0	0.000000 *	0/10	* CumulativeSums
9	0	0	0	0	1	0	0	0	0	0.000000 *	1/10	* Runs
2	1	0	2	1	0	1	0	3	0	0.350485	10/10	LongestRun
0	0	1	5	2	2	0	0	0	0	0.004301	10/10	Rank
4	4	1	0	0	0	0	0	1	0	0.004301	8/10	FFT
3	0	1	1	2	0	0	2	0	1	0.350485	8/10	NonOverlappingTemplate
3	0	1	1	2	1	1	0	1	0	0.534146	10/10	OverlappingTemplate
0	0	1	1	2	2	1	3	0	0	0.350485	10/10	Universal
3	4	0	0	2	1	0	0	0	0	0.017912	9/10	ApproximateEntropy
0	0	0	0	0	0	0	0	0	0	----	7/7	RandomExcursions
0	0	0	0	0	0	0	0	0	0	----	7/7	RandomExcursionsVariant
0	1	3	1	1	1	0	0	1	2	0.534146	10/10	Serial
1	3	0	0	1	1	1	0	2	1	0.534146	10/10	Serial
1	1	1	1	2	0	0	1	1	2	0.911413	10/10	LinearComplexity

generator is ⟨data/1.pi⟩

C1	C2	C3	C4	C5	C6	C7	C8	C9	C10	P-VALUE	PROPORTION	STATISTICAL TEST
0	2	0	0	3	0	2	0	1	2	0.213309	10/10	Frequency
0	1	0	3	0	1	1	1	2	1	0.534146	10/10	BlockFrequency
0	1	1	2	1	0	1	2	2	0	0.739918	10/10	CumulativeSums
1	1	1	1	0	1	2	2	1	0	0.911413	10/10	CumulativeSums
1	0	0	5	0	0	1	1	0	2	0.008879	10/10	Runs
1	1	0	1	2	1	0	1	2	1	0.911413	10/10	LongestRun
0	1	1	1	1	2	1	0	0	3	0.534146	10/10	Rank
2	3	1	2	1	1	0	0	0	0	0.350485	10/10	FFT
1	1	1	0	1	3	0	1	1	1	0.739918	10/10	NonOverlappingTemplate
0	0	1	1	1	1	4	1	1	0	0.213309	10/10	OverlappingTemplate
1	0	0	0	1	2	1	0	3	2	0.350485	9/10	Universal
0	1	1	1	0	3	0	0	4	0	0.035174	10/10	ApproximateEntropy
0	0	0	2	0	1	2	0	0	2	----	7/7	RandomExcursions
0	0	0	1	1	0	0	2	2	1	----	7/7	RandomExcursionsVariant
2	0	1	3	0	0	1	0	1	2	0.350485	10/10	Serial
1	1	1	0	1	2	0	0	1	3	0.534146	10/10	Serial
0	1	0	3	2	0	2	2	0	0	0.213309	10/10	LinearComplexity

Fig. 7. Final analysis report of NIST test suite; (Left side) Bernardi et al. [5], (Right side) proposed method.

To see the pattern of the generated bit stream, we converted the bit stream into the form of a bitmap. Figure 8 shows the result of visualizing the generator output before and after training. The two images are different. The results before training are repeated with regular patterns. However, as the inner state changes in the training process, it learns not to produce a predictable repetitive pattern, which improves the randomness of generator output in comparison to that produced before training.

4.2 Comparison with Existing PRNGs

The exitings PRNGs were measured on an Intel Core i5-8259 CPU@2.30 GHz x 8, 16 GB RAM and Ubuntu 18.04.4LTS environment. The proposed method was measured by connecting the Edge TPU to the embedded processor. When an operation not supported by the TPU is performed, the CPU performs subsequent operations. The proposed method performs multiplication using only fully connected layers supported by the Edge TPU. So, it's the result of inference from the TPU. Table 3 is the experimental result. The proposed method was slower than the PRNGs in the desktop environment. However, Xorshift, Mersenne Twister decreases the speed when measured on STM32F4 as in [20]. In addition, in [21], lightweight PRNG shows a speed of 0.16 GB/s. As mentioned

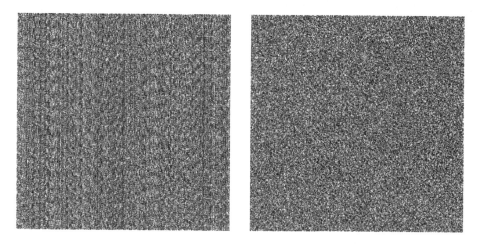

Fig. 8. Visualization of random number generated by the generator. (left) before training and (right) after training.

earlier, the proposed method is the result of using TPU on embedded. Considering this point, we think that the proposed method has a high speed as a PRNG for an embedded processor.

Table 3. Comparison with existing PRNGs.

	Throughput	Method	Machine
Xorshift128+	8.3 GB/s	XOR, Shift	Desktop
Xoroshiro128+	8.5 GB/s	XOR, Shift	Desktop
PCG64	4.3 GB/s	LCG	Desktop
MT19937-64	2.9 GB/s	Twisted GFSR	Desktop
MPCG [21]	0.16 GB/s	PCG	Embedded processors
This work	1.0 GB/s	GAN (Deep Learning)	Embedded processors

4.3 Next Bit Test

CSPRNG must satisfy the next bit test and be resistant to state compromise extensions attacks. The next bit test is that given a bit stream of m-bits, the $m+1th$ bit should not be predicted. As explained earlier, we train to be unpredictable which bit stream will come after some bit stream. So, if the proposed PRNG has been trained to reduce its losses sufficiently, the next bit test is satisfied. Figure 9 shows the losses of the generator and predictor. The loss values of the generator and predictor decrease in a similar pattern. And it can be seen that the value is decreasing to very close to 0.00.

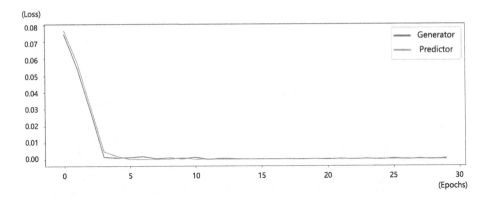

Fig. 9. Loss of a generator and a predictor.

4.4 State Compromise Attack Resistance

In general, CSPRNG uses an unpredictable random seed from the operating system, special hardware, or external sources. And it is not safe to use the built-in random function. In this work, entropy sources that can be collected in an embedded environment are used as random seed. It is necessary to reseed the CSPRNG to recover from potential state compromise [20]. In our proposed method, we generate $1,099,200$-bits with one seed and then reseeded. Reseed from a new entropy source to generate a more unpredictable bit stream. As the entropy of the random seed increases, the performance as a CSPRNG will improve to ensure the security of the cryptographic algorithm. Considering these features of GAN-PRNG, it is faster than other lightweight PRNG and satisfies the requirements of CSPRNG.

5 Conclusion

In this paper, we presented a novel GAN-based PRNG for embedded processors. The new GAN model was designed for embedded processors. The model was successfully ported to the Edge TPU. Finally, the random number sequence produced by this method passed the NIST test suite. In our future work we will applying other GAN models for high randomness and efficiency. Depending on the GAN model, the quality of random number sequences is totally different. In addition, we plan to reduce the random seed length for a resource-constrained environment.

Acknowledgement. This work was partly supported by the National Research Foundation of Korea (NRF) grant funded by the Korea government (MSIT) (No. NRF-2020R1F1A1048478) and this work was partly supported by Institute for Information & communications Technology Promotion (IITP) grant funded by the Korea government(MSIT) (No. 2018-0-00264, Research on Blockchain Security Technology for IoT Services).

References

1. Dabal, P., Pelka, R.: FPGA implementation of chaotic pseudo-random bit generators. In: Proceedings of the 19th International Conference Mixed Design of Integrated Circuits and Systems, MIXDES 2012, pp. 260–264. IEEE (2012)
2. Pande, A., Zambreno, J.: A chaotic encryption scheme for real-time embedded systems: design and implementation. Telecommun. Syst. **52**(2), 551–561 (2013)
3. Azzaz, M., Tanougast, C., Sadoudi, S., Dandache, A.: Real-time FPGA implementation of Lorenz's chaotic generator for ciphering telecommunications. In: 2009 Joint IEEE North-East Workshop on Circuits and Systems and TAISA Conference, pp. 1–4. IEEE (2009)
4. de la Fraga, L.G., Torres-Pérez, E., Tlelo-Cuautle, E., Mancillas-López, C.: Hardware implementation of pseudo-random number generators based on chaotic maps. Nonlinear Dyn. **90**(3), 1661–1670 (2017). https://doi.org/10.1007/s11071-017-3755-z
5. De Bernardi, M., Khouzani, M.H.R., Malacaria, P.: Pseudo-random number generation using generative adversarial networks. In: Alzate, C., et al. (eds.) ECML PKDD 2018. LNCS (LNAI), vol. 11329, pp. 191–200. Springer, Cham (2019). https://doi.org/10.1007/978-3-030-13453-2_15
6. Shannon, C.E.: Communication theory of secrecy systems. Bell Syst. Tech. J. **28**(4), 656–715 (1949)
7. Nyquist, H.: Thermal agitation of electric charge in conductors. Phys. Rev. **32**(1), 110 (1928)
8. Stipčević, M.: Fast nondeterministic random bit generator based on weakly correlated physical events. Rev. Sci. Instrum. **75**(11), 4442–4449 (2004)
9. Figotin, A., et al.: Random number generator based on the spontaneous alpha-decay. US Patent 6,745,217, 1 June 2004
10. Stefanov, A., Gisin, N., Guinnard, O., Guinnard, L., Zbinden, H.: Optical quantum random number generator. J. Mod. Opt. **47**(4), 595–598 (2000)
11. Vincent, C.: The generation of truly random binary numbers. J. Phys. E: Sci. Instrum. **3**(8), 594 (1970)
12. Schneier, B., Kohno, T., Ferguson, N.: Cryptography Engineering: Design Principles and Practical Applications. Wiley, Hoboken (2013)
13. Abadi, M., et al.: Tensorflow: a system for large-scale machine learning. In: 12th USENIX Symposium on Operating Systems Design and Implementation (OSDI 2016), pp. 265–283 (2016)
14. Sengupta, J., Kubendran, R., Neftci, E., Andreou, A.G.: High-speed, real-time, spike-based object tracking and path prediction on google edge TPU. In: AICAS, pp. 134–135 (2020)
15. Goodfellow, I., et al.: Generative adversarial nets. In: Advances in Neural Information Processing Systems, pp. 2672–2680 (2014)
16. Schawinski, K., Zhang, C., Zhang, H., Fowler, L., Santhanam, G.K.: Generative adversarial networks recover features in astrophysical images of galaxies beyond the deconvolution limit. Monthly Not. R. Astron. Soci. Lett. **467**(1), L110–L114 (2017)
17. Wang, X., et al.: ESRGAN: enhanced super-resolution generative adversarial networks. In: Leal-Taixé, L., Roth, S. (eds.) ECCV 2018. LNCS, vol. 11133, pp. 63–79. Springer, Cham (2019). https://doi.org/10.1007/978-3-030-11021-5_5
18. Schuster, M., Paliwal, K.K.: Bidirectional recurrent neural networks. IEEE Trans. Signal Process. **45**(11), 2673–2681 (1997)

19. Hong, S.L., Liu, C.: Sensor-based random number generator seeding. IEEE Access **3**, 562–568 (2015)
20. Kietzmann, P., Schmidt, T., Wählisch, M.: A guideline on pseudorandom number generation (PRNG) in the IoT, July 2020
21. Paul, B., Khobragade, A., Javvaji Sai, S., Goswami, S.S.P., Dutt, S., Trivedi, G.: Design and implementation of low-power high-throughput PRNGs for security applications. In: 2019 32nd International Conference on VLSI Design and 2019 18th International Conference on Embedded Systems (VLSID), pp. 535–536 (2019)

A RDBMS-Based Bitcoin Analysis Method

Hyunsu Mun, Soohyun Kim, and Youngseok Lee

Chungnam National University, Daejeon, Republic of Korea
{munhyunsu,shkim95,lee}@cnu.ac.kr

Abstract. Due to the proliferation of Bitcoin and Ethereum, over 1000 cryptocurrencies have appeared in the market. As hundreds of thousands of cryptocurrency transactions are taken place per day, cryptocurrency exchange, service operators, or government agencies have to observe user transaction activities for legal concerns or economic purposes. Cryptocurrencies generally use the blockchain structure where every transaction is connected in the linked list with the public key hash function. In order to examine specific transaction, address or a group of addresses, called a cluster, we need an efficient cryptocurrency data analyzer and scalable storage. Though a few studies on cryptocurrency analysis tools have been presented, they do not generally satisfy all the requirements. In this paper, we propose an extensible and user-friendly Bitcoin analysis software based on RDBMS. From extensive Bitcoin experiments, we demonstrate that RDBMS queries are useful to perform analysis of Bitcoin transaction, cluster and graph. In addition, we show that the indexed SQLite3 database provides quick response time and the extensible Bitcoin storage. This study contributes to a method of analyzing Bitcoin blockchain data using an easy-to-use RDBMS.

Keywords: Bitcoin · Cryptocurrency · Relational database system · Data analysis

1 Introduction

As over 1000 cryptocurrencies, including Bitcoin and Ethereum, have been popular in the market across the world[1], the cryptocurrency exchange, service providers, and government agencies have to monitor cryptocurrency transaction activities and user trends for various purposes. Cryptocurrency exchanges such as Huobi, Binance, Bithumb, and UPBit provide services to deposit, withdraw, and transfer the cryptocurrency. On the other hand, in the deep web,

---

[1] https://coinmarketcap.com/all/views/all/.

This work was supported by Institute of Information & communications Technology Planning & Evaluation (IITP) grant funded by the Korea government (MSIT) (No. 2020-0-00901, Information tracking technology related with cyber crime activity including illegal virtual asset transactions).

D. Hong (Ed.): ICISC 2020, LNCS 12593, pp. 235–253, 2021.
https://doi.org/10.1007/978-3-030-68890-5_13

online markets such as Silkroad and Dream market allow a customer to purchase illegal goods with the cryptocurrency. Transactions in the deep web are often used for crimes related with weapons, drugs, child pornography, and ransomware. Government agencies should collect and analyze the cryptocurrency of illegal transactions for tracking and detecting abnormal transactions early.

Though the Bitcoin software itself is an open-source project, there is no functional or manageable analysis platform for cryptocurrency exchange operators or government agencies. Recently, there have been a few Bitcoin analysis tools such as BlockSci, GraphSense, BTCSpark and Bitcoin-Abe in open-source projects. BlockSci provides a Python wrapper and C++ core engine for the Bitcoin analysis. As BlockSci focuses on high performance, it requires dedicated LevelDB and data structure. GraphSense is a visual analyzer of Bitcoin transactions and it needs a distributed filesystem called Cassandra. BTCSpark is an Apache Spark-based Bitcoin analyzer platform, which has not been updated recently. Bitcoin-Abe is a MySQL-based platform that has a web interface. So it is not suited for general purpose processing and analyzing Bitcoin data.

Overall, Bitcoin data analysis are classified into three functions. First, we need to retrieve the Bitcoin transaction and its address information and to search for specific transactions with time or date conditions. Second, we have to examine the transactions among a group of addresses, called a cluster. For this purpose, we have to build a set of addresses into a tagged cluster, which is often carried out by heuristic algorithms. Then, we have to investigate the wallets in the cluster and transactions between clusters. Third, we perform graph analysis algorithms on the Bitcoin graph consisting of Bitcoin addresses and transactions.

When designing a Bitcoin analysis method, we have to meet the following challenges.

- Scalable Bitcoin data ingestion and storage: Bitcoin data includes transaction information consisting of public key and hash function. Bitcoin addresses and transaction data should be ingested into the storage which should support scalable and efficient manipulation functions of a large amount of blockchain data.
- Easy interface of analytics: Bitcoin analysis often needs plain questions that can be expressed with user-friendly interface for the data analyzers. For this purpose, we should build transaction information and related metadata.
- Compatibility of software integration: Bitcoin analysis requires integration with the related software such as Bitcoin core, data storage, visualization and web servers.

In this paper, we propose an extensible and scalable Bitcoin analysis based on relational database. We harness the relational database to ingest and store Bitcoin data. For scalability, we present three-layer databases with seven Bitcoin tables which can store Bitcoin transaction and address information. With the relational database such as SQLite3 and MariaDB, we support easy analytics in queries. We present representative queries on Bitcoin analytics, cluster analytics, and graph analytics. As we use SQLite3 database, our analysis method can be easily integrated with the web or visualization software. This study contributes to a method of analyzing Bitcoin blockchain data using an easy-to-use RDBMS.

2 Related Work

As Bitcoin software[2] is an open source project, the Bitcoin P2P network is available to the public. Bitcoin software provides only the access function to the P2P network and it does not have convenient and useful analysis modules. There are various analysis tools with their own databases as well as Bitcoin block parsers that extract information from Bitcoin block byte files (Table 1).

Kalodner et al. [1] proposed a Bitcoin data analysis software named BlockSci[3] that uses an append-only data structure instead of atomicity, consistency, isolation, and durability (ACID) features of the database. It is known that BlockSci has the improved performance with C++ for cluster analysis. Bitcoin data can be expressed as an address-transaction graph or a graph that uses an address and a transaction as nodes. Masarah et al. [2] proposed a software called GraphSense[4] that expresses and analyzes Bitcoin transactions in a graph, and analyzed the relationship of Bitcoin addresses used in 35 ransomwares from 2013 to 2017.

BTCSpark[5], proposed by Rubin et al. [3], is a software that allows a user to analyze the Bitcoin blockchain on the Apache Spark that uses a distributed cluster. In 2011, an open source project called Bitcoin-Abe[6] allowed data storage using a SQL database such as PostgreSQL, MySQL's InnoDB engine, and SQLite. Since Bitcoin-Abe was created to provide the analysis function of Bitcoin block data through the web, it is suitable for examining bitcoin blockchain, transaction, and address information, but not for performing Bitcoin address clustering and graph analysis.

Bitcoin and Ethereum have different structures, making it difficult to analyze them all with one tool. Massimo et al. [4] proposed an API tool named BlockAPI[7] for analyzing the blockchain, demonstrating that Bitcoin and Ethereum data can be stored and analyzed in MongoDB, MySQL, and PostgreSQL.

It is essential to analyze the addresses owned by cryptocurrency exchanges to understand Bitcoin transactions. Zhen et al. [5] found addresses of 10 famous cryptocurrency exchanges such as Huobi and Binance using Bitcoin address clustering heuristics, and classified the addresses of those exchanges into user addresses, hot wallets, and cold wallets. They suggest 20 features of a cold wallet that stores Bitcoins, a hot wallet that is used for transactions, and a user address.

Since users can have multiple Bitcoin addresses, in order to know the owner of the Bitcoin address, it is necessary to cluster Bitcoin addresses through Bitcoin

Table 1. Comparison of Bitcoin analysis tools: our method vs. previous studies.

	Ours	BlockSci	GraphSense	BTCSpark	Bitcoin-Abe	BlockAPI
Cryptocurrency	Bitcoin Ethereum	Bitcoin Bitcoin-like	Bitcoin	Bitcoin	Bitcoin	Bitcoin Ethereum
Language	SQL	C++ Python 3	Scala	Scala	C++	JAVA Scala
Dependency	-	LevelDB	BlockSci Casandra	Spark	LevelDB MySQL	LevelDB
Bitcoin Analytics	○	○	○	○	○	○
Cluster Analytics	○	○	×	×	×	×
Graph Analytics	○	×	○	×	×	×
Response Time	Low	Low	High	High	High	High

address heuristics or graph analysis. Mikkel Alexander *et al.* [6] use the Bitcoin address cluster data provided by Chainalysis to classify the Bitcoin address cluster into 10 categories such as exchange, gambling, ransomware, scam, and tor market by machine learning. Lee *et al.* [7] analyzed the flow of Bitcoin addresses collected at Tor darknet and proposed a method to track criminals' behavior using exchanges to exchange Bitcoins. Baokun *et al.* [8] proposed a method to create a Bitcoin address cluster using the Bitcoin address heuristic and the improved the Louvain community detection algorithm. Zhen *et el.* [9] performed a Bitcoin address cluster by combining the Bitcoin-address tag collected from the surface web and Bitcoin graph analysis, and then found the entity of the cluster.

3   RDMBS-Based Bitcoin Analysis

3.1   Architecture

For the RDBMS-based Bitcoin analysis, we designed a Bitcoin data collector that ingests Bitcoin information from `Bitcoin Core` as shown in Fig. 1, The Bitcoin data collector imports transaction data through `JSON-RPC` into the temporary JSON files. In order to explain the address and transaction information, we collect tag data corresponding addresses from web, like WalletExplorer[8], and Tor-based deep web. We store the Bitcoin data in the `SQLite3` consisting of three databases and seven data tables. The Bitcoin analyzer provides analysis queries on the `SQLite3` tables. The web and visualization modules like a Neo4j[9], Matplotlib[10], NetworkX[11] can be connected to the `SQLite3` database APIs. We make the database construction script, detailed schema, and example queries available on Github project[12].

---

[8] https://www.walletexplorer.com/.
[9] https://neo4j.com/.
[10] https://matplotlib.org/.
[11] https://networkx.github.io/.
[12] https://github.com/munhyunsu/BitcoinAnalysis/.

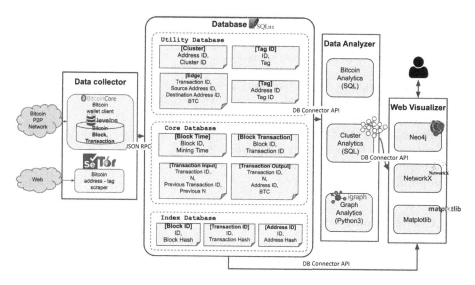

Fig. 1. RDBMS-based Bitcoin Analysis Architecture

3.2   Database Schema

Although `Bitcoin Core` provides transaction data on the blockchain, it will take a long time to retrieve addresses or transactions for the given time period or specific cluster. In order to minimize the access time to transactions across the long blockchain, we create the transaction tables for the input and output indexed by block id, transaction id, and address id (Appendix A). We also create the cluster table where each cluster consists of a set of addresses with the help of crowd-sourced tag information and the active probe method.

We present three-layer databases of index, core, and utility as follows. Index database has key identifier tables for blocks, transactions and addresses with the integer type. The index database is useful for reducing the long Bitcoin hash value to the integer variable so we can decrease the size of Bitcoin core tables by 65% from 906 GB to 314 GB. The core database maintain the transaction information tables through `JSON-RPC` by iterating all transactions along the blockchain. Due to the core database, we can improve the response time of transaction analysis including specific addresses for input or output, and clustering heuristics. We also retain utility database for the fast response time of joining tables.

3.3   Bitcoin Analytics

Based on the Bitcoin database, we can easily retrieve Bitcoin address, transaction, and cluster information with date, time, and clusters' conditions. For example, if a user wants to perform analysis on transactions and addresses generated during a specific period, the user must know the transaction hash mined during that period (Listing B2). By using the proposed analysis method, the query can be analyzed as a conditional statement. The representative examples of Bitcoin analytics queries are as follows.

– Retrieving a number of transactions for a time period
– Finding a top addresses in the order of Bitcoin balance
– Retrieving a transaction fee
– Retrieving UTXO addresses
– Calculating withdrawing count and volume
– Calculating Transfer count and volume
– Computing Address balance

3.4 Cluster Analytics

A Bitcoin cluster is defined as a set of addresses belonging to a single organization such as an cryptocurrency exchange. For instance, an exchange will issue an account and related addresses for a customer. A marketplace or ransomware operators often maintain their own addresses, which form a cluster. Therefore, we need to group a set of addresses for a specific cluster. As Bitcoin has the anonymous feature, we cannot explicitly know which address belongs to which organization. For this purpose, we employ Bitcoin address clustering heuristics.

In our analysis method, we implemented two representative Bitcoin address clustering heuristics. The first one is the multi-input heuristic which groups the input addresses for a single transaction, because they will usually belong to a single owner (Listing B4). Though the multi-input heuristic is not perfect, it is still useful for finding the seed of Bitcoin address clusters. The other one is the one-time change heuristic that groups newly generated addresses not used by any future transaction. After applying the address clustering step, we put the appropriate tag to the cluster by using the crowd-sourcing data such as http:// www.walletexplorer.com. Then, we provide queries for the cluster analytics such as intra- or inter-cluster address or transaction information. A sample of cluster analysis queries are as follows.

– Grouping a set of addresses into a cluster with the given heuristic (e.g., multi-input or one-time change)
– Retrieving a set of addresses belonging to a cluster
– Retrieving transaction information between clusters
– Sorting the addresses within a given cluster (e.g., hot wallet or cold wallet addresses of a cluster)

3.5 Graph Analytics

As Bitcoin addresses and transactions can be modeled as a graph, we can issue graph analysis queries as follows.

– Calculating indegree and outdegree of an address node
– Computing PageRank of address nodes
– Finding shortest paths between addresses or clusters
– Finding maximum flow between two clusters

With queries on the Bitcoin graph, we can examine the graph characteristics. In addition, it is possible to analyze important addresses among 7 million Bitcoin addresses by using the PageRank algorithm (Listing B5). We can use various graph analysis tools such as NetworkX, iGraph, or Neo4j for analyzing the Bitcoin address transaction graph after exporting the csv file from the Bitcoin tables (Listing B7). From RDBMS tables, we can generate Bitcoin address transaction edges. Listing B6 is a query that lists all bitcoin address transaction graph information. We can also extract a subgraph with conditions such as duration, block height, address tag or cluster.

4 Analysis Results with Queries

4.1 Bitcoin Data

For the experiment, we import all blocks the height from 0 to 644,806 into the database as shown in Table 2. We extract input transaction, output transaction, and cluster tables from raw block data for efficient query performance. For the continuous update of Bitcoin addresses and transactions, we append the newly created Bitcoin block data to the existing table.

Table 2. Bitcoin data information for database construction.

Block height	0–644,806
Date	2009-01-03 06:15:05 (UTC)–2020-08-22 06:24:55 (UTC)
Transactions	560,882,950
Addresses	704,688,729
Database volume	357 GB

4.2 Which Address Has the Largest Amount of Bitcoin?

As Bitcoin is based on unspent transaction output (UTXO), we must add all the UTXO BTC values of the address to find the balance of a specific address. For the balance query operation, we add the deposits to the target address from every block, which takes long response time even with high performance tools like BlockSci. In order to know the balance for each address, we create a UTXO table up to the last block and calculate the amount of BTC of the address by performing a GROUP BY operation on each address (Listing B3).

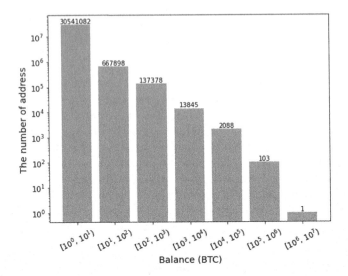

Fig. 2. Histogram of Bitcoin balance by address on 2020-08-22 (Block height: 644,805)

Bitcoin is generated every time a block is mined and 18467457.8422912 BTC has been generated as of 2020-08-22 06:24:55 (UTC). 31,362,395 addresses hold this Bitcoin, and the distribution is shown in Fig. 2. The Bitcoin address 35hK24tcLEWcgNA4JxpvbkNkoAcDGqQPsP has 355 UTXOs and 1.14% of the generated Bitcoins as shown in Table 3. According to https://bitcointalk.org/, this address is known as the cold wallet of the cryptocurrency exchange Huobi.com.

Table 3. Top 10 Bitcoin holdings and the number of UTXOs on 2020-08-22 (Block height: 644,805)

Rank	Bitcoin address	Balance (% of Bitcoins)	# of UTXOs
1	35hK24tcLEWcgNA4JxpvbkNkoAcDGqQPsP	247,502 (1.34)	355
2	37XuVSEpWW4trkfmvWzegTHQt7BdktSKUs	94,505 (0.51)	70
3	34EiJfy4jGF32M37aQ2ZobupwiRQWa1Siy	92,857 (0.50)	7
4	1FeexV6bAHb8ybZjqQMjJrcCrHGW9sb6uF	79,957 (0.43)	336
5	34xp4vRoCGJym3xR7yCVPFHoCNxv4Twseo	73,436 (0.40)	45
6	3D8qAoMkZ8F1b42btt2Mn5TyN7sWfa434A	70,000 (0.38)	171
7	1HQ3Go3ggs8pFnXuHVHRytPCq5fGG8Hbhx	69,370 (0.38)	212
8	37tRFZw7n94Jddq6TfVs3MbCXmDX6eMfeY	68,101 (0.37)	5
9	3JurbUwpsAPqvUkwLM5CtwnEWrNnUKJNoD	65,236 (0.35)	9
10	bc1qgdjqv0av3q56jvd82tkdjpy7gdp...	60,000 (0.32)	5

4.3   Clustering Bitcoin Addresses with Heuristics

We have performed a multi-input heuristic for all Bitcoin transactions on more than 641,039 blocks. For improving the performance of heuristic, we save the temporary cluster information on the memory. The final clustered addresses are recorded in the table. Figure 3 shows the cluster size distribution for the multi-input and one-time change heuristics.

With our own multi-input heuristics, we processed 691,806,723 addresses and grouped 418,493,080 addresses into 61,918,407 clusters. We observed that 95.7% of clusters are small, with less than 10 addresses in Fig. 3a. On the other hand, we found super-clusters with more than 10 million addresses.

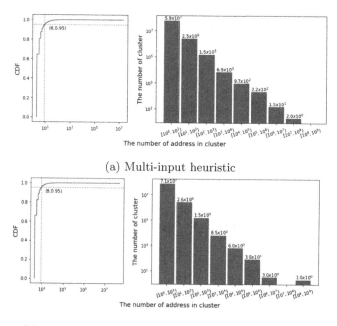

(a) Multi-input heuristic

(b) Multi-input heuristic and one-time change heuristic

Fig. 3. The cluster size distribution for 690 million addresses on 641,039 blocks.

Multi-input heuristic cannot cluster UTXOs, and addresses that are not used together are not clustered into the same cluster. To solve this problem, we clustered the Bitcoin addresses using both the one-time change heuristic and the multi-input heuristic. As a result, one-time change heuristic grouped 464,473,950 addresses into 74,339,475 clusters. We clustered 4,598,0870 more addresses compared to using only multi-input heuristic, but super cluster with 177 million addresses clustered into one was created as Fig. 3b. Because of this heuristic inaccuracy, we presume that the Bitcoin address heuristic should compensated with machine learning or graph analysis.

4.4   Identify Clusters with Address-Tag Information

We find out the entity of the cluster by using the Bitcoin address-tag information collected by https://www.walletexplorer.com. From the result of the multi-input heuristic, 137,733 (0.2%) clusters were identified, which contained 86,278,071 (20.62%) addresses. Multiple tags such as cryptocurrency exchange, ransomware, mixing, and gambling were attached to a single cluster.

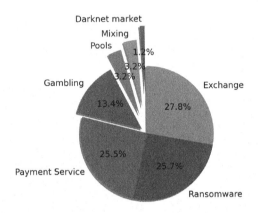

Fig. 4. The cluster size distribution of the multi-input heuristic for 690 million addresses on 641,039 blocks.

As shown in Fig. 4, We identify 86,278,071 addresses into exchanges, ransomware, payment services, and gambling. 92.4% of the addresses are included in the top 4 categories. Bter.com, NoobCrypt, Epay.info, and SecondsTrade.com are entities with the most weight in each category.

4.5   What Is the Amount and Count of Transactions to CryptoLocker Addresses?

After applying the Bitcoin address clustering heuristic, we identify the owner of the cluster through Bitcoin address-tag information. Using our database, we can perform intra- or inter-cluster analysis. The representative inter-cluster analysis example is to examine the transaction history between clusters. Ransomware victims usually pay attackers through exchanges. Therefore, in order to analyze the amount of ransomware fee, it is necessary to investigate transactions transferred from the exchange cluster to the ransomware cluster.

Figure 5 shows the number of Bitcoin transactions and BTC amount transferred to Cryptolocker from September 5, 2013 to May 2014. According to Wikipedia and news reports, Cryptolocker has broken out and continued until May 2014. However, through the Cryptolocker inter-cluster analysis, we observe that no transaction has occurred since February 2014. A total of 2,789.15 BTC was transferred, and the number of transactions was 1,839 with the largest trading volume in October 2013 when Cryptolocker was widely spread out.

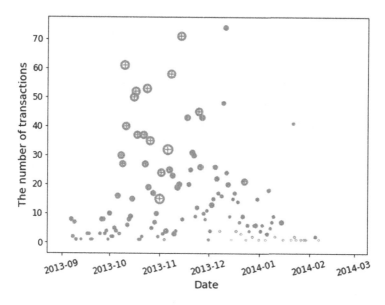

Fig. 5. The number of daily Bitcoin transactions and Bitcoin amount sent to Cryptolocker (Circle size: BTC amount, 2013-09-01–2014-03-31)

4.6   List the Hot Wallet Addresses of a Korean Exchange A

Cryptocurrency exchange operators not only exchange real and virtual currencies, but also issue Bitcoin addresses for a customers and perform transactions on behalf of customers. In general, for this purpose, exchanges use three types of wallet addresses: a user wallet that is distributed to users; a hot wallet used to collect cryptocurrencies deposited at the user's address and to send cryptocurrency to outside the exchange; a cold wallet that stores cryptocurrency for storage and security. Among these three wallets, it is essential to find a hot wallet in order to identify and analyze the wallet address of the exchange because the hot wallet is the center of operation of the exchange.

We use the high indegree or outdegree of the address node as the feature for finding exchange's hot wallets that deposit Bitcoins to the user's address and transfer money to other addresses. On the Bitcoin graph, we performed intra-cluster analysis to find hot wallet in the Korean exchange A. Table 4 shows the top 12 degree addresses in Korean exchange A, with high transaction count (degree). From this experiment, we observe that 12 of the 3,545 addresses of the exchange A has a degree of more than 1000.

4.7   Graph Analysis Algorithm Using Graph Tools

Bitcoin blockchain can be represented as a graph because the input address and the output address are connected with a transaction. After building a graph with Bitcoin address nodes and transaction edges, we carry out graph analysis.

Table 4. Top 12 hot wallet list of Korean exchange A

Rank	Bitcoin address	Degree	Value
1	1En5ErLPzF9RMeP8z8hjna3..........	2,870	21,198
2	1JeyZBDbJTz5d1rfSkGqywzw..........	2,520	122
3	19iGtbDzXSASmcyJFbdgCiFi..........	2,082	102
4	1KHFeyp2Sb4xXg1rjnNRi4c8..........	2,076	28,549
5	1Gxd9c2VuLcQjee1tubhHSJS..........	2,072	108
6	19Ls2qFMEztRVgSYyFFtFReE..........	1,516	14
7	1CsTzASjqs8f63pzcw5f9LJo..........	1,477	128
8	181acE6XdV4JToqMFRNmmKDq..........	1,402	70
9	1GoxkdmiZzFKwndzGihHcW82..........	1,352	33
10	1QJ13PRLkWBF4s1XGKUMr9Ab..........	1,162	12,051
11	18rWxfA3Qv6uFKwKexGtskPx..........	1,130	4,842
12	1AvGPjBB3PcdhLYwy7twKFrP..........	1,116	50

We extract a graph of Silkroad, which is the famous black market from 2011 to 2013, and investigate the main address with the PageRank algorithm. As the PageRank algorithm can reveal the importance of nodes in the weighted graph, we employ this algorithm in iGraph for analyzing the core of Bitcoin addresses. We used iGraph[13] to generate and analyze the extracted edge data (Listing B6,B7).

After finding the most important address of the Silkroad cluster with the PageRank algorithm, we looked at the transaction change of the address over time. Figure 6 is a graph showing the transaction volume over time. After the Silkroad was arrested by the FBI in 2013[14], its operation was stopped. However, we can observe that the recently occurred transactions.

4.8   Community Detection Algorithm on Korean Exchange B Cluster

We use the Bitcoin address heuristic in order to find a Bitcoin address cluster. However, as the heuristics do not provide the accurate results, we harness the alternative clustering algorithm with graph community detection algorithms.

We built a Bitcoin address-transaction graph of a Korean exchange B, and found clusters with the Leiden community detection algorithm as shown in Fig. 7. From this the community detection algorithm experiment, we can observe that the largest community was a cluster of 51,714 out of 275,952 addresses. There were 7 communities over 10,000 addresses.

---

[13] https://igraph.org/.
[14] https://en.wikipedia.org/wiki/Silk_Road_(marketplace).

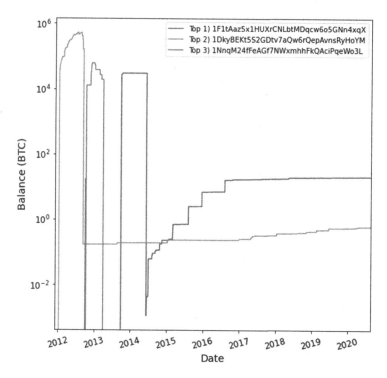

Fig. 6. The balance of top 3 PageRank address in Silkroad marketplace graph

4.9   Performance

For the Bitcoin analysis method, both the initial database construction performance and the quick response time of a query are important. In particular, when analyzing large-scale blockchain data, we have to construct the Bitcoin database within acceptable time and process queries quickly. In SQLite3, we often meet the slow response time of analysis queries without database optimization. For instance, we increase `PRAGMA cache_size` of `SQLite3`. This is the size of the cache used by the database, and has a great impact on the database construction performance. When building the database with the initial value of 8 KB, it takes about 40 h. However, with 300 GB, it takes only about 7 h.

Indexing is a major technology that accelerates search performance in RDBMS and is important in our proposed analysis method. For example, when performing a task of generating a Bitcoin address edge list, it takes only 15 min with index. However, without index, the address-edge generation time increases to over 24 h. In addition, our three-layer table structure has a great impact on improving performance. For example, when creating the edge table (TxID, SrcAddrID, DstAddrID, BTC) in advance, the Bitcoin address edge list query job was completed only in 2 s.

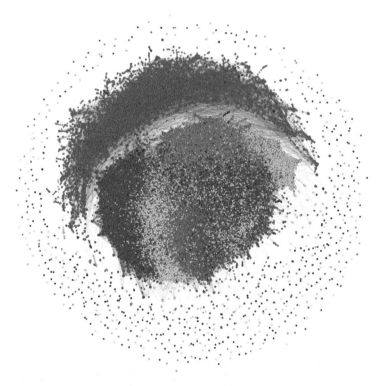

Fig. 7. A result of Leiden community detection on B address graph

5   Conclusion

The problem of RDBMS Bitcoin analysis is difficult to solve because we have to deal with large-scale data storage construction and complicated Bitcoin analysis jobs. In this paper, we propose an extensible and user-friendly Bitcoin analysis software based on RDBMS. We present three-layer databases with seven Bitcoin tables which can store Bitcoin transaction and address information. We show that the proposed tool supports Bitcoin analytics, cluster analytics, and graph analytics through an example of analyzing Bitcoin addresses, clusters, and graphs. The proposed method can be used for various applications with scalability. In the future, we plan to enhance Bitcoin address cluster analysis, machine learning, and graph analysis functions.

Appendix

A   Three-Layer Table and Index Creation Query

Listing A1. Bitcoin Index Tables

```
CREATE TABLE IF NOT EXISTS BlkID (
    id INTEGER PRIMARY KEY,
    blkhash TEXT NOT NULL UNIQUE);
CREATE TABLE IF NOT EXISTS TxID (
    id INTEGER PRIMARY KEY,
    txid TEXT NOT NULL UNIQUE);
CREATE TAblE IF NOT EXISTS AddrID (
    id INTEGER PRIMARY KEY,
    addr TEXT NOT NULL UNIQUE);
```

Listing A2. Bitcoin Core Tables and Indices

```
CREATE TABLE IF NOT EXISTS BlkTime (
    blk INTEGER PRIMARY KEY,
    unixtime INTEGER NOT NULL);
CREATE TABLE IF NOT EXISTS BlkTx (
    blk INTEGER NOT NULL,
    tx INTEGER NOT NULL,
    UNIQUE (blk, tx));
CREATE TABLE IF NOT EXISTS TxIn (
    tx INTEGER NOT NULL,
    n INTEGER NOT NULL,
    ptx INTEGER NOT NULL,
    pn INTEGER NOT NULL,
    UNIQUE (tx, n));
CREATE TABLE IF NOT EXISTS TxOut (
    tx INTEGER NOT NULL,
    n INTEGER NOT NULL,
    addr INTEGER NOT NULL,
    btc REAL NOT NULL,
    UNIQUE (tx, n, addr));
CREATE INDEX idx_BlkTime_2 ON BlkTime(unixtime);
CREATE INDEX idx_BlkTx_2 ON BlkTx(tx);
CREATE INDEX idx_TxIn_3_4 ON TxIn(ptx, pn);
CREATE INDEX idx_TxOut_3 ON TxOut(addr);
```

Listing A3. Bitcoin Utility tables and Indices

```
CREATE TABLE IF NOT EXISTS Edge (
    tx INTEGER NOT NULL,
    src INTEGER NOT NULL,
    dst INTEGER NOT NULL,
    btc REAL NOT NULL,
    UNIQUE (tx, src, dst, btc));
```

```
CREATE TABLE IF NOT EXISTS Cluster (
    addr INTEGER PRIMARY KEY,
    cluster NOT NULL);
CREATE TABLE IF NOT EXISTS TagID (
    id INTEGER PRIMARY KEY,
    tag TEXT UNIQUE);
CREATE TABLE IF NOT EXISTS Tag (
    addr INTEGER NOT NULL,
    tag INTEGER NOT NULL,
    UNIQUE (addr, tag));
CREATE INDEX idx_Edge_1 ON Edge(tx);
CREATE INDEX idx_Edge_2 ON Edge(src);
CREATE INDEX idx_Edge_3 ON Edge(dst);
CREATE INDEX idx_Cluster_2 ON Cluster(cluster);
```

B   Example Queries (Code Block)

Listing B1. (CB:AddrID) Address ID by Address hash

```
SELECT DBINDEX.AddrID.id
FROM DBINDEX.AddrID
WHERE DBINDEX.AddrID.addr = 'ADDRHASH';
```

Listing B2. (CB:TxTime) Transaction hash for a specific period

```
SELECT BlkTx.tx
FROM BlkTx
INNER JOIN BlkTime ON BlkTime.blk = BlkTx.blk
WHERE (
  SELECT STRFTIME('%s', '2020-06-01T00:00:00+00:00')) <=
    BlkTime.unixtime
AND BlkTime.unixtime <= (
  SELECT STRFTIME('%s', '2020-06-31T23:59:59+00:00'));
```

Listing B3. (CB:Balance) Increase or decrease the balance of the address for a specific period

```
SELECT Income.value-Outcome.value AS Balance
FROM
(SELECT SUM(btc) AS value
 FROM TxOut
 WHERE TxOut.addr = ([CB:AddrID]) AND
       TxOut.tx IN ([CB:TxTime]) AS Income,
(SELECT SUM(btc) AS value
 FROM TxIn
 INNER JOIN TxOut ON TxIn.ptx = TxOut.tx AND
            TxIn.pn = TxOut.n
 WHERE TxOut.addr = ([CB:AddrID]) AND
       TxIn.tx IN ([CB:TxTime])) AS Outcome;
```

Listing B4. (CB:MultiInput) Multi input Bitcoin address heuristic

```
SELECT TxOut.addr AS addr
FROM TxIn
INNER JOIN TxOut ON TxIn.ptx = TxOut.tx AND TxIn.pn = TxOut.n
WHERE txIn.tx IN (SELECT TxIn.tx
                  FROM TxIn
                  INNER JOIN TxOut ON TxIn.ptx = TxOut.tx AND
                                      TxIn.pn = TxOut.n
                  WHERE addr = ([CB:AddrID]))
GROUP BY addr;
```

Listing B5. (CB:Edge) Address-Transaction graph edge

```
SELECT TXI.tx, TXI.addr, TXO.addr, TXO.btc
FROM (
  SELECT TxIn.tx AS tx, TxIn.n AS n,
         TxOut.addr AS addr, TxOut.btc AS btc
  FROM TxIn
  INNER JOIN TxOut ON TxOut.tx = TxIn.ptx AND
             TxOut.n = TxIn.pn) AS TXI
INNER JOIN (
  SELECT TxOut.tx AS tx, TxOut.n AS n,
         TxOut.addr AS addr, TxOut.btc AS btc
  FROM TxOut) AS TXO ON TXO.tx = TXI.tx;
```

Listing B6. (CB:ExportGraph) Export csv file for address transaction graph

```
.header on
.mode csv
.once edge.csv
SELECT SRC.addr AS saddr, DST.addr AS daddr,
       SUM(Edge.btc) AS btc, Edge.src AS saddr_id,
       Edge.dst AS daddr_id, COUNT(Edge.tx) AS cnt
FROM Edge
INNER JOIN TxID ON TxID.id = Edge.tx
INNER JOIN AddrID AS SRC ON SRC.id = Edge.src
INNER JOIN AddrID AS DST ON DST.id = Edge.dst
WHERE Edge.src in (
  SELECT Cluster.addr
  FROM Cluster WHERE Cluster.cluster IN (
    SELECT Cluster.cluster
    FROM Cluster WHERE Cluster.addr IN (
      SELECT Tag.addr
      FROM Tag WHERE Tag.tag = (
        SELECT TagID.id
        FROM TagID WHERE TagID.tag = 'TAG'))))
AND    Edge.dst in (
  SELECT Cluster.addr
  FROM Cluster WHERE Cluster.cluster IN (
    SELECT Cluster.cluster
```

```
   FROM Cluster WHERE Cluster.addr IN (
      SELECT Tag.addr
      FROM Tag WHERE Tag.tag = (
         SELECT TagID.id
         FROM TagID WHERE TagID.tag = 'TAG'))))
GROUP BY Edge.src, Edge.dst;
```

Listing B7. PageRank algorithm using iGraph

```
## Export edge to csv using query (file: edge.csv)
df = pd.read_csv('./edge.csv')
vertices = set()
edges = list()
weights = list()
for index, row in df.iterrows():
    if row['saddr'] not in vertices:
        vertices.add(row['saddr'])
    if row['daddr'] not in vertices:
        vertices.add(row['daddr'])
    edges.append((row['saddr'], row['daddr']))
    weights.append((row['cnt']))
vertices = list(vertices)
g = igraph.Graph()
g.add_vertices(vertices)
g.add_edges([(x[0], x[1]) for x in edges])
g.es['weight'] = weights
# Comunity detection
partition = g.community_leiden(
               objective_function='modularity')
# Page rank
pagerank = g.pagerank(weights=weights)
```

References

1. Kalodner, H., Goldfeder, S., Chator, A., Möser, M., Narayanan, A.: BlockSci: design and applications of a blockchain analysis platform. arXiv preprint arXiv:1709.02489 (2017)
2. Paquet-Clouston, M., Haslhofer, B., Dupont, B.: Ransomware payments in the bitcoin ecosystem. J. Cybersecur. **5**(1), tyz003 (2019)
3. Rubin, J.: BTCSpark: scalable analysis of the bitcoin blockchain using spark. Dec **16**, 1–14 (2015)
4. Bartoletti, M., Lande, S., Pompianu, L., Bracciali, A.: A general framework for blockchain analytics. In: Proceedings of the 1st Workshop on Scalable and Resilient Infrastructures for Distributed Ledgers, pp. 1–6 (2017)
5. Li, Z., Zheng, Y., Li, Q., Ming, W., Peng, K.: Dragnet: a method for tagging bitcoin addresses of exchanges (2020)
6. Harlev, M.A., Yin, H.S., Langenheldt, K.C., Mukkamala, R., Vatrapu, R.: Breaking bad: de-anonymising entity types on the bitcoin blockchain using supervised machine learning. In: Proceedings of the 51st Hawaii International Conference on System Sciences (2018)

7. Lee, S., et al.: Cybercriminal minds: an investigative study of cryptocurrency abuses in the dark web. In: Network and Distributed System Security Symposium, pp. 1–15. Internet Society (2019)
8. Zheng, B., Zhu, L., Shen, M., Du, X., Guizani, M.: Identifying the vulnerabilities of bitcoin anonymous mechanism based on address clustering. Sci. China Inf. Sci. **63**(3), 1–15 (2020). https://doi.org/10.1007/s11432-019-9900-9
9. Zhang, Z., Zhou, T., Xie, Z.: Bitscope: scaling bitcoin address deanonymization using multi-resolution clustering. In: Proceedings of the 51st Hawaii International Conference on System Sciences (2018)

Fault and Side-Channel Attack

Federated Learning in
Side-Channel Analysis

Huanyu Wang$^{(\boxtimes)}$ and Elena Dubrova

KTH Royal Institute of Technology, Stockholm, Sweden
{huanyu,dubrova}@kth.se

Abstract. Recently introduced federated learning is an attractive framework for the distributed training of deep learning models with thousands of participants. However, it can potentially be used with malicious intent. For example, adversaries can use their smartphones to jointly train a classifier for extracting secret keys from the smartphones' SIM cards without sharing their side-channel measurements with each other. With federated learning, each participant might be able to create a strong model in the absence of sufficient training data. Furthermore, they preserve their anonymity. In this paper, we investigate this new attack vector in the context of side-channel attacks. We compare the federated learning, which aggregates model updates submitted by N participants, with two other aggregating approaches: (1) training on combined side-channel data from N devices, and (2) using an ensemble of N individually trained models. Our first experiments on 8-bit Atmel ATxmega128D4 microcontroller implementation of AES show that federated learning is capable of outperforming the other approaches.

Keywords: Federated learning · Side-channel attack · AES · Power analysis

1 Introduction

Federated learning (FL) is a new paradigm in machine learning that can help meet regulatory requirements (GDPR [31], HIPAA [1]) and mitigate privacy concerns while taking advantage of massive distributed data [15,16,23]. FL allows its participants to collaboratively train a global model without sharing participant's local training data. At every communication round, each participant trains a local model based on his/her training data and submits the model updates to the server. The server employs a secure aggregation [3] to build a global model by averaging the local models' weights. Motivating applications for FL include image classifiers for self-driving cars, keyboard next-word predictors, and personalized product recommendation services [18].

However, as any great scientific discovery, FL can potentially be used with malicious intent. Since FL preserves not only training data confidentiality, but also participant's anonymity, its setting is very appealing to adversaries. Furthermore, an adversary who does not have enough training data might still be

© Springer Nature Switzerland AG 2021
D. Hong (Ed.): ICISC 2020, LNCS 12593, pp. 257–272, 2021.
https://doi.org/10.1007/978-3-030-68890-5_14

able to create a strong deep-learning model by training in a FL framework. For example, adversaries can use their smartphones to jointly train a classifier for extracting secret keys from the smartphones' SIM cards without sharing their local side-channel measurements with each other. At each round, every participant independently trains a local model update based on traces captured from his/her profiling device and uploads it to the aggregator, where the submitted updates are combined to construct a global model. The aggregator can be either a participant, or a third party.

In this paper, we investigate this new attack vector in the context of Deep-Learning Side-Channel Attacks (DL-SCAs). DL-SCA is one of the most powerful attacks against implementations of cryptographic algorithms at present [26]. During the execution of a cryptographic algorithm, physical implementations tend to leak side-channel information which is related to the secret key. An adversary first trains a deep-learning model on power traces captured from profiling devices which he/she controls, and then applies the trained model to recover the key of a victim device. Using more than one device for profiling (called *multi-source training*), is known to reduce the negative effect of inter-chip variation, which is prominent in advanced technologies, and helps generalization [9,33,35].

Another known technique for reducing generalization error in machine learning is *bootstrap aggregating*, or *bagging* [4]. In bagging, several different, separately trained models are used in an ensemble to vote on the output results. Since different models usually do not make the same errors on the test set, on average, an ensemble of N models is expected to perform better than its members [11]. Bagging has been successfully applied to power analysis of hardware implementations of Advanced Encryption Standard (AES) [34]. The attack presented in [34] uses an ensemble of three Convolutional Neural Networks (CNNs) trained on different attack points.

While it is obvious that a DL-SCA in FL framework will outperform a DL-SCA based on a single classifier trained on a single profiling device, the outcome of a competition between FL (model-level aggregation), bagging (output-level aggregation), and multi-source training (data-level aggregation) methods is not evident. We present such an evaluation in this paper. We apply FL, bagging, and multi-source training aggregation methods to power analysis of a microcontroller implementation of AES. Our first experiments show that FL is capable of outperforming the other two approaches.

2   Background

This section reviews the background, including AES-128 and the general concept of DL-SCA.

2.1   AES-128

The AES [8] is a symmetric encryption algorithm standardized by NIST in FIPS 197 and included in ISO/IEC 18033-3. AES-128 takes a 128-bit block of plaintext

Algorithm 1. Pseudo-code of the AES-128 algorithm [8, 33]

```
// AES-128 Cipher
// in: 128 bits (plaintext)
// out: 128 bits (ciphertext)
// Nr: number of rounds, Nr = 10 for AES-128
// Nb: number of columns in a state, Nb = 4
// kₑ: expanded key K, Nb * (Nr + 1) = 44 words, (1 word = Nb bytes)
state = in;
AddRoundKey(state, kₑ[0, Nb − 1]);
for round = 1 step 1 to Nr − 1 do
    SubBytes(state); // Point of attack in round 1
    ShiftRows(state);
    MixColumns(state);
    AddRoundKey(state, kₑ[round * Nb, (round + 1) * Nb − 1]);
end for
SubBytes(state);
ShiftRows(state);
AddRoundKey(state, kₑ[Nr * Nb, (Nr + 1) * Nb − 1]);
out = state;
```

and a 128-bit key K as input and computes a 128-bit block of ciphertext as output. In this section, we describe AES-128 algorithm whose implementation is used in our experiments. Its pseudo-code is shown in Algorithm 1. AES-128 performs encryption iteratively, in 10 rounds for the 128-bit key. Each round except the last repeats the four steps: non-linear substitution, transposition of rows, mixing of columns, and round key addition. The last round does not mix columns. The non-linear substitution is implemented by the function $SubBytes()$ which applies the 8-input 8-output substitution box (S-Box) to $state$ byte-by-byte. As any block cipher, AES can be used in several modes of operation. In our experiments we use *Electronic Codebook* (ECB) mode, in which the message is divided into blocks and each block is encrypted separately.

An attack point is a selected intermediate state which can be used to describe the power consumption. We use the output of the S-Box procedure in the first round as the attack point in our following experiments. Since its value needs to be loaded from a memory onto a data bus, which is known to be the dominant fraction of the total power consumed by a software implementation of AES.

2.2 Deep Learning Side-Channel Attacks

Side-channel attacks were pioneered by Paul Kocher in his seminal paper on *timing analysis* [14] where he has shown that non-constant running time of a cipher can leak information about its key. Kocher has also introduced *power analysis* [13] which exploits the fact that circuits typically consume differing amounts of power based on their input data. The power consumption remains one of the most successfully exploited side-channels today. We focus on power analysis in this paper.

Table 1. Local model's architecture summary.

Layer type	Output shape	Parameter #
Input (Dense)	(None, 200)	19400
Dense 1	(None, 200)	40200
Dense 2	(None, 200)	40200
Dense 3	(None, 200)	40200
Dense 4	(None, 200)	40200
Output (Dense)	(None, 256)	51456
Total parameters: 231,656		

The target of a side-channel attack is to recover the 128-bit key $K \in \mathcal{K}$ of AES, where $\mathcal{K}$ is the set of all possible keys. To recover the key, the attacker uses of a set of known input data (e.g. the plaintext) and a set of the physical measurements (e.g. power consumption). Usually a divide-and-conquer strategy is used in which the key K is divided into 8-bit *subkeys* K_k, and the subkeys K_k are recovered independently, for $k \in \{1, 2, \ldots, 16\}$.

With advances in deep learning, side-channel attacks become several orders of magnitude more effective since deep-learning techniques are good at finding correlations in raw data [4,11,37]. Deep learning can be used in side-channel analysis in two settings: profiling and non-profiling. *Profiling* attacks [22] first learn a leakage profile of the cryptographic algorithm under attack, and then attack. *Non-profiling* attacks [30] recover the secret keys directly, as the traditional Differential Power Analysis [13] or Correlation Power Analysis (CPA) [5]. In non-profiling attacks, deep neural networks are commonly used as pre-processing techniques. In this paper, we focus on profiling attacks.

A profiling deep-learning side-channel attack is done in two stages.

1. At the *profiling* stage, the selected type of deep-learning model is trained to learn a leakage profile of the cryptographic algorithm under attack for all possible values of the sensitive variable. The sensitive variable is typically a subkey. The training is done by using a large set of side-channel data captured from profiling device(s) which are labeled according to the selected attack point and the leakage model.
2. At the *attack* stage, the trained model is used to classify the side-channel data captured from a victim device.

A *leakage model* describes the leakage of a device at attack point during the execution of the algorithm. Common leakage models for power analysis are the identity, the Hamming weight, and the Hamming distance. In this paper, we use the identity model which assumes that the power consumption is proportional to the value of the data processed at the attack point.

3 Training of Local Models

In this section we describe how local models are trained.

3.1 Choice of Neural Network Type

Previous work investigated which type of deep neural networks is suitable for various side-channel analysis scenarios. For example, CNNs have been successfully applied to bypass the trace misalignment and to overcome jitter-based countermeasures [6]. CNNs were also used to break protected AES [10,12,21,26] and the convolutional layer has been applied to handle the noise in both EM [36] and power traces [38]. However, when traces are well-synchronized and less noisy, Multiple Layer Perception (MLP) seems to be a more suitable choice in side-channel context [32,33], since MLPs are more computational efficient during the profiling stage. MLPs are shown successful in extracting keys from both software [2,9,19,21,33] and hardware [17] implementations of AES.

In our experiments, the target board contains an 8-bit microcontroller which is programmed to a standard version of AES-128 without masks. In [20], we know that the leakage in software implementations is time-dependent and less noisy since instructions are executed sequentially. This makes deep-learning models easier to learn features from traces [34]. For each subkey, the model only needs to learn the features within a specific part of traces. Furthermore, we capture traces by using ChipWhisperer [25], which assures a perfect trace alignment. For these reasons, we use MLPs.

3.2 Training Process

Given a set of power traces $\mathcal{T} = \{ T_1, T_2, ..., T_m \}$, where T denotes a single trace and m represents the number of traces in the dataset, the objective is to classify a trace T to its label $l(T) \in \mathcal{L}$ and derive the subkey K_k from the recovered label, where $\mathcal{L}$ is the set of intermediate data processed at the attack point.

The process of training a neural network can be viewed as to build a function $\mathcal{M} : \mathbb{R}^n \rightarrow \mathbb{I}^{|\mathcal{L}|}$, which maps a trace T to a *score* vector $S = \mathcal{M}(T) \in \mathbb{I}^{|\mathcal{L}|}$ whose element s_i represents the probability of the label with value $i \in \mathcal{L}$, where $\mathbb{I} = \{x \in \mathbb{R} \mid 0 \leq x \leq 1\}$. We use the *categorical cross-entropy loss* to quantify the classification error of the network. To minimize the loss, the gradient of the loss with respect to the score vector S is computed and back-propagated through the network to tune model's internal parameters according to the *RMSprop* optimizer [28]. This is repeated for a chosen number of iterations called *epochs*.

Once the network is trained, to classify a trace T captured from the victim device with an unknown label $l(T)$, we determine the most likely label $\tilde{l}$ among all $|\mathcal{L}|$ candidates as

$$\tilde{l} = \arg\max_{i \in |\mathcal{L}|} s_i$$

If $\tilde{l} = l(T)$, the classification is successful and the attacker is able to derive the key from the obtained intermediate value.

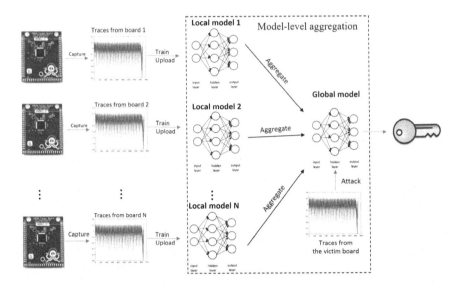

Fig. 1. Model-level aggregation in the deep-learning side-channel analysis context.

3.3 Choice of Neural Network Architecture

The architecture of MLP networks used in our experiments is shown in Table 1. The network contains an input layer, four hidden layers and an output layer. The input size is set to 96, which is corresponding to the number of data samples for one S-box execution in a trace, see Fig. 5. By using the *identity* power model, the set of intermediate data processed at the attack point can be set as $\mathcal{L} = \{0, 1, ..., 255\}$. Thus, the output size of the model is $|\mathcal{L}| = 256$.

4 Aggregation Methods

This section describes model-, output- and data-level aggregation methods in the side-channel analysis context.

4.1 Model-Level Aggregation

Figure 1 illustrates the model-level aggregation for SCA which utilizes the horizontal federated learning framework [23,29]. There are N participants (clients) jointly constructing a federated deep-learning model. Suppose each client j has n_j traces captured from his/her private profiling device, for $j \in \{1, \ldots, N\}$. The total number of training traces of N clients is denoted by n, with $n = \sum_{j=1}^{N} n_j$.

At the beginning of the training process, a typical model structure is initialized by an aggregator (server) and sent to each client. At each communication round t, a random fraction $\eta_t \in [0, 1]$ of N clients is selected by the aggregator to independently update local models based on their private data and upload

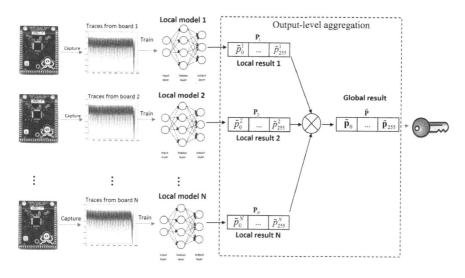

Fig. 2. Output-level aggregation in the deep-learning side-channel analysis context.

the updates to the aggregator. In our experiment, we set $\eta_t = 1$, which means that all clients contribute to the global model in each communication round. For each client, local updates are typically done by using the *Stochastic Gradient Descent* (SGD) taken on the private data based on the weights ω_0^t of the shared global model:

$$\omega_j^{t+1} = \omega_0^t - \alpha \nabla \phi(\omega_j^t)$$

where α is the learning rate, $\nabla \phi$ is gradient of the classification loss ϕ, and ω_j^t are weights of current local model of the client j.

A typical aggregation approach of federated learning is *averaging*. The aggregator computes the weights ω_0^{t+1} of the global model by averaging the weights of submitted local models:

$$\omega_0^{t+1} = \sum_{j=1}^{N} \frac{n_j}{n} \omega_j^{t+1}$$

At the end of communication round t, the aggregator sends the global model with the weights ω_0^{t+1} back to each client.

All clients can use the global model to classify the data samples captured from a victim device.

4.2   Output-Level Aggregation

Figure 2 shows an overview of the output-level aggregation approach with N participants, which is inspired by a machine learning algorithm called *bootstrap aggregating* [4]. In this approach, N different classifiers are trained on traces

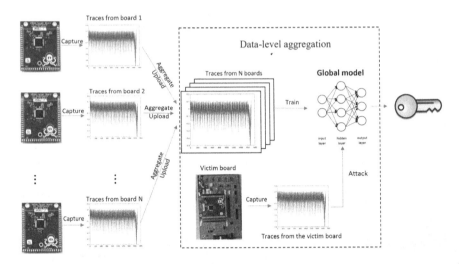

Fig. 3. Data-level aggregation in the deep-learning side-channel analysis context.

captured from N different devices independently. During the attack stage, these trained models are used to classify the same trace captured from the victim board and then the aggregator obtains the final classification result by multiplying all guess vectors generated by local classifiers. Suppose each participant j uses n_j private traces captured from his/her profiling device to train the local model j, and obtains the corresponding score vector S_j by classifying the trace captured from the victim device. Thus, the final classification result (cumulative guess vector) **S** can be defined as:

$$\mathbf{S} = \sum_{j=1}^{N} S_j$$

Combining the outputs of several weak classifiers are expected to perform better than local classifiers since different models usually do not make the same errors on the test set. This can avoid overfitting and reduce the generalization error [27].

4.3   Data-Level Aggregation

Figure 3 illustrates the data-level aggregation. The basic idea is to train a model on traces captured from different devices, which is proved to be an efficient way to mitigate the effect caused by the board diversity.

In a data-level aggregation approach, each participant j uploads n_j private traces captured from his/her profiling device to the aggregator, for $j \in \{1, \ldots, N\}$. The aggregator combines the local data into a global data set with $\mathbf{n} = \sum_{j=1}^{N} n_j$ traces and trains the global model on this mixed dataset. After the training, the aggregator sends the model back to each participant and all clients can use this model to classify the trace captured from the victim device.

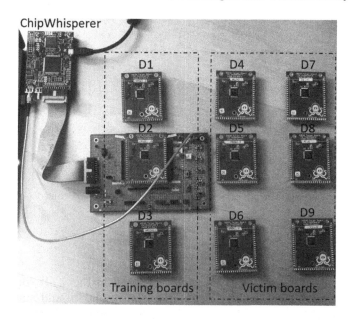

Fig. 4. Equipment for power analysis. In our experiments, three XMEGA boards are used as the profiling boards and others are for testing.

5 Assumptions

Profiling side-channel attacks assume that:

1. The attacker has at least one device, called the *profiling* device, which is similar to the device under attack and runs the same implementation of AES.
2. The attacker has a full control over the profiling device (can apply chosen plaintext, program chosen keys, and do physical measurements).
3. The attacker has a physical access to the victim device to measure some side-channel signals during the execution of AES.

In addition, in this paper we assume that only a single power trace from a victim device is available to the attacker. Such attacks are called the *single-trace attack*. Single-trace attacks are particularly threatening because they can recover the key even if the key is changed for every session.

6 Experimental Setup

This section describes our experimental setup, including the equipment we used and how we capture the power trace during the execution of AES.

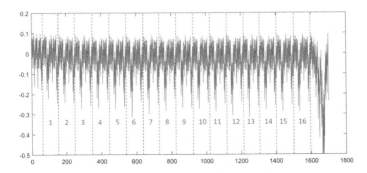

Fig. 5. Segment of a power trace from an 8-bit ATxmega128D4 microcontroller representing 16 executions of S-box.

6.1   Equipment for Power Analysis

The equipment we use for power analysis is shown in Fig. 4. It consists of the ChipWhisperer-Lite board, the CW308 UFO mother board and nine CW308T-XMEGA target boards. In the sequel, we refer to these target boards as $D_1, D_2, \ldots, D_9$. The ChipWhisperer is a hardware security evaluation toolkit based on a low-cost open hardware platform and an open source software [24]. The ChipWhisperer-Lite can be used to measure power consumption with the maximum sampling rate of 105 MS/s.

The CW308 UFO board is a generic platform for evaluating multiple targets [7]. The target board is plugged in a dedicated U connector. The CW308T-XMEGA target board contains an 8-bit ATxmega128D4 microcontroller. We programmed the microcontrollers to the same version of AES-128 in Electronic codebook (ECB) mode of operation. In our experiments, D_1, D_2 and D_3 are used for profiling and others are the victim boards.

6.2   Power Trace Acquisition

To collect training data, 300K power traces are captured from each profiling device during the execution of AES for randomly selected plaintexts and keys. We also 1 K power traces from each victim device for testing. Figure 5 shows the segment of a power trace from an 8-bit ATxmega128D4 microcontroller, representing 16 executions of S-box in the 1st encryption round. The S-box is a 8×8 invertible mapping. AES-128 executes S-box 16 times in each round. One can see the distinct shape of each S-box execution.

6.3   Estimation Metrics

In our experiments, the adversary has a strictly limited access to the victim device, which means there is only 1 trace can be captured by the adversary and classified by the trained deep-learning model. We term this test as single-trace

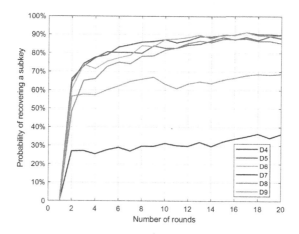

Fig. 6. Probability of recovering a subkey from a single trace captured from devices D_4–D_9 by using the global model generated from each communication round in the model-level aggregation.

test. In this case, we test the model on a single trace captured from the victim device. If the correct subkey has the highest probability in the generated score vector, the attack is successful. Otherwise, the attack fails. The estimation metrics used in our experiment is called single-trace key recovery rate [33], which represents the success probability of recovering a subkey from only **1** trace captured from the victim board.

7 Evaluation Results

In this section, we apply model-, output- and data-level aggregation methods to power analysis of AES-128 and compare their results. In all experiments, we simulate a scenario that $N = 3$ participants have the same number of training traces.

7.1 Results of Model-Level Aggregation

In the first experiment, three participants jointly build a global MLP model by using the federated leanring framework and each participant has $n_j = 300K$ traces captured from D_j for training his/her local model, for $j \in \{1, 2, 3\}$. All local models are trained for 40 epochs with a learning rate $\alpha = 0.0001$ and the local minimum batch size is set to 128. We have 20 communication rounds in total and at the end of each round, the global model is sent back to all participants to further update the model based on their local traces.

We test the generated global model of each round on a single trace captured from the victim device. We run 1,000 single-trace tests for each model and each victim device to obtain the average. Figure 6 shows the average probability of

Table 2. Probability of recovering a subkey from a single trace using aggregated models.

Device	Aggregation method		
	Model-level approach	Output-level approach	Data-level approach
D_4	89.8%	64.5%	74.6%
D_5	91.2%	76.0%	83.0%
D_6	91.4%	66.0%	73.6%
D_7	35.5%	18.4%	37.5%
D_8	88.5%	68.3%	62.3%
D_9	69.6%	58.8%	81.5%
Average	77.7%	58.7%	68.8%

these models recovering a subkey from a single trace captured from different victim devices. In Fig. 6, one can see that the federated model generated from the 17th communication round has the highest average success probability over all rounds. The 2nd column in Table 2 shows the probability of recovering a subkey from a single trace captured from different victim devices by using this model. We can see that the average is 77.7%.

7.2   Results of Output-Level Aggregation

In the output-level aggregation, three participants first train their local MLP models independently. Each participant j has $n_j = 300K$ traces captured from D_j 60K traces set aside for validation, for $j \in \{1, 2, 3\}$. We denote these local classifiers as local model 1, 2 and 3, respectively. Afterwards, the trained local models are used to classify the trace captured from the victim device and the aggregator applies an element-wise multiplication to all local models' classification results. Such an approach is known to work well for power analysis of hardware implementations of AES [34].

The 3rd column of Table 2 shows the probability of recovering a subkey from a single trace by using the output-level aggregation approach. The average probability of recovering a subkey from a single trace captured from the victim device is 58.6%. Note that we use the same attack point (S-box output in the 1st round) to create labels for local models. In [34], different attack points are used for training different local models. This might have negatively affected the ensemble's results.

For a comparison, We further test each local models and Table 3 shows the results for local model 1, 2 and 3. We see that success probabilities of these local models vary a lot for different devices. This is because different pairs of devices have different amounts of variability. Some devices are less different, some are more different. For example, on the one hand, local model 1 and 2 can recover a subkey from a single trace captured from D_5 in 48.4% and 63.8% of cases, respectively. Contrary, for local model 3, the subkey recovery rate for D_5 is only

Table 3. Probability of recovering a subkey from a single trace by using local models in the output-level aggregation (averaged by 1,000 tests).

Device	Local model 1	Local model 2	Local model 3
D_4	29.1%	42.6%	40.8%
D_5	48.4%	63.8%	21.8%
D_6	38.3%	33.6%	39.7%
D_7	6.8%	10.4%	57.9%
D_8	27.3%	36.1%	50.0%
D_9	33.9%	51.8%	35.4%
Average	34.9%	41.3%	40.9%

21.8%. Probably D_1 and D_2 are less different from D_5 than D_3. On the other hand, local model 3 significantly outperforms local model 1 and 2 on D_7 (57.9% vs 6.8% and 10.4%, respectively). Probably D_3 is similar to D_7, while D_1 and D_2 are very different from D_7. High dissimilarly of D_1 and D_2 from D_7 might be the reason why the global model in the federated learning case performs so poorly on D_7. From Table 3 we can also conclude that D_3 is very different from D_9, which explains the worse result of the global model in the federated learning case for D_9 as compared to the results for $D4$–$D6$ and D_8.

7.3 Results of Data-Level Aggregation

In this experiment, each participant 300K training traces captured from D_1, D_2 and D_3, respectively, and upload the training data to the aggregator. The aggregator combines theses sets into a global training set of 900K and sets 270K traces for validation. Then, the aggregator trains a MLP model on this mixed training dataset, which is a data-level aggreagtion for combining different profiling devices.

The 4th column of Table 2 shows the probability of recovering a subkey from a single trace using the data-level aggregation approach. We can see that the average is 68.8%.

8 Conclusion

We compared the federated learning approach to two other aggregating approaches - on data and on output levels. Our first results show that federated learning is capable of outperforming the other approaches. This is quite surprising.

Intuitively, it should be more difficult to train a global model in a federated learning framework due to challenges related to training on distributed data while keeping these data private. Apparently, in the case of power analysis, moves due to averaging of weights of local models take neural network parameters into the region of function space which is beneficial for the optimization of the objective function.

We plan to further investigate this phenomena by training new models using other combinations of profiling devices.

Acknowledgment. This work was supported in part by the research grant 2018-04482 from the Swedish Research Council.

References

1. Atchinson, B.K., Fox, D.M.: From the field: the politics of the health insurance portability and accountability act. Health Affairs **16**(3), 146–150 (1997)
2. Benadjila, R., Prouff, E., Strullu, R., Cagli, E., Dumas, C.: Study of deep learning techniques for side-channel analysis and introduction to ASCAD database. ANSSI, France & CEA, LETI, MINATEC Campus, France, vol. 22 (2018). https://eprint.iacr.org/2018/053.pdf
3. Bonawitz, K., et al.: Practical secure aggregation for privacy-preserving machine learning. In: Proceedings of the 2017 ACM SIGSAC Conference on Computer and Communications Security, pp. 1175–1191 (2017)
4. Breiman, L.: Bagging predictors. Mach. Learn. **24**(2), 123–140 (1996)
5. Brier, E., Clavier, C., Olivier, F.: Correlation power analysis with a leakage model. In: Joye, M., Quisquater, J.-J. (eds.) CHES 2004. LNCS, vol. 3156, pp. 16–29. Springer, Heidelberg (2004). https://doi.org/10.1007/978-3-540-28632-5_2
6. Cagli, E., Dumas, C., Prouff, E.: Convolutional neural networks with data augmentation against jitter-based countermeasures. In: Fischer, W., Homma, N. (eds.) CHES 2017. LNCS, vol. 10529, pp. 45–68. Springer, Cham (2017). https://doi.org/10.1007/978-3-319-66787-4_3
7. CW308 UFO Target. https://wiki.newae.com/CW308_UFO_Target
8. The Design of Rijndael. ISC. Springer, Heidelberg (2020). https://doi.org/10.1007/978-3-662-60769-5_9
9. Das, D., Golder, A., Danial, J., Ghosh, S., Raychowdhury, A., Sen, S.: X-deepsca: cross-device deep learning side channel attack. In: Proceedings of the 56th Annual Design Automation Conference 2019, pp. 1–6 (2019)
10. Gilmore, R., Hanley, N., O'Neill, M.: Neural network based attack on a masked implementation of AES. In: 2015 IEEE International Symposium on Hardware Oriented Security and Trust (HOST), pp. 106–111. IEEE (2015)
11. Goodfellow, I., Bengio, Y., Courville, A.: Deep Learning. MIT Press, Cambridge (2016). http://www.deeplearningbook.org
12. Jin, M., Zheng, M., Hu, H., Yu, N.: An enhanced convolutional neural network in side-channel attacks and its visualization. arXiv preprint arXiv:2009.08898 (2020)
13. Kocher, P., Jaffe, J., Jun, B.: Differential power analysis. In: Wiener, M. (ed.) CRYPTO 1999. LNCS, vol. 1666, pp. 388–397. Springer, Heidelberg (1999). https://doi.org/10.1007/3-540-48405-1_25
14. Kocher, P.C.: Timing attacks on implementations of Diffie-Hellman, RSA, DSS, and other systems. In: Koblitz, N. (ed.) CRYPTO 1996. LNCS, vol. 1109, pp. 104–113. Springer, Heidelberg (1996). https://doi.org/10.1007/3-540-68697-5_9
15. Konečný, J., McMahan, H.B., Ramage, D., Richtárik, P.: Federated optimization: Distributed machine learning for on-device intelligence. arXiv preprint arXiv:1610.02527 (2016)

16. Konečnỳ, J., McMahan, H.B., Yu, F.X., Richtárik, P., Suresh, A.T., Bacon, D.: Federated learning: strategies for improving communication efficiency. arXiv preprint arXiv:1610.05492 (2016)
17. Kubota, T., Yoshida, K., Shiozaki, M., Fujino, T.: Deep learning side-channel attack against hardware implementations of AES. In: 2019 22nd Euromicro Conference on Digital System Design (DSD), pp. 261–268. IEEE (2019)
18. Li, Q., Wen, Z., Wu, Z., Hu, S., Wang, N., He, B.: A survey on federated learning systems: Vision, hype and reality for data privacy and protection (2019)
19. Maghrebi, H.: Deep learning based side channel attacks in practice. Technical Report, IACR Cryptology ePrint Archive 2019, vol. 578 (2019)
20. Maghrebi, H., Portigliatti, T., Prouff, E.: Breaking cryptographic implementations using deep learning techniques. In: Carlet, C., Hasan, M.A., Saraswat, V. (eds.) SPACE 2016. LNCS, vol. 10076, pp. 3–26. Springer, Cham (2016). https://doi.org/10.1007/978-3-319-49445-6_1
21. Martinasek, Z., Dzurenda, P., Malina, L.: Profiling power analysis attack based on MLP in DPA contest v4. 2. In: 2016 39th International Conference on Telecommunications and Signal Processing (TSP), pp. 223–226. IEEE (2016)
22. Martinasek, Z., Malina, L., Trasy, K.: Profiling power analysis attack based on multi-layer perceptron network. In: Mastorakis, N., Bulucea, A., Tsekouras, G. (eds.) Computational Problems in Science and Engineering. LNEE, vol. 343, pp. 317–339. Springer, Cham (2015). https://doi.org/10.1007/978-3-319-15765-8_18
23. McMahan, H.B., Moore, E., Ramage, D., Hampson, S., et al.: Communication-efficient learning of deep networks from decentralized data. arXiv preprint arXiv:1602.05629 (2016)
24. NewAE Technology Inc.: Chipwhisperer. https://newae.com/tools/chipwhisperer
25. O'Flynn, C., Chen, Z.D.: ChipWhisperer: an open-source platform for hardware embedded security research. In: Prouff, E. (ed.) COSADE 2014. LNCS, vol. 8622, pp. 243–260. Springer, Cham (2014). https://doi.org/10.1007/978-3-319-10175-0_17
26. Perin, G., Ege, B., van Woudenberg, J.: Lowering the bar: deep learning for side-channel analysis (white-paper). In: Proceedings of BlackHat, pp. 1–15 (2018)
27. Polikar, R.: Ensemble learning. In: Ensemble Machine Learning, pp. 1–34. Springer, Heidelberg (2012). https://doi.org/10.1007/978-1-4419-9326-7_1
28. Robbins, H., Monro, S.: A stochastic approximation method. Ann. Math. Stat. **22**, 400–407 (1951)
29. Smith, V., Chiang, C.K., Sanjabi, M., Talwalkar, A.S.: Federated multi-task learning. In: Advances in Neural Information Processing Systems, pp. 4424–4434 (2017)
30. Timon, B.: Non-profiled deep learning-based side-channel attacks. IACR Cryptol. ePrint Arch. **2018**, 196 (2018)
31. Voigt, P., von dem Bussche, A.: The EU General Data Protection Regulation (GDPR). A Practical Guide. Springer, Cham (2017). https://doi.org/10.1007/978-3-319-57959-7
32. Wang, H.: Side-Channel Analysis of AES Based on Deep Learning. Master's thesis, KTH, School of Electrical Engineering and Computer Science (EECS) (2019)
33. Wang, H., Brisfors, M., Forsmark, S., Dubrova, E.: How diversity affects deep-learning side-channel attacks. In: 2019 IEEE Nordic Circuits and Systems Conference (NORCAS): NORCHIP and International Symposium of System-on-Chip (SoC), pp. 1–7. IEEE (2019)
34. Wang, H., Dubrova, E.: Tandem deep learning side-channel attack against FPGA implementation of AES. Cryptology ePrint Archive, Report 2020/373 (2020). https://eprint.iacr.org/2020/373

35. Wang, H., Forsmark, S., Brisfors, M., Dubrova, E.: Multi-source training deep learning side-channel attacks. IEEE 50th International Symposium on Multiple-Valued Logic (2020)

36. Wang, R., Wang, H., Dubrova, E.: Far field em side-channel attack on AES using deep learning. Cryptology ePrint Archive, Report 2020/1096 (2020). https://eprint.iacr.org/2020/1096

37. Wu, Y., Shen, K., Chen, Z., Wu, J.: Automatic measurement of fetal cavum septum pellucidum from ultrasound images using deep attention network. In: 2020 IEEE International Conference on Image Processing (ICIP), pp. 2511–2515. IEEE (2020)

38. Yang, G., Li, H., Ming, J., Zhou, Y.: CDAE: towards empowering denoising in side-channel analysis. In: International Conference on Information and Communications Security, pp. 269–286. Springer (2019)

Differential Fault Based Key Recovery Attacks on TRIAD

Iftekhar Salam$^{(\boxtimes)}$ ⬛, Kim Young Law, Luxin Xue, and Wei-Chuen Yau ⬛

School of Electrical and Computer Engineering, Xiamen University Malaysia,
Sepang, Malaysia
{iftekhar.salam,swe1704069,cst1609025,wcyau}@xmu.edu.my

Abstract. We demonstrate two fault based key recovery attacks – a bit-flipping fault attack, and a random fault attack, on the authenticated encryption stream cipher TRIAD. The attacks discussed in this paper are applicable mainly due to the simplicity of the TRIAD keystream function during the first few hundred rounds. We investigated the algebraic normal form (ANF) of the first 160 output keystream bits of TRIAD. The ANF of these output keystream bits shows specific patterns that are used in our fault attacks. We first use these patterns with a bit-flipping fault model combined with solving a system of low degree algebraic equations that requires 85 faults to recover the secret key of TRIAD, with a data complexity of $2^{7.43}$. We then relax our assumptions by using a random fault model. The random fault model described in this paper is also combined with solving a system of low degree algebraic equations and requires on average 170 faults to recover the secret key of TRIAD with a data complexity of $2^{8.01}$. For both of the attacks, the complexity of solving the low degree algebraic equations is negligible. We have not performed experiments on the actual hardware implementation, but we have verified all the attacks using simulation on TRIAD software implementation.

Keywords: Fault attack · Random fault · TRIAD · NIST LWC project · Key recovery

1 Introduction

The Lightweight Cryptography (LWC) Project [1] is an ongoing project organized by the National Institute of Standards and Technology (NIST) to evaluate and standardize lightweight cryptographic algorithms to be used in a resource-constrained environment. Some applications of the LWC include automobiles, smart home appliances, Internet of Things (IoT) devices, medical sensors. Fifty-six candidates were chosen as the Round 1 candidates of the NIST LWC project, where TRIAD [2] is one of the candidates. In this paper, we analyze the security of TRIAD against several fault attack models.

TRIAD is an authenticated encryption (AE) stream cipher algorithm. The construction of TRIAD is inspired by the design of TRIVIUM [3], which was a

© Springer Nature Switzerland AG 2021
D. Hong (Ed.): ICISC 2020, LNCS 12593, pp. 273–287, 2021.
https://doi.org/10.1007/978-3-030-68890-5_15

candidate in the eStream project [4]. TRIAD adopts a new structure without changing Trivium's global structure and increases the security level to 112 bits. Few cryptanalysis models have been investigated against TRIAD. He et al. [5] demonstrated a key recovery attack on a reduced round (521-round) initialization phase of TRIAD using a cube attack with a cube size of 16. With a cube size of 30, it led to a key recovery attack for the 565-round TRIAD. The complexity for these cube attacks on the 521-round TRIAD and 565-round TRIAD are $2^{24.81}$ and 2^{48}, respectively. Kesarwani et al. [6] also performed a cube tester on TRIAD. The authors provided heuristics to obtain useful cubes for the cube tester. With a cube size of 34, it can distinguish 550 initialization rounds of TRIAD. All the existing key recovery attacks on TRIAD are performed on the reduced-round variants of the cipher.

In this work, we demonstrate two fault based key recovery attacks on the full version of TRIAD. To the best of our knowledge, there are currently no existing fault attacks that have been applied to TRIAD. The motivation of this research is to determine the applicability of fault attacks to TRIAD. This study's findings contribute to the understanding of the security level provided by TRIAD against the bit-flipping fault attack and the random fault attack. We show that both of these attacks can recover the secret key of TRIAD with practical complexity. The work in this paper may also contribute to the consideration of the evaluation of TRIAD.

2   Fault Attack

Fault attack exploits the output of a cryptographic implementation after an error is induced in the cryptographic scheme. Each stream cipher responds to faults differently, whether the fault is intentional or accidental. An adversary may recover information regarding the secret key or the state by comparing the faulty and fault-free outputs. The main goal of the fault attacks is to recover the secret key or the internal state. The fault attack is an active attack that requires an adversary to access the cipher's physical implementation.

2.1   Fault Attack Models

The adversarial models of a fault attack are defined based on the assumptions made on an adversary's capability. These fault attack models consider different criteria, including the modification type of the fault, the number of bits affected by the fault, the precision of the fault, and the duration of the fault.

The modification type defines the impact of the fault in the target location. Depending on the modification type, fault models are categorized into three types: stuck-at-fault, bit-flipping fault, and random fault.

– Stuck-at-fault: sets the faulty bit(s) to a specific value of either zero or one. Stuck-at-fault is also known as a set-to-zero fault or a set-to-one fault when the fault injection sets the target location to a value of either zero or one, respectively.

- Bit-flipping fault: complement the contents of the target fault location.
- Random fault: the target location is set to either zero or one with an equal probability. That is, the probability of the bit in the target location may remain the same or may get complemented with a probability of 0.5.

The number of bits affected by the fault can also be considered a factor in modeling the fault attacks. Depending on the model, fault injection may affect a single bit, a few bits, one byte, or multiple bytes.

Based on the precision of affecting the target location and the timing of the fault injection, fault models can be categorized into three types: precise control, moderate control, and random control. In the precise control model, the adversary is assumed to have access to fault injection equipment that can inject faults at a precise time and precise location of the internal components. In the moderate control model, the adversary has reduced control over injecting faults to the target location at any given time. For the random control model, the adversary does not have any control over the fault target location and the timing; that is, the adversary can inject a fault that may affect a random location at a random time.

Fault injection can be either transient or permanent, depending on the duration that the faults remain active. The transient fault remains active only for a specific duration, typically a single clock cycle; whereas, the permanent fault remains active for the entire duration of the operation.

2.2 Fault Injection Techniques

The fault injection techniques may fall under one of the three categories: invasive, non-invasive, or semi-invasive. The invasive technique requires an adversary to tamper with the physical device, which may cause harm to the device. The non-invasive technique will not cause any physical damage to the device. The semi-invasive technique will fall in between the invasive and non-invasive techniques.

The fault injection can be performed using inexpensive methods such as voltage glitch [7], temperature variations [8], clock glitch [9], electromagnetic pulses [10], or laser beam [11]. Among these methods, voltage glitch and clock glitch have low precision; whereas, the laser beam induced fault has a comparatively high precision on the fault location. The faults also may have a different impact, such as skipping an instruction, complementing a bit, destroying a memory cell, depending on the method/equipment used to inject the faults.

The fault attacks presented in this paper require an adversary to inject faults at a specific location and time. Regarding the impact of the fault, we discuss two different attacks – one with a bit-flipping fault and the other with a random fault. Skorobogatov and Anderson [11] have already demonstrated fault attacks that use a focused flash light to set specific memory cells to a value of zero or one. With the advances in the fault attack techniques, it was also demonstrated that multiple laser beams could be used to complement multiple memory cells simultaneously [12], as well as to complement the same memory cell multiple times [13]. So, we consider the assumptions made in this paper are feasible to apply the fault attacks on TRIAD.

3   Description of TRIAD

TRIADv1 is a family of lightweight authenticated encryption stream cipher, designed by Subhadeep Banik, Takanori Isobe, Willi Meier, Yosuke Todo, and Bin Zhang [2]. TRIADv1 consists of two main parts. The first part describes an authenticated encryption mode named TRIAD-AE. The second part describes a hash function named TRIAD-HASH. The authenticated encryption mode TRIAD-AE uses a stream cipher algorithm called TRIAD-SC to ensure the message's confidentiality and a message authentication code (MAC) algorithm called TRIAD-MAC to detect the message's integrity. Our work mainly focuses on the confidentiality component, and so TRIAD-SC is the primary focus of our investigation.

TRIAD-AE uses a 128-bit key, $K \in \{0,1\}^{128}$ and a 96-bit nonce $N \in \{0,1\}^{96}$. The cipher accepts an arbitrary length byte-array plaintext M and an arbitrary length byte array for associated data A. The output of TRIAD-AE consists of a ciphertext $C \in \{0,1\}^{\ell_m}$, where ℓ_m is the length of original plaintext M. TRIAD-AE also outputs a 64-bit tag T.

3.1   TRIAD-AE Component Functions

TRIAD has a 256-bit internal state, consisting of three feedback shift registers (FSR) $a, b,$ and c. The registers a, b, c are of length 80, 88 and 88 bits, respectively and their contents are denoted as $a = (a_0||a_1|| \cdots ||a_{79})$, $b = (b_0||b_1|| \cdots ||b_{87})$, and $c = (c_0||c_1|| \cdots ||c_{87})$.

Keystream Generation Function. TRIAD-AE consists of a stream cipher TRIAD-SC. The keystream is generated by combining a set of state bits. At time t, the keystream generation of TRIAD-AE is computed as:

$$z_t = a_{67}^t \oplus a_{79}^t \oplus b_{63}^t \oplus b_{87}^t \oplus c_{67}^t \oplus c_{87}^t \oplus b_{84}^t c_{84}^t \tag{1}$$

Feedback Functions. TRIAD-AE has three non-linear feedback functions f_1, f_2 and f_3. These feedback functions at time t are defined as:

$$f_1^t = a_{67}^t \oplus a_{79}^t \oplus b_{84}^t c_{84}^t \oplus a_{72}^t a_{78}^t \oplus b_{65}^t \tag{2}$$

$$f_2^t = b_{63}^t \oplus b_{87}^t \oplus b_{64}^t b_{86}^t \oplus c_{83}^t \tag{3}$$

$$f_3^t = c_{67}^t \oplus c_{87}^t \oplus c_{76}^t c_{86}^t \oplus a_{73}^t \tag{4}$$

State Update Function. The state of the TRIAD-AE needs to be updated after the generation of the keystream z so that the same keystream is not repeated at the next time instant. TRIAD-AE uses a state update function TriadUpd to update the 256-bit internal state. The contents of registers a, b and c at time $t+1$ are updated as per Eqs. (5), (6), and (7), respectively.

$$a_i^{t+1} = \begin{cases} f_3^t \oplus msg^t & \text{for } i = 0 \\ a_{i-1}^t & \text{for } i = 1, \cdots, 79 \end{cases} \tag{5}$$

$$b_i^{t+1} = \begin{cases} f_1^t \oplus msg^t & \text{for } i = 0 \\ b_{i-1}^t & \text{for } i = 1, \cdots, 87 \end{cases} \tag{6}$$

$$c_i^{t+1} = \begin{cases} f_2^t \oplus msg^t & \text{for } i = 0 \\ c_{i-1}^t & \text{for } i = 1, \cdots, 87 \end{cases} \tag{7}$$

As shown in Eqs. (5), (6), and (7), all the registers in a, b and c are updated by shifting, except that registers a_0, b_0 and c_0 are updated by combining the output of specific non-linear feedback functions, f_1, f_2 and f_3 with the input message, msg. Depending on the operating mode of the cipher, the input message, msg, could be either the plaintext, or the associated data, or a fixed constant. Note that for TRIAD-SC, the input message, msg, is set to zero, that is, $msg = 0$.

3.2   Operation Phases of TRIAD-AE

TRIAD-AE has mainly four operation phases: initialization, encryption, tag generation, and decryption & verification. For encryption and decryption, TRIAD-AE uses a stream cipher algorithm TRIAD-SC, whereas, for the tag generation, TRIAD-AE uses a separate message authentication code (MAC) algorithm called TRIAD-MAC. Both of these algorithms, TRIAD-SC and TRIAD-MAC, need to go through the initialization process. Note that TRIAD-SC and TRIAD-MAC use two different initialization functions for their initialization process. Here, we briefly discuss the initialization and keystream generation process of TRIAD-SC as we apply a fault attack in the TRIAD-SC. Interested readers may refer to the original specification of TRIAD [2] for more details on the other phases of operation.

The initialization process of TRIAD-SC uses a 128-bit key K, a 96-bit nonce N, and a fixed 32-bit constant as the inputs to the internal state of the stream cipher. First, the inputs are loaded into the registers a, b, and c in a pre-defined format. The internal state is then updated using the state update function TriadUpd for 1024 iterations without producing any keystream bits. That is, for these 1024 iterations the contents of registers a, b and c are updated using Eqs. (5), (6), and (7), respectively. During these updates, the external message, msg is set to zero. The state obtained after these updates is the initial internal state of TRIAD-SC. Once the initialization phase of TRIAD-SC is completed, the cipher iteratively computes the keystream z_t using Eq. (1), and XORs the keystream with the corresponding plaintext to compute the ciphertext. Simultaneously, at each iteration t, the state is updated accordingly.

4   Fault Attacks on TRIAD

We applied two different fault attacks to TRIAD – a bit-flipping fault attack and a random fault attack. In this section, we describe these attacks. First, we describe some theoretical observations on the application of the fault attacks on TRIAD. Next, we implement these attacks to verify our theoretical observations. The techniques used in this paper are adapted from the work of Dey et al. [14], Salam et al. [15] and Bartlett et al. [16].

4.1 Algebraic Normal Form (ANF) of the Keystream Function

Our fault attacks on TRIAD first require to generate a set of keystream equations, and then identify specific patterns in the equations. These equations have certain structure which may help us to identify the fault target locations. To understand the structure of the keystream function, we first compute the algebraic normal form (ANF) of successive keystream bits constructed in terms of the initial state of TRIAD. Let $S = \{s_0, \cdots, s_{255}\}$ denotes the initial internal state of TRIAD. These 256 bits of the initial state is formed by concatenating the initial state of registers a^t, b^t, and c^t, at time $t = 1024$; that is, $S = (s_0, \cdots, s_{79}, s_{80}, \cdots, s_{167}, s_{168}, \cdots, s_{255}) = (a_0^{1024}, \cdots, a_{79}^{1024}) \| (b_0^{1024}, \cdots, b_{87}^{1024}) \| (c_0^{1024}, \cdots, c_{87}^{1024})$. Using the variables from the initial state S, we can use Sage [17] to generate the algebraic normal form (ANF) of successive keystream bits of TRIAD. Based on Eq. (1), we implemented the TRIAD keystream function in Sage to compute the keystream equation, z_t, for different time instant, t. Examples of the ANF of the first five keystream equations $z_0, \cdots, z_4$ are shown below.

$$z_0 = s_{164}s_{252} \oplus s_{67} \oplus s_{79} \oplus s_{143} \oplus s_{167} \oplus s_{235} \oplus s_{255} \tag{8}$$

$$z_1 = s_{163}s_{251} \oplus s_{66} \oplus s_{78} \oplus s_{142} \oplus s_{166} \oplus s_{234} \oplus s_{254} \tag{9}$$

$$z_2 = s_{162}s_{250} \oplus s_{65} \oplus s_{77} \oplus s_{141} \oplus s_{165} \oplus s_{233} \oplus s_{253} \tag{10}$$

$$z_3 = s_{161}s_{249} \oplus s_{64} \oplus s_{76} \oplus s_{140} \oplus s_{164} \oplus s_{232} \oplus s_{252} \tag{11}$$

$$z_4 = s_{160}s_{248} \oplus s_{63} \oplus s_{75} \oplus s_{139} \oplus s_{163} \oplus s_{231} \oplus s_{251} \tag{12}$$

In our fault attacks, we use the following observations from the ANF of the generated equations.

i. The equations have a clear pattern, consisting of a quadratic term and several other linear terms. The variables involved in the quadratic terms are unique, as they do not appear anywhere else in that equation. This particular structure forms the basis of our bit-flipping fault attack to identify the target fault register(s) and the corresponding register bit(s) that may be recovered with the fault injection.

ii. There are several linear terms (register s_j) involved, which are also appearing uniquely in each of these equations. Using such unique linear terms, we can compute the value of a random fault induced in the register s_j. This observation, combined with observation (i), forms the basis of our random fault attack. We first use the equations with the unique linear terms to identify the target register's faulty value and then use the equations that contain the corresponding unique quadratic term to recover particular register bit(s).

We analyze the ANF of the first 160 keystream bits to identify equations consisting of such unique patterns. More details on the usage of such equations to apply fault attacks are provided in the later sections.

For the rest of the paper, we use the notation z'_{i,s_j} to denote the ith bit of faulty keystream for a fault injection at register s_j, where $i \geq 0$ and $0 \leq j \leq 255$.

We use the notation e_j to indicate a random fault that has been injected into a target register s_j. For all the attacks, we limit our fault injections only to the register c^{1024}, that is, initial state bits $s_{168}, \cdots, s_{255}$.

4.2 Bit-Flipping Fault Attack on TRIAD

We first attempt to recover the initial states of the cipher using a bit-flipping fault. The bit-flipping fault exploits the differences in the faulty and fault-free outputs when a bit is complemented in a specific target register. We make several assumptions to apply the bit-flipping fault attack, including:

i. An adversary has access to the first 157 bits of the keystream.
ii. An adversary can complement the bit in a specific target register.

We first take a theoretical approach towards the bit-flipping attack. The first step in the theoretical approach is to generate the equations in terms of the initial states. In this case, we used SageMath to generate the algebraic normal from (ANF) of the keystream equations of TRIAD in terms of the initial state variables, as discussed in Sect. 4.1. After that, we carefully examined the structures of the equations to identify quadratic terms with unique variables in the equations.

We take Eq. (8) as an example to illustrate the theoretical observations on the bit-flipping fault attack on TRIAD. In Eq. (8), notice that there is a quadratic term involving two initial state register bits: s_{252} and s_{164}. Also, these two bits do not appear anywhere else in the equation. Suppose the adversary introduces a bit-flipping fault in the state bit s_{252}. Based on the fault at register s_{252}, a faulty equation can be produced and written as:

$$z'_{0,s_{252}} = s_{164}\overline{s_{252}} \oplus s_{67} \oplus s_{79} \oplus s_{143} \oplus s_{167} \oplus s_{235} \oplus s_{255} \qquad (13)$$

where $z'_{0,s_{252}}$ indicates the 0th bits of the faulty keystream when a fault is applied to s_{252} and $\overline{s_{252}}$ represents the complement of s_{252}. Applying XOR operation on Eqs. (8) and (13), the value of s_{164} can be recovered as shown in Eq. (14).

$$z_0 \oplus z'_{0,s_{252}} = s_{164}(s_{252} \oplus \overline{s_{252}})$$
$$= s_{164} \qquad (14)$$

A similar observation can also be applied where an adversary can apply bit-flipping fault in s_{164} to recover the register bit s_{252}. Our analysis shows that the patterns in the first 157 keystream equations are sufficient to recover all the initial state bits of TRIAD, except s_{79}, s_{167}, and s_{255}. However, this approach requires a comparatively large number of faults (253 faults) and may require to inject the faults in all the three registers $a^{1024}, b^{1024}, c^{1024}$.

Instead of recovering all the bits by injecting faults, we used fault injections only to recover certain parts of the initial state. The remaining bits can be recovered by solving a system of equations. The goal is to recover the initial state with a minimal amount of faults. Particularly, we used the bit-flipping faults to recover the initial register bits $s_{80}, \cdots, s_{164}$ and $s_{168}, \cdots, s_{254}$, that is, $b_0^{1024}, \cdots, b_{84}^{1024}$ and $c_0^{1024}, \cdots, c_{86}^{1024}$, respectively.

In our fault attack, we aim to minimize the number of faults required to recover these register bits. We notice that the same fault can recover multiple register contents simultaneously by considering the faulty keystream equations at different time instants. For example, consider the 82nd keystream bit z_{82} as below.

$$z_{82} = s_{54}s_{60} \oplus s_{130}s_{152} \oplus s_{82}s_{170} \oplus s_{146}s_{234} \oplus s_{230}s_{240} \oplus s_{242}s_{252} \oplus s_{49} \oplus s_{59} \oplus s_{61}$$
$$\oplus s_{71} \oplus s_{85} \oplus s_{127} \oplus s_{129} \oplus s_{153} \oplus s_{173} \oplus s_{221} \oplus s_{233} \oplus s_{237} \oplus s_{241} \oplus s_{253} \qquad (15)$$

Notice that in both keystream bits z_0 and z_{82} (Eqs. (8) and (15)), the register bit s_{252} appears in quadratic terms together with s_{164} and s_{242}, respectively. Therefore, with a single fault injection at s_{252} we can compute $z_0 \oplus z'_{0,s_{252}} = s_{164}$ and $z_{82} \oplus z'_{82,s_{252}} = s_{242}$. In our application, we identify such registers as the target locations to minimize the number of faults.

Table 1 shows the patterns identified from the equations that can be used to apply a fault attack with minimal faults. As shown in Table 1, an adversary applies faults in registers $s_{168}, \cdots, s_{252}$ (i.e., $c_0^{1024}, \cdots, c_{84}^{1024}$) to recover the register bits $s_{80}, \cdots, s_{164}$ and $s_{168}, \cdots, s_{254}$.

Based on these patterns from Table 1, Algorithm 1 illustrates the bit-flipping fault injection targets in $s_{168}, \cdots, s_{252}$ to recover the register bits $s_{80}, \cdots, s_{164}$ and $s_{168}, \cdots, s_{254}$. In Algorithm 1, we use:

- faulty keystream bits $z'_{0,s_{252}}, \cdots, z'_{84,s_{168}}$ to recover $s_{164}, \cdots, s_{80}$,
- faulty keystream bits $z'_{82,s_{252}}, \cdots, z'_{156,s_{178}}$ to recover $s_{242}, \cdots, s_{168}$, and
- faulty keystream bits $z'_{68,s_{244}}, \cdots, z'_{79,s_{233}}$ to recover $s_{254}, \cdots, s_{243}$.

In total, Algorithm 1 requires an adversary needs to apply 85 bit-flipping faults to recover 172 register bits of the initial state. The process for recovering the remaining initial state bits are explained in Sects. 4.4.

4.3   Random Fault Attack on TRIAD

The random fault attack model extends the concept of the bit-flipping fault attack model described in Sect. 4.2. As compared to the bit-flipping fault attack model, the random fault model is more practical. The assumptions are more realistic as the fault generated is random, and the effect of the fault is unknown to the adversary. The injection of a random fault may not affect the target register's contents or may complement the target register's contents with an equal probability. So, unlike the bit-flipping model, the random fault model does not guarantee to recover the contents of the target register with just a single fault. Multiple faults may be needed in order to recover each register bit. We make the following assumptions for our random fault attack on TRIAD.

Table 1. Useful equation patterns to recover $s_{80},\cdots,s_{164}$ and $s_{168},\cdots,s_{254}$ by injecting faults at registers $s_{168},\cdots,s_{252}$

Fault target, s_j	Required faulty keystream, z'_{i,s_j}	Recovered bit
s_{252}	$z'_{0,s_{252}}$	s_{164}
	$z'_{82,s_{252}}$	s_{242}
s_{251}	$z'_{1,s_{251}}$	s_{163}
	$z'_{83,s_{251}}$	s_{241}
$\vdots$	$\vdots$	$\vdots$
s_{246}	$z'_{6,s_{246}}$	s_{158}
	$z'_{88,s_{246}}$	s_{236}
s_{245}	$z'_{7,s_{245}}$	s_{157}
	$z'_{89,s_{245}}$	s_{235}
s_{244}	$z'_{8,s_{244}}$	s_{156}
	$z'_{90,s_{244}}$	s_{234}
	$z'_{68,s_{244}}$	s_{254}
s_{243}	$z'_{9,s_{243}}$	s_{155}
	$z'_{91,s_{243}}$	s_{233}
	$z'_{69,s_{243}}$	s_{253}
$\vdots$	$\vdots$	$\vdots$
s_{179}	$z'_{73,s_{179}}$	s_{91}
	$z'_{155,s_{179}}$	s_{169}
	$z'_{133,s_{179}}$	s_{189}
s_{178}	$z'_{74,s_{178}}$	s_{90}
	$z'_{156,s_{178}}$	s_{168}
	$z'_{134,s_{178}}$	s_{188}
s_{177}	$z'_{75,s_{177}}$	s_{89}
	$z'_{135,s_{177}}$	s_{187}
s_{176}	$z'_{76,s_{176}}$	s_{88}
	$z'_{136,s_{176}}$	s_{186}
$\vdots$	$\vdots$	$\vdots$
s_{169}	$z'_{83,s_{169}}$	s_{81}
	$z'_{143,s_{169}}$	s_{179}
s_{168}	$z'_{84,s_{168}}$	s_{80}
	$z'_{144,s_{168}}$	s_{178}

i. An adversary has access to the first 157 bits of the keystream.
ii. An adversary can inject a random fault several times in a specific target register.

Algorithm 1: Bit-flipping fault attack on TRIAD

Input: Fault target location(s): $s_{168}, \cdots, s_{252}$
Output: Initial state bits: $s_{80}, \cdots, s_{164}$, and $s_{168}, \cdots, s_{254}$

1 Initialise with the K and N
2 Generate and store the first 157 bits of the fault-free keystream $z_0, \cdots, z_{156}$
3 **for** $j \leftarrow 252$ **to** 168 **do**
4 | Re-initialise the cipher with the K, N
5 | Inject bit-flipping fault to s_j
6 | Compute the faulty keystream z'_{252-j,s_j}
7 | Output $s_{j-88} = z_{252-j,s_j} \oplus z'_{252-j,s_j}$
8 | **if** $j \geq 178$ **then**
9 | | Compute the faulty keystream z'_{334-j,s_j}
10 | | Output $s_{j-10} = z_{334-j,s_j} \oplus z'_{334-j,s_j}$
11 | **if** $j \geq 233$ AND $j \leq 244$ **then**
12 | | Compute the faulty keystream z'_{312-j,s_j}
13 | | Output $s_{j+10} = z_{312-j,s_j} \oplus z'_{312-j,s_j}$

We take the keystream bits z_0 and z_3 (Eqs. (8) and (11)) as an example to illustrate the theoretical considerations for applying the random fault attack on TRIAD. Notice that the keystream bit z_0 in Eq. (8) consists of a quadratic monomial $s_{252}s_{164}$. These two variables s_{252} and s_{164} appears together only in the quadratic term in Eq. (8). Carefully looking into the patterns, we notice that the keystream bit z_3 in Eq. (11) also contains both of these variables s_{252} and s_{164}, but in this case, these variables appear as unique linear terms. Using these observations, we can use keystream bit z_3 to determine whether the injected random fault has complemented the target registers s_{252}, and s_{164}. That is, we can determine the effect of the random fault by such equations where the target register appears as a unique linear term in the corresponding keystream polynomial. For example, suppose a random fault e_{252} is applied to s_{252}. Let s'_{252} denotes the corresponding faulty register, where $s'_{252} = s_{252} \oplus e_{252}$. The corresponding faulty keystream bit $z'_{3,s_{252}}$ can be written as:

$$z'_{3,s_{252}} = s_{161}s_{249} \oplus s_{64} \oplus s_{76} \oplus s_{140} \oplus s_{164} \oplus s_{232} \oplus s'_{252} \tag{16}$$

XOR-ing the faulty and fault free keystream $z'_{3,s_{252}}$ and z_3 we get

$$z'_{3,s_{252}} \oplus z_3 = s_{252} \oplus s'_{252}$$
$$= s_{252} \oplus s_{252} \oplus e_{252}$$
$$= e_{252} \tag{17}$$

Equation (17) allows an adversary to identify whether the target register bit has been complemented or not. The fault value $e_{252} = 0$ means that the fault did not complement register bit s_{252}, while $e_{252} = 1$ means that the fault has complemented the bit in the target location s_{252}. Now, if the register bit s_{252}

Table 2. Unique linear terms in the equations to identify the fault value e_j

Fault Target, s_j	Required faulty keystream, z'_{i,s_j}	Recovered faulty value, e_j
s_{252}	$z'_{3,s_{252}}$	e_{252}
s_{251}	$z'_{4,s_{251}}$	e_{251}
s_{250}	$z'_{5,s_{250}}$	e_{250}
$\vdots$	$\vdots$	$\vdots$
s_{169}	$z'_{86,s_{169}}$	e_{169}
s_{168}	$z'_{87,s_{168}}$	e_{168}

has been complemented, then similar to Sect. 4.2, an adversary is able to use keystream equations z_0 and $z'_{0,s_{252}}$ to recover the register bit s_{164}. As the fault is random, the target register bit may not get complemented with single fault injection, and multiple faults may need to be injected until the desired outcome is achieved. Analysis of the equation patterns reveals that this concept of the random fault can be used to identify the fault impact on all the required target registers and hence can be used to recover the 253 state bits of TRIAD.

However, to reduce the number of faults, as similar to Sect. 4.2, we only aim to recover the register bits $s_{80}, \cdots, s_{164}$ and $s_{168}, \cdots, s_{254}$ using such random faults. We can apply the same process from Sect. 4.2, except that the injected fault is random, and the value of the injected fault needs to be determined first using the process shown in Eq. (17). To recover the register bits specified in Table 1 with a random fault model, we need to inject random faults in registers $s_{168}, \cdots, s_{252}$, and also need to determine the value of the random fault $e_{168}, \cdots, e_{252}$ using suitable equations. We found that there are enough suitable equations among the first 160 keystream functions. Table 2 lists the necessary patterns in the keystream bits where these registers $s_{168}, \cdots, s_{252}$ appear as a unique linear term and can be used to recover the corresponding fault value $e_{168}, \cdots, e_{252}$, injected into those registers.

Using the keystream bits from Table 2, we need to repeat the fault injection to the same target register s_j until the value of the injected fault is one, i.e., $e_j = 1$. Once the random fault e_j has complemented the corresponding target register s_j, we can use the same approach from Sect. 4.2 to recover the resulting state bit. Algorithm 2 shows the process of the random fault attack on TRIAD. The difference between Algorithm 1 and 2 is that we used a random fault in the latter algorithm, and the fault may be required to inject several times to the same target register.

4.4 Recovering the Remaining Register Bits

The fault attacks discussed in Sects. 4.2 and 4.3 recovers the register bits $s_{80}, \cdots, s_{164}$ and $s_{168}, \cdots, s_{254}$. The register bits $s_0, \cdots, s_{79}$, $s_{165}, \cdots, s_{167}$ and

---

Algorithm 2: Random fault attack on TRIAD

Input: Fault target location(s): $s_{168}, \cdots, s_{252}$
Output: Initial state bits: $s_{80}, \cdots, s_{164}$, and $s_{168}, \cdots, s_{254}$

1   Initialise with the K and N
2   Generate and store the first 157 bits of the fault-free keystream $z_0, \cdots, z_{156}$
3   **for** $j \leftarrow 252$ **to** 168 **do**
4   Re-initialise the cipher with the K, N
5   Inject a random fault to s_j
6   Compute the faulty keystream z'_{255-j,s_j}
7   **if** $z'_{255-j,s_j} \oplus z_{255-j,s_j} = 0$ **then**
8   Go back to step 4 and repeat
9   **else**
10   Output $s_{j-88} = z_{252-j,s_j} \oplus z'_{252-j,s_j}$
11   **if** $j \geq 178$ **then**
12   Compute the faulty keystream z'_{334-j,s_j}
13   Output $s_{j-10} = z_{334-j,s_j} \oplus z'_{334-j,s_j}$
14   **if** $j \geq 233$ AND $j \leq 244$ **then**
15   Compute the faulty keystream z'_{312-j,s_j}
16   Output $s_{j+10} = z_{312-j,s_j} \oplus z'_{312-j,s_j}$

---

s_{255} are not recovered directly using the faults. In this section we discuss the recovery of these remaining bits by solving a set of equations with low algebraic degree.

The equation system used in the fault attack covers keystream bits up to the first 157 rounds. During these 157 rounds, the degree of the output keystream function ranges from 2 to 6. However, we can reduce this degree further by substituting the variables recovered using fault attacks. So, we first substitute the values from registers $s_{80}, \cdots, s_{164}$ and $s_{168}, \cdots, s_{254}$ into keystream bits $z_0, \cdots, z_{156}$. The substitution of the known (recovered using faults) values into the equation system resulted in a comparatively low degree equation system. In fact, after substituting the recovered variables, a majority of these keystream equations are turned into linear equations with respect to the unknown variables. We use Buchberger's algorithm [18] to compute Gröbner bases to solve these low degree equations, which resulted in the recovery of all the remaining unknown register bits. Therefore, in general, with the fault attacks from Sects. 4.2 and 4.3, an adversary is able to recover all the initial state bits of TRIAD.

4.5 Experimental Analysis and Discussions

The implementations of the experiments are conducted in SageMath on a standard laptop with 16GB of memory. We implemented Algorithm 1 to verify the theoretical observations from Sect. 4.2. The physical injection of the fault is simulated through the code in SageMath, assuming that we can complement

Table 3. Comparison of bit-flipping and random fault attacks on TRIAD

Fault type	Total number of required faults	Data complexity	Nonce reuse
Bit-flipping	85	$2^{7.43}$	$2^{6.41}$
Random	170	$2^{8.01}$	$2^{7.41}$

the contents of the target register. We assume that the attacker can access the selected keystream outputs and inject faults in the targeted locations at a chosen time. The experiment is repeated multiple times with 1,000 random keys and nonces to recover the selected register bits from Table 1 by computing the differentials of the corresponding faulty and fault-free keystream outputs. We compared the recovered bits with the actual bits for each of these experiments, and all these experiments can successfully retrieve the respective register bits. In our experiments, a total of 85 bit-flipping faults are applied to recover the state bits from Table 1.

We also experimentally verified the theoretical observations on the random fault attack as described in Algorithm 2. The physical injection of the random fault e_j is simulated using the built-in random function in Python. To verify the random fault attack, we first experimentally determined the average number of random faults required to complement the bit in the respective target register s_j, i.e., to get $e_j = 1$. We found that, on average, two random faults are required to complement the contents of a single target register s_j. That is, a total of $85 \times 2 = 170$ faults are expected to recover the initial register bits from Table 1. We also tested Algorithm 2 using 1,000 random keys and nonces, and all the results from these experiments verify our expected outcome.

We also verified the recovery of the remaining state bits for each of the bit-flipping fault and the random fault attack experiments with random keys and nonces. We substitute the recovered register values $s_{80}, \cdots, s_{164}, s_{168}, \cdots, s_{252}$ in the keystream bits $z_0, \cdots, z_{156}$, and then solve the resulting low degree equations using Gröbner bases. Experimental results confirm the recovery of TRIAD's initial state for all the experiments with different random keys and nonces. The computational complexity of solving these equations is negligible.

Table 3 provides an overall comparison between the bit-flipping fault attack and the random fault attack. As shown in Table 3, the bit-flipping fault attack requires an adversary to apply 85 bit-flipping faults and observe $2^{7.43}$ keystream bits to recover TRIAD's initial state. On the other hand, the random fault attack requires an adversary to apply in average 170 random faults and observe $2^{8.01}$ keystream bits to recover TRIAD's initial state.

Note that the state update function of TRIAD is bijective. Therefore, a state recovery essentially means a key recovery, as we can clock backward from the initial state to the loaded state ($t = 0$) to recover the secret key, K.

5   Conclusion

This paper describes two fault-based key recovery attacks on TRIAD under the nonce-reuse scenario. We demonstrated a bit-flipping fault attack on TRIAD that can recover the initial state with 85 faults. We also demonstrated a random fault attack that can recover the initial state of TRIAD with about 170 faults on average. The recovery of TRIAD's initial state leads to a key recovery by clocking backward to the beginning state; hence, both of these attacks can recover the secret key without any additional faults. These attacks do not require injecting faults in all the registers of TRIAD; instead, it targets the fault injection in only one of the feedback shift register, c. Both of these key recovery attacks can be performed with practical complexity. The random fault attack requires a slightly large number of faults, but this attack is comparatively more practical as the random fault model's assumption is less stringent.

We have not performed these attacks on a physical device but have simulated the attacks on TRIAD's software implementation. We note that these types of fault injections have been demonstrated in other hardware devices [11–13], and so we consider these approaches to be feasible. The results confirm the importance of having adequate physical protection of the device to prevent an adversary from using these vulnerabilities.

Acknowledgements. This research is supported by Xiamen University Malaysia Research Fund (Grant No: XMUMRF/2019-C3/IECE/0005).

References

1. NIST Lightweight Cryptography Project (2019). https://csrc.nist.gov/projects/lightweight-cryptography
2. Banik, S., Isobe, T., Meier, W., Todo, Y., Zhang, B.: TRIAD v1 - A Lightweight AEAD and Hash Function based on Stream Cipher, NIST Lightweight Cryptography (LWC) Project (2019). https://csrc.nist.gov/CSRC/media/Projects/Lightweight-Cryptography/documents/round-1/spec-doc/TRIAD-spec.pdf
3. De Cannière, C., Preneel, B.: Trivium: A stream cipher construction inspired by block cipher design principles. In: Katsikas, S.K., López, J., Backes, M., Gritzalis, S., Preneel, B. (eds.) Information Security - ISC 2006, LNCS, vol. 4176, pp. 171–186. Springer, Heidelberg (2006). https://doi.org/10.1007/11836810_13
4. eSTREAM: the ECRYPT Stream Cipher Project. https://www.ecrypt.eu.org/stream/. Accessed 11 Sep 2020
5. He, Y., Wang, G., Li, W., Ren, Y.: Improved cube attacks on some authenticated encryption ciphers and stream ciphers in the Internet of Things. IEEE Access **8**, 20920–20930 (2020). https://doi.org/10.1109/ACCESS.2020.2967070
6. Kesarwani, A., Sarkar, S., Venkateswarlu, A.: Some cryptanalytic results on TRIAD. In: Hao, F., Ruj, S., Sen Gupta, S. (eds.) Progress in Cryptology - INDOCRYPT 2019, LNCS, vol. 11898, pp. 160–174. Springer, Cham (2019). https://doi.org/10.1007/978-3-030-35423-7_8
7. Schmidt, J., Herbst, C.: A practical fault attack on square and multiply. In: 5th Workshop on Fault Diagnosis and Tolerance in Cryptography, pp. 53–58. IEEE, Washington, DC (2008). https://doi.org/10.1109/FDTC.2008.10

8. Hutter, M., Schmidt, J.: The temperature side channel and heating fault attacks. In: Francillon, A., Rohatgi, P. (eds.) Smart Card Research and Advanced Applications - CARDIS 2013, LNCS, vol. 8419, pp. 219–235. Springer, Cham (2014). https://doi.org/10.1007/978-3-319-08302-5_15

9. Amiel, F., Clavier, C., Tunstall, M.: Fault analysis of DPA-resistant algorithms. In: Breveglieri, L., Koren, I., Naccache, D., Seifert, JP. (eds.) Fault Diagnosis and Tolerance in Cryptography - FDTC 2006, LNCS, vol. 4236, pp. 223–236. Springer, Heidelberg (2006). https://doi.org/10.1007/11889700_20

10. Barenghi, A., Breveglieri, L., Koren, I., Naccache, D.: Fault injection attacks on cryptographic devices: theory, practice, and countermeasures. Proc. IEEE **100**(11), 3056–3076 (2012). https://doi.org/10.1109/JPROC.2012.2188769

11. Skorobogatov, S.P., Anderson, R.J.: Optical fault induction attacks. In: Kaliski, B.S., Koç, K., Paar, C. (eds.) Cryptographic Hardware and Embedded Systems - CHES 2002, LNCS, vol. 2523, pp. 2–12, Springer, Heidelberg (2003). https://doi.org/10.1007/3-540-36400-5_2

12. Selmke, B., Heyszl, J., Sigl, G.: Attack on a DFA protected AES by simultaneous laser fault injections. In: 2016 Workshop on Fault Diagnosis and Tolerance in Cryptography (FDTC), pp. 36–46. IEEE, Santa Barbara (2016). https://doi.org/10.1109/FDTC.2016.16

13. Trichina, E., Korkikyan, R.: Multi fault laser attacks on protected CRT-RSA. In: 2010 Workshop on Fault Diagnosis and Tolerance in Cryptography, pp. 75–86. IEEE, Santa Barbara (2010). https://doi.org/10.1109/FDTC.2010.14

14. Dey, P., Rohit, R.S., Sarkar, S., Adhikari, A.: Differential fault analysis on Tiaoxin and AEGIS family of ciphers. In: Mueller, P., Thampi, S., Alam, B.M., Ko R., Doss, R., Alcaraz, C.J. (eds.) Security in Computing and Communications - SSCC 2016, CCIS, vol. 625, pp. 74–86, Springer, Singapore (2016). https://doi.org/10.1007/978-981-10-2738-3_7

15. Salam, I., Mahri, H.A., Simpson, L., Bartlett, H., Dawson, E., Wong, K.K.: Fault attacks on Tiaoxin-346. In: Proceedings of the the Australasian Computer Science Week - ASCW 2018, pp. 1–9. ACM Digital Library, New York (2018). https://doi.org/10.1145/3167918.3167940

16. Bartlett, H., Dawson, E., Mahri, H.A., Salam, M.I., Simpson, L., Wong, K.K-H.: Random fault attacks on a class of stream ciphers, security and communication networks, vol. 2019, Article ID 1680263, 12 pages (2019). https://doi.org/10.1155/2019/1680263

17. The Sage Developers. SageMath, The Sage Mathematics Software System (Version 9.0) (2020). https://www.sagemath.org

18. Buchberger, B.: Gröbner-bases: an algorithmic method in polynomial ideal theory. In: Bose, N.K. (ed.) Multidimensional Systems Theory, pp. 184–232. Reidel Publishing Company, Dodrecht (1985)

Author Index

Printed in the United States
By Bookmasters